Beyond measure:

Modern physics, philosophy, and the meaning of quantum theory

Beyond measure:
Modern physics, philosophy, and the meaning of quantum theory

Jim Baggott

OXFORD
UNIVERSITY PRESS

This book has been printed digitally and produced in a standard specification
in order to ensure its continuing availability

OXFORD
UNIVERSITY PRESS

Great Clarendon Street, Oxford OX2 6DP
United Kingdom

Oxford University Press is a department of the University of Oxford.
It furthers the University's objective of excellence in research, scholarship,
and education by publishing worldwide. Oxford is a registered trade mark of
Oxford University Press in the UK and in certain other countries

British Library Cataloguing in Publication Data
Data available

Library of Congress Cataloging in Publication Data
Data available

ISBN 978-0-19-852536-3

Printed and bound by CPI Group (UK) Ltd, Croydon, CR0 4YY

To Judy

Happy anniversary

KERNER: Now we come to the exciting part. We will watch the bullets of light to see which way they go. This is not difficult, the apparatus is simple. So we look carefully and we see the bullets one at a time, and some hit the armour plate and bounce back, and some go through one slit, and some go through the other slit, and, of course, none go through both slits.

BLAIR: I knew that.

KERNER: You knew that. Now we come to my favourite bit. The wave pattern has disappeared! It has become particle pattern, just like with real machine-gun bullets.

BLAIR: Why?

KERNER: Because we looked. So, we do it again, exactly the same except now without looking to see which way the bullets go; and the wave pattern comes back. So we try again while looking, and we get particle pattern. Every time we don't look we get wave pattern. Every time we look to see how we get wave pattern, we get particle pattern. The act of observing determines the reality.

Tom Stoppard, *Hapgood*

HEISENBERG: I mean the Copenhagen Interpretation. The Copenhagen Interpretation works. However we got there, by whatever combination of high principles and low calculation, of most painfully hard thought and most painfully childish tears, it works. It goes on working.

Michael Frayn, *Copenhagen*

Contents

Foreword

No other theory of the physical world has caused such consternation as quantum theory, for no other theory has so completely overthrown the previously cherished concepts of classical physics and our everyday apprehension of reality. For philosophers, it has been a romping ground of epistemological adventure or pessimism about science's ability to expose ultimate truth. For physicists, it has required a confrontation with the nature of physical reality and a heady inhalation of new attitudes. For all scientists and technologists, it has been the key to advances in all fields of endeavour, from genetics to superconductivity.

The extraordinary feature of quantum theory is that although we do not understand it, we can apply the rules of calculation it inspires, and compute properties of matter to unparalleled accuracy, in some cases with a precision that exceeds that currently obtained from experiment. It has been said that around 30 per cent of manufacturing economy stems from the application of quantum mechanics: that is not bad for a theory that we do not understand and suggests that there would be an extraordinary surge in the economy should we ever understand the theory properly, for understanding always enhances application.

The trouble has always been to find a guide to edge us towards that understanding. We need a guide who can build up an appreciation of the content of quantum mechanics, showing how experiments done over a century ago gradually impelled brilliant minds towards the selective discarding of their intellectual foundations. Then, with the confidence that the new theory of matter and radiation works, and works extraordinarily well, our guide must lead us into the darkened back rooms of the subject, and show us just why the theory is so perplexing. Then, we need to be shown that 'obvious' ways of recovering our pre-conditioning do not work. We might wonder perhaps that there may be mechanisms below the level that quantum mechanics considers, the so-called hidden variables, which actually guide particles in the more homely manner we have been trained to expect, and we need to be shown that such possibly graspable classical straws can be ruled out not from our armchairs but by experiment. Then, once we accept that the world is far more bizarre than we would ever have thought, we need to be pointed in the direction in which experts are currently thinking.

Jim Baggott is an extraordinary guide, who does the job I have described with great skill. As well as being steeped in the problems of interpretation that he has found fascinating for years, his writing is pellucid and well informed. He leads us from the nineteenth century to the twenty-first, from the puzzlement of a century ago to the deeper puzzlement of the present. Few who read this book with the care and attention it deserves will come away not convinced that there is something very odd indeed in the nature of the universe, and such is the style of writing that they will be able to adjust their level of comprehension to the exposition. Be prepared, then, in these pages to meet the new duality, the new complementarity of enlightenment and bewilderment.

Peter Atkins
Oxford, January 2003

Preface

It is now over 11 years since *The meaning of quantum theory* was first published. The original purpose of that book was to introduce undergraduate and postgraduate students of physical science to the fundamental conceptual and philosophical problems of quantum theory at a reasonably rigorous level, without reaching for mathematical apparatus more challenging than a little vector algebra. Most students are taught about quantum theory as though the conceptual and philosophical problems do not exist or are irrelevant to their understanding. Either by design or default they are fed the orthodox 'Copenhagen' interpretation of quantum theory, originally developed by Niels Bohr, Werner Heisenberg, Wolfgang Pauli and their colleagues in the 1920s and 1930s. When faced with the theory's inherent non-understandability under this interpretation, students are likely to blame themselves for failing to come to terms with what is one of the most important theoretical foundations of modern physical science. This is a great pity, because this non-understandability can, in fact, be traced to the anti-realism of the Copenhagen interpretation. The theory is, quite simply, not *meant* to be understood.

In the time since its publication, little has happened to undermine the original book's fundamental purpose. However, experimental tests of quantum non-locality and complementarity have become evermore sophisticated. In all cases quantum theory has been vindicated. Today the theory remains a mysterious black top hat from which white rabbits continue to be pulled. Students are usually advised not to ask how this particular conjuring trick is done.

There are by now many popular presentations of the inherent weirdness of the quantum world that are light on jargon and contain no mathematics. Some of these are well written and provide genuine insight into the underlying problems. In combination, however, these different presentations serve to create the impression that there are two theories – the serious one with its abstract mathematical formalism that all students of physical science must learn how to apply without worrying overmuch about what it all means and the weird one guaranteed to provide much pointless debate for the less serious or downright foolish and naïve.

Beyond measure represents a complete re-writing and updating of the original work, and attempts to bridge the gulf between these presentations. I have grounded my discussion of the theory's profound problems directly in a much simplified version of its mathematical formalism, so that the quantum weirdness can be seen to be directly attributable to this structure. Moreover, I have tried to do this in a way that most undergraduate students should be able to follow. I believe the need for this approach remains very valid.

The basic structure of *The meaning of quantum theory* remains, but *Beyond measure* contains more history, more philosophy, and attempts to bring the reader reasonably up to date with the results of fundamental experiments that have been successfully performed in the time since publication of the original book, as well as recent thinking on alternative interpretations and the frontiers of quantum cosmology, quantum gravity and what might be loosely termed as potential applications of the phenomenon of quantum entanglement in computing, cryptography and teleportation. In his book *Schrödinger's kittens*, popular science writer John Gribbin commended *The meaning of quantum theory* to his readers, 'if you read around the equations'. I like John and trust his judgement, and have therefore chosen in *Beyond measure* to excise the mathematics and confine it to a long series of appendices which, I hope, improves the flow and readability of the main text whilst still providing interested readers with the opportunity to delve into the mathematics should they wish to do so.

I owe a debt of thanks to Alastair Rae, who undertook to review the entire manuscript and make innumerable suggestions to improve the book's clarity, eliminate my misconceptions and remove errors. My thanks also go to Peter Atkins for a similar review of the manuscript and for his flattering Foreword. The misconceptions and errors that remain are, of course, entirely of my own making. I am again indebted to Oxford University Press for their indulgence.

Quantum theory is a subject that generates considerable interest, speculation and debate both within and outside the communities of practising theorists and experimentalists. Since publication of *The meaning of quantum theory*, I have been very conscious of the many letters I have received from readers that have, sadly, gone unanswered. I offer my sincere apologies to those who should have been better served. In compensation, I have set up a website devoted to *Beyond measure* and its subject matter, which includes a forum with separate discussion topics covering the history, formalism and philosophy, and interpretation of quantum theory, together with experimental tests and alternative interpretations. The site is something of an experiment itself, and will be maintained for as long as the interest is sustained. You can visit it at www.meaningofquantumtheory.com.

Jim Baggott
Reading, July 2003

Part I

Discovery

1
An act of desperation

A scientist in the late nineteenth century could be forgiven for believing that the major elements of physics were built on unshakeable foundations and effectively established for all time. The efforts of generations of scientists, philosophers, and mathematicians had culminated in Isaac Newton's grand synthesis in the late seventeenth century. Newton's work had been shaped by a further 200 years of theoretical and experimental science into a marvellous structure that we now call classical physics. This physics appeared to explain almost every aspect of the physical world: the interplay of force and motion in the dynamics of moving objects, thermodynamics, optics, electricity, magnetism, and gravitation. Its scope was vast: from the objects of everyday experience on earth to the furthest reaches of the visible universe. So closely did theory agree with and explain experimental observations that there could be no doubt about its basic correctness—its essential 'truth'. Admittedly, there were a few remaining problems, but these seemed to be trivial compared with the fundamentals—a matter of dotting a few is and crossing some ts.

And yet within 30 years, these trivial problems had turned the world of physics upside down and, as we will see, completely subverted our notions of physical reality. When extended to the microscopic world of atoms, the foundations of classical physics were shown to be not only shakeable but built on sand. The emphasis changed. The physics of Newton was mechanistic, deterministic, logical, and certain—there appeared to be little room for any doubt about what it all meant. In contrast, the new quantum physics was to be characterized by its indeterminism, illogicality, and uncertainty; nearly 80 years after its discovery, its meaning remains far from clear.

It seems incredible that we should willingly trade certainty for quantum confusion and doubt. But, make no mistake, despite its simplicity, its appealing visual images and its resonance with our common understanding, classical physics failed. The quantum description was built amidst the ruins of the structure that preceded it, and it is therefore appropriate that we begin our journey from within this classical landscape.

Newton's legacy

Newton's was a synthesis unsurpassed in the history of science. There is no doubt that he drew heavily on the work of his predecessors: Nicolas Copernicus, Johannes Kepler, René Descartes, and Galileo Galilei, to name but a few. But Newton did not simply repackage knowledge already established—he radically extended the very ambition of physical science and transformed the way that we should seek to interpret its mathematical principles. He gave physics a considerably deeper level of meaning.

The real power of Newton's mechanics is manifest in his second law of motion. This is now commonly expressed through the mathematical relation: $F = ma$, force equals mass times acceleration.[1] Stop and reflect on this equation for a moment. This is a mathematical statement, as powerful as it is simple and beguiling. It introduces the somewhat abstract concept of force, but in this case, abstract does not necessarily mean unfamiliar, as anyone who has spent even a short time in a world of objects in motion can testify. Force might be abstract but we can feel it; we can exert it; we can directly experience its effects. 'Physics', the Austrian philosopher Karl Popper once exclaimed, 'is that!' as he grabbed a book from a table and slammed it down.[2]

Mass enters Newton's equations of motion as *inertial mass*—a measure of an object's resistance to a change in its motion. To accelerate an object with twice the mass, I need to apply twice the force. If I apply no force, I get no acceleration. This does not mean that when I apply no force, I get no motion at all. In fact, the result of applying no force is Newton's first law, also known as the principle of inertia: an object not subjected to any external force moves in a straight line with constant velocity. Our over-familiarity with the consequences of the first law tends to hide the fact that this is, in fact, all a bit counter-intuitive and certainly contradicted the 'common sense' physics of the ancient Greek philosopher Aristotle, which had provided the basis for understanding the physical world for 2000 years and which said that when I remove the motive force, the object stops moving.

On the surface, you might think this all seems pretty basic stuff, good for understanding the accelerated motions of bronze balls rolling on inclined planes, or objects dropped from the masts of ships. But Newton was chasing a much, much bigger prize. At first sight, the prospect of extending the physical principles underlying the motions of terrestrial objects and applying them to celestial objects didn't look too promising. It had in fact been Kepler who had first introduced the concept of inertia, but he used it to argue that it is an inherent 'inertness' of matter that makes it necessary for a 'mover' to make it move. Take the mover away, and the movement stops (as Aristotle had said it would). Galileo argued that it is inertia that keeps the planets going in the absence of any force, but going around in *circles*. Descartes was the first to say that inertia keeps objects moving in *straight lines* in the absence of force (a bit of a problem for Descartes, though, as straight lines did not feature at all in the motions of celestial objects).

Newton's solution was the law of universal gravitation. A fundamental force of attraction is exerted between two masses along the line drawn between their centres, varying in

---

[1] Actually, this famous equation did not appear in quite this form in Newton's *Philosophiae naturalis principia mathematica* when it was published in 1687.

[2] Horgan, John. (1992). *Scientific American*, November.

magnitude with the inverse square of the distance between the centres. When combined with a counterbalancing centrifugal force, Newton's equations can be applied to planetary motion and solved to give Kepler's empirical 'laws' as a natural consequence of the underlying physical principles. With the law of gravitation in place, Newton had completed his synthesis—all objects, from apples on earth to planets and stars, obeyed the same universal principles of mechanics expressed through just three laws of motion and a law of gravitation. Amazing.

Prior to Newton, the 'laws' of physics were perhaps little more than numerical or mathematical relationships between experimental observations. Newton transformed the scientists' perspective. He established his laws as deep, underlying physical principles that govern the behaviour of all objects, all mass, all motion, throughout the entire universe.

Newton had demonstrated that all the material objects we see around us, including all that we can see in a night sky, obey the same set of fundamental principles. These principles describe effects that are somewhat counter-intuitive, if we inform our intuition only through the observation of everyday objects and apply our common sense. These principles are written in the language of mathematics, and the symbols that form the content of this language represent somewhat abstract concepts. However, there is no doubting the meaning of these symbols; the concepts are familiar and relevant to our direct experience, and their interpretation appears straightforward.

But Newton had made some sacrifices. He needed an absolute space and an absolute time to provide an ultimate inertial 'frame of reference' against which all motion could in principle be measured. Much more worrisome, however, was the means by which the force of gravity was meant to exert its influence. In all of Newton's mechanics, force is a physical phenomenon exerted through *contact* between one object and another. Gravity appeared to be an influence felt through mutual action-at-a-distance between objects, exerted instantaneously, with no intervening medium to provide a mechanism for making contact other than a very hypothetical, all-pervading, tenuous form of matter called the *ether*, that was thought to fill the void. Descartes had already rejected this approach. Newton himself was extremely uncomfortable with it. And thus is was that Newton's grand design carried within it the seeds of its own destruction.

Light at the turn of the century

Newton was also keen to extend the scope of his mechanics into the domain of optics, and concluded on the basis of relatively scant evidence that light consists of tiny particles, or corpuscles, which should in principle be subject to the same laws of motion. Such was Newton's stature in the scientific community that his pronouncements tended to be accepted without much question, despite the fact that reasonably well-developed wave theories of light had been expounded by his contemporaries Robert Hooke and Christiaan Huygens.

In a series of papers read to the Royal Society between 1801 and 1803, nearly 80 years after Newton's death, Thomas Young revived the wave theory as the only possible explanation of light diffraction and interference phenomena. In 1807, he described an experiment that will forever be associated with his name and which will return constantly to haunt the pages of this book. He demonstrated that the passage of light through two narrow, closely spaced holes or slits (often called Young's slits) produces a pattern of bright and dark fringes (see Fig. 1.1). These are readily explained in terms of a wave theory in which

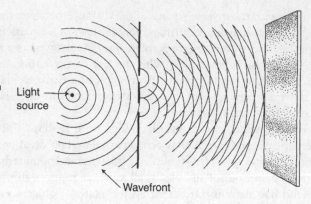

Fig. 1.1 Light interference in a double-slit apparatus. In this picture, successive wave crests are identified by the continuous lines shown to be emanating from the source and spreading out beyond the two slits. These are *wavefronts*. The overlapping of the wavefronts gives rise to constructive and destructive interference, manifested by observation of a series of light and dark fringes.

Light source

Wavefront

the peaks and troughs of the waves from the two slits start out 'in step' (or in phase). Where a peak of one wave is coincident with a peak of the other, the two waves add and reinforce (called constructive interference), giving rise to a bright fringe. Where a peak of one wave is coincident with a trough of another, the two waves cancel (called destructive interference), giving a dark fringe. Despite the apparent logic of his explanation, Young's views were roundly rejected by the physics community at the time, with some condemning his explanation as 'destitute of every species of merit'. Newton's corpuscular theory had dominated physics for a century and had become something of a dogma: arguments against it were not readily accepted.

But the wave theory of light just would not lie down and die. It attracted more adherents, including a young French physicist called Augustin Fresnel, who improved on many of Young's experiments and started to develop a detailed mathematical description of diffraction and interference effects. When, in 1817, the French Academy of Sciences decided to resolve the issue one way or the other by offering a prize for the best experimental study and associated interpretation in terms of a theory of light, Fresnel's was one of only two submissions.

The results of Fresnel's work have become the stuff of legend. He provided a theoretical interpretation of the results of a very simple experiment. Place a small object such as a ball or a disk in the path of a larger diameter beam of light, and the light bends around the edges of the object to form interference patterns at the edges of the shadow cast by the object. This is all a little difficult to explain in terms of a corpuscular theory. Fresnel provided an explanation in terms of waves but struggled with the precise mathematical formulation of the problem.

Unfortunately for Fresnel, sitting on the panel of judges was the noted mathematician Siméon Poisson, a defender of Newton against the wave theory who used to remark that there are only two things that make life worth living: doing mathematics and teaching it. Poisson decided to undertake a sophisticated analysis of Fresnel's wave problem, using mathematics that was beyond Fresnel. He concluded that whilst the wave theory provides an adequate explanation of the interference effects at the edges of the shadow, it also predicts that right in the darkest centre of the shadow should be a bright spot—itself the result of interference—with an intensity rising to 100 per cent of the intensity of the light beam. Absurd! Poisson was about to reject Fresnel's work (and with it the wave theory) when he brought his conclusion to the attention of the chairman of the panel of judges, the experimentalist François Arago. Arago suggested that, before rejecting the work, he actually set up the experiment and look for the spot. Against all expectations, he found it (it is today still called Poisson's spot). Fresnel won the prize, and the wave theory of light turned the tide against Newton's corpuscles.

Perhaps the most conclusive evidence in favour of a wave theory of light came in the 1860s from James Clerk Maxwell's work on electricity and magnetism. The connection between electricity and magnetism had been established for some time, most notably through the extraordinary experimental work of Michael Faraday. In 1831, Faraday had experimented with a soft iron ring, around which two wires were wound, one on each side of the ring. He found that by connecting and disconnecting an electric battery to one of the wires, he could induce an electric current to flow in the other. As the two wires were clearly insulated from each other, Faraday concluded that the current in the second wire must result from an influence of some kind of force generated by passing a current along the first. A few months later, he succeeded in generating an electric current in a helical coil of wire by moving a permanent bar magnet along the central axis inside the helix. He had discovered the principle of electromagnetic induction, thereby laying the scientific foundations of the electrical industry.

The prevailing ideas regarding these effects hinted at the same kind of action at a distance that appeared to lie at the heart of Newton's gravity, though of very different strength. But Faraday's experiments on induction led him to think differently. If you try to push the north poles of two permanent bar magnets together, you can feel the resistance between them literally in midair. The force seems to act outside the material. Faraday speculated that the space in between matter was somehow responsible for the forces and went on to develop the idea of a *force field*. Induction worked because of the influence of the changing electric or magnetic force field on the wire in which the current was induced. In fact, one of Faraday's favourite demonstrations was to reveal the 'lines of force' of the force field by sprinkling iron filings on a sheet of paper held over a bar magnet.

As Newton had done, so Maxwell surveyed what was known about electricity and magnetism as he sat down to construct a theoretical description. Much qualitative and quantitative information was available already in several 'laws', such as Charles Augustin de Coulomb's inverse square law for the force between two charged objects, Faraday's laws of induction, Karl Friedrich Gauss's laws for the electric and magnetic fields, André Ampere's law for electrical circuits, and the work of Jean Biot and Felix Savart on the quantitative connection between electric currents and magnets. Drawing heavily on analogies with fluid mechanics, Maxwell proposed the existence of electromagnetic fields whose properties are described by a theoretical structure consisting of four basic equations (see Appendix 1). These equations provoke two kinds of reactions among contemporary scientists, mathematicians, and commentators. There are those (the majority) who find their elegant simplicity and visual beauty irresistible, and there are those who condemn the equations as signalling the triumph of mathematical formalism over explanatory power. The mathematics is indeed hard to digest, but after all, the problem is not a simple one. Maxwell was wrestling with closely interconnected, time-varying, three-dimensional electric and magnetic fields which are properly described in terms of vectors—they have properties of both magnitude and direction. The result is perhaps inevitably somewhat esoteric for those unfamiliar with the language of vector calculus (as, indeed, was Maxwell himself).

Whilst Maxwell's equations are beautifully simple, they are not symmetrical in their treatment of the interconnected electric and magnetic fields. Magnetic fields are generated

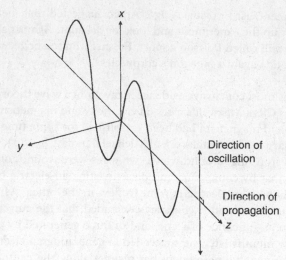

Fig. 1.2 A plane wave—the wave oscillates at right angles to the direction of propagation.

by the movement of electric charge *and* by time-varying electric fields. Electric fields are generated by time-varying magnetic fields only. As we have no evidence for the existence of magnetic charge (magnetic monopoles), it is, of course, impossible to make Maxwell's equations symmetrical by generating an electric field through the movement of magnetic charge.[3]

Maxwell made no assumptions about how these fields move through space. Nevertheless, when Maxwell's equations are recast to express the interdependence of the electric and magnetic fields propagating through free space, they point unambiguously to a wave-like motion (see Appendix 1). In 1845, Faraday had demonstrated that a magnetic field exerted an influence over light as it passed through different media. In an impromptu public discussion some months later,[4] he speculated that light itself was a wave disturbance in a force field. Nineteen years later, Maxwell found that his equations not only implied wave-like motion but also suggested that the speed of the waves was the speed of light itself.

Furthermore, for one-dimensional plane waves—waves constrained to oscillate in only one dimension—Maxwell's equations do not allow the field to vary in the direction of propagation. In other words, plane electromagnetic waves (and hence plane polarized light waves) are *transverse* waves; they oscillate at right angles to the direction in which they are moving, as Young had proposed about 40 years earlier. An example of such a plane wave is shown in Fig. 1.2.

A few difficulties remained, however. For example, all wave motion requires a medium to support it, and the ether was again called to duty as the favoured medium for light waves. Faraday had rejected the notion of the ether, but Maxwell had leaned heavily on

[3] Permanent magnets produce magnetic fields as a result of *internal* electric currents within their constituent atoms.

[4] Legend has it that Charles Wheatstone (of 'Wheatstone bridge' fame) was due to speak at the Royal Institution on this occasion, but he had succumbed to stage fright moments before he was due to start his lecture and had disappeared. To this day, invited speakers are kept in a locked room 30 min before they are due to appear to make sure they do not 'escape'.

the concept in developing his equations, and as late as 1878, he had 'no doubt that the interplanetary and interstellar spaces are occupied by a material substance or body'.

But if the existence of the ether was accepted, certain physical consequences had to follow. The earth's motion through a motionless ether should give rise to a drag effect, and hence there should be measurable differences in the speed of light depending on the direction it is travelling relative to the earth. This idea was put to its most stringent test by Albert Michelsen and Edward Morley in 1887. They found no evidence for a drag effect and hence no evidence for relative motion between the earth and the ether. This is one of the most important 'negative' experiments ever performed and led to the award of the 1907 Nobel prize in physics to Michelsen.

The ether had become something of an intractable problem. However, despite this, there was much to be satisfied with. As the century turned, Newton's grand mechanical design remained unassailable. Scientists were either willing to forgive gravitational action at a distance or quietly forget that this was a problem because the structure worked so wonderfully well, and it was so obviously *right*. Newton had been shown to be fallible on the question of light, but it was now clear how a wave theory of light fitted into the equally wonderful structure created by Maxwell to describe electromagnetism. We should, perhaps, forgive our late-nineteenth-century scientist his general smugness and sense of security. He probably felt that the problems of action at a distance and the ether would be resolved eventually. He was not to know that it would be another, seemingly innocuous problem involving light that would bring the whole structure crashing down. This was the problem of black-body radiation, and solving it led to the development of quantum theory.

Black-body radiation and the ultraviolet catastrophe

When we heat an object to very high temperatures, it gains energy and emits light. We use phrases such as 'red hot' or 'white hot' to describe this effect. By the middle of the nineteenth century, it was pretty clear that the radiation was caused by mechanical oscillations of electric charge within the object's material. A higher temperature implied greater oscillation, in terms of both amplitude and frequency, as it resulted in an increase in the intensity of radiation emitted and a shift towards the blue end of the visible spectrum. Theoreticians conceived the notion of a 'black body', so called because the same rules apply to the absorption of energy by a completely non-reflecting object (which would therefore be black). This is a model object intended to serve as a good approximation for the properties of real objects but which is theoretically easier to describe. A black body absorbs and emits radiation 'perfectly'; that is, it does not favour any particular range of radiation frequencies over another. Thus, the intensity of radiation emitted by a black body is directly related to the amount of energy in the body when it is in thermal equilibrium with its surroundings.

The theory of black-body radiation has a fascinating history, not only because it encompasses the discovery of quantum theory but also because its development is so typical of the frequently tortuous paths scientists follow to sometimes new and unexpected destinations. Theoretical physicists realized that they could probe the properties of a black body by studying the properties of radiation trapped inside a cavity with perfectly absorbing (i.e. black) walls. Such a cavity is simply a box with insulating walls that can be heated and which is punctured with a small pinhole through which radiation can enter and leave. The radiation trapped inside behaves to all intents and purposes like a gas. Its properties (such as its intensity and frequency) can be measured by observing the small proportion of the radiation that leaks through the pinhole (in itself insufficient to disturb the equilibrium), and the temperature of the radiation can be obtained from the temperature of the box.

In the winter of 1859–1860, the German physicist Gustav Kirchoff came to a remarkable conclusion. He was able to demonstrate that the ratio of emitted to absorbed energy depends only on the frequency of the radiation and the temperature inside the cavity, and not on shape of the cavity, the shape of its walls, or the nature of the material from which the cavity is made. This ratio is measurable as a property called the spectral (or radiation) density, which is effectively the intensity of radiation trapped in the cavity per unit volume per unit frequency interval, measured at a specific temperature. Kirchoff's result implied that the spectral density is something fundamental, an aspect of the underlying physics of the radiation itself, and challenged the scientific community to discover the functional form of its dependence on both frequency and temperature.

The theoreticians devised models for black-body radiation based on vibrations or oscillations of the electromagnetic field trapped inside the cavity. These vibrations were assumed to be caused by the interaction between the electromagnetic field and a set of oscillators of a largely imaginary and unspecified nature, their primary purpose being to ensure that the energy was properly distributed among the accessible frequencies through 'recycling' the radiation by re-absorption and re-emission. Today, we would identify these oscillators as highly excited electrons within the atoms of the cavity material. Energy is released from the material in the form of light (ultraviolet, visible, and infrared, depending on the temperature), and the oscillators help to bring the radiation trapped inside the cavity to equilibrium—a dynamic balance between energy absorption and emission. (Remember that in the latter half of the nineteenth century, there was still much uncertainty about the reality of atoms and molecules, and J. J. Thompson's experiments confirming the existence of electrons were not performed until 1897.)

It was imagined that as the external temperature of a cavity is increased, so the distribution of the frequencies of the oscillators shifts to higher ranges. This in turn causes vibrations in the electromagnetic field of increasingly higher frequencies, with a certain oscillator frequency giving rise to the same frequency of vibration of the field. The vibrations were visualized as standing waves; waves which 'fit' exactly in the space between the walls of the cavity and which were reinforced by constructive interference.

Towards the end of the nineteenth century, breakthroughs in the experimental study of infrared (heat) radiation emitted from a radiation cavity allowed the models developed by the theoreticians to be stringently tested. For example, in 1896, Wilhelm Wien had used a simple model (and had made some unjustified assumptions) to derive a mathematical expression that linked the spectral density to the cube of the frequency, moderated by a term that decreased exponentially with the ratio of frequency to temperature (see Appendix 2). This seemed to be quite acceptable and was supported by the experiments of Friedrich Paschen in 1897. However, new experimental results obtained by Otto Lummer and Ernst Pringsheim reported in 1900 showed that Wien's formula failed in the low frequency (long wavelength) infrared region, as illustrated in Fig. 1.3(a).

In June 1900, Lord Rayleigh (William Strutt) published details of a theoretical model based on the 'modes of ethereal vibration' in a radiation cavity. Each mode possessed a specific frequency, and could take up and give out energy continuously. Rayleigh assumed a classical distribution of energy over these modes. Such a distribution requires that, at

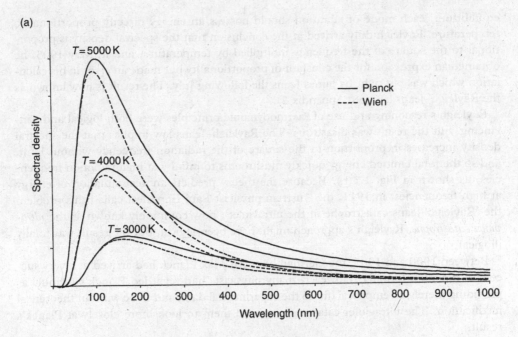

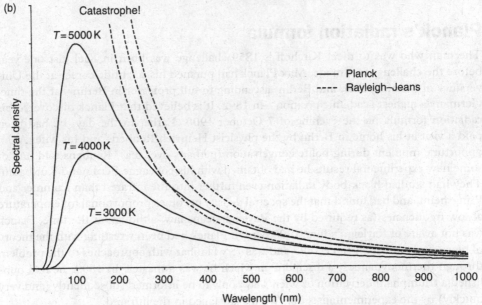

Fig. 1.3 (a) Comparison of the predictions of Planck's radiation law and Wien's law for three different temperatures, $T = 3000$ K, 4000 K, and 5000 K. Wien's law accurately reproduces the behaviour of black-body radiation at very high frequencies (short wavelengths) but fails at lower frequencies (longer wavelengths). The discrepancy is most noticeable at higher temperatures. (b) Comparison of the predictions of Planck's radiation law and the Rayleigh–Jeans law for the same three temperatures. The Rayleigh–Jeans law approaches the behaviour of black-body radiation at very low frequencies (very long wavelengths) but gives catastrophic results in the ultraviolet.

equilibrium, each mode of vibration should possess an energy directly proportional to temperature. Rayleigh duly arrived at the conclusion that the spectral density is proportional to the square of the frequency multiplied by temperature, and in May 1905, he obtained an expression for the constant of proportionality but made an error in his calculation which was put right by James Jeans the following July. The result is now known as the Rayleigh–Jeans law (see Appendix 2).

Rayleigh's reasoning and use of thermodynamic principles were both logical and convincing, but the result was disastrous. The Rayleigh–Jeans law implies that the spectral density increases in proportion to the square of the radiation frequency without limit, and so the total emitted energy quickly mushrooms to infinity at high radiation frequencies, as shown in Fig. 1.3(b). Because the theory predicts an accumulation of energy at high frequencies, in 1911, the Austrian physicist Paul Ehrenfest called this problem the 'Rayleigh–Jeans catastrophe in the ultraviolet', now commonly known as the *ultraviolet catastrophe*. Rayleigh's approach might have been logical, but the result was totally illogical.

Between 1900 and 1905, the German physicist Max Planck had arrived at a very successful radiation formula. However, many physicists had regarded Planck's formula as providing merely an empirical 'fit' to the experimental data, and to be without theoretical justification. The ultraviolet catastrophe caused them to look more closely at Planck's result.

Planck's radiation formula

The man who was to meet Kirchoff's 1859 challenge was born in Kiel just one year before the challenge was made. Max Planck had pursued his scientific career at the Universities of Munich, Kiel, and Berlin, ascending to full professor in Berlin—at the time, Germany's highest academic position—in 1892. It is believed that Planck discovered his radiation formula on the evening of 7 October 1900. Earlier in the day, he had been paid a visit at his home in Berlin by the physicist Heinrich Reubens and his wife. At an opportune moment during polite conversation (perhaps over tea), Reubens told him of some new experimental results he had obtained with his colleague Ferdinand Kurlbaum. They had studied black-body radiation even further into the infrared than Lummer and Pringsheim, and had found that the spectral density becomes proportional to temperature at low frequencies (as required by the Rayleigh–Jeans law, although at the time, Planck was not aware of Rayleigh's June 1900 paper). Planck had been wrestling with the theory of black-body radiation since 1895 and was very familiar with approaches to the problem through thermodynamics, a discipline in which he was a master. In 1899, he had published a triumphant derivation of Wien's law, only to be informed subsequently (and very quickly) by the experimentalists that Wien's law failed in the infrared.

Planck used the information given to him by Reubens to adapt his derivation by making some rather *ad hoc* assumptions. The result was an expression which fitted all the available experimental data (see Appendix 2). At a meeting of the German Physical Society on 19 October, Planck discussed his derivation of Wien's law, drew immediate attention to its shortcomings, and proposed his new formula. If he was a little cagey about its derivation, it was because he had obtained the result somewhat circuitously, and he wanted first to invest it with a little more physical meaning. The next day, Reubens compared his experimental results with Planck's formula and found the agreement to be 'completely satisfactory'. It was clear that, despite its weak theoretical foundations, Planck's formula had captured the experimental truth.

Planck's concern to discover the proper physical basis of his new radiation formula was to lead to 'a few weeks of the most strenuous work of my life'.[5] He again chose to approach the problem through thermodynamics, drawing on the analogy between radiation trapped inside a cavity and the properties of a gas confined within the walls of a container. The link between energy and temperature can be made via the basic thermodynamic quantity known as entropy. Entropy is a somewhat abstract quantity that we tend to interpret as the amount of 'disorder' in a system, and it had been known for some time that in a closed system (i.e. one prevented from exchanging energy with the outside world), entropy will increase spontaneously to a maximum. This is the second law of thermodynamics, the subject of Planck's 1879 doctoral thesis.

Using basic thermodynamics, Planck derived an expression for the entropy of an oscillator[6] in terms of its internal energy and its frequency of oscillation that is consistent with his radiation formula and therefore consistent with experiment (see Appendix 2). Thus, the physical basis of the radiation formula would be established if a second, theoretical, expression for the entropy could be derived more directly from the underlying physical properties of the oscillators themselves.

At the time that Planck was struggling to find an alternative derivation, the Austrian physicist Ludwig Boltzmann had long advocated a new approach to the calculation of thermodynamic quantities using statistics. Planck had not liked Boltzmann's statistical approach at all, but over the previous 5 years or so, he had gradually been won over to its benefits, and he was forced to use it now. As he later explained in a letter to Robert Williams Wood[7]:

what I did can be described as simply an act of desperation...A theoretical interpretation [of the radiation formula]...had to be found at any cost, no matter how high.

Ludwig Boltzmann was an adherent of a strictly mechanical approach to interpreting and understanding physical phenomena. In 1877, he developed his own entirely mechanical interpretation of the second law of thermodynamics. Entropy, he argued, is simply a measure of the probability of finding a mechanical system composed of discrete atoms or molecules in a particular 'state'. The second law is therefore a general statement that a system with a low probability (low entropy) will evolve in time into a state of higher probability (higher entropy). The equilibrium state of a system is the one of highest probability and therefore maximum entropy, that is, it is the most likely state.

Boltzmann's approach to calculating the entropy of a gas was to assume that its total available energy could be thought of as being distributed over all its molecules in a series of different states or *complexions*. For example, one possible (though very unlikely) complexion would put all the energy in one molecule only and none in the remainder. Other, more likely, complexions would have a more even distribution of energy over the molecules, but the important point was to take account of *all* the possible combinations. In going

---

[5] The quotation comes from Planck's address to the Swedish Academy; see Kuhn, Thomas S. (1978). *Black-body theory and the quantum discontinuity 1894–1912*, Oxford University Press, Oxford, p. 97.

[6] Planck's early work on black-body radiation referred in fact to an elaborate theory involving 'resonators'. By 1909, he had accepted that the special properties of resonators were not required to support the theoretical basis of his radiation formula, and he reverted to using 'oscillators'.

[7] Planck, Max, letter to Wood, Robert Williams, 7 October 1931.

through this process, it was necessary for Boltzmann to assume that the energy itself could be broken down into finite elements. It was not necessary under this assumption for the energy elements to have a fixed size or to be constrained in any way to discrete values. Boltzmann simply needed to break up the energy into elements so that he could turn the problem into one of working out the number of possible permutations of the energy elements over the molecules, which is easy using the mathematics of combinatorials. It did not imply that the energies of individual molecules were restricted in anything other than a classical sense.

A complexion is therefore built by restricting the energies of the molecules to integer multiples of some notional energy element, and then counting the number of molecules with energies *in the range* zero to one element, one element to two elements, two to three, and so on. From this it is apparent that energy remains continuously variable in this analysis, and all that Boltzmann had done was divide it up into 'buckets' so that he could count the number of molecules in each bucket, and hence the different possible permutations of molecules and energy buckets. He reasoned that the most probable state of the gas would be the one with the highest number of possible permutations, representing maximum entropy (or 'disorder'). By equating the maximum number of possible permutations to the most probable energy distribution, it was a relatively simple step to the calculation of the entropy itself.

In applying Boltzmann's ideas to the theory of black-body radiation, Planck followed Boltzmann's reasoning and assumed that the total energy could be split up into a collection of energy elements, which were then statistically distributed over a large number of oscillators. Planck almost certainly worked backwards from the result that he was aiming for, because he chose to make the energy elements *indistinguishable*, directly related to the frequency of the oscillators according to his now famous relation: $\varepsilon = h\nu$—energy element equals a constant of proportionality (now called Planck's constant) times frequency—and fixed in size as integer multiples of $h\nu$. In making these choices, he was following a very different path from Boltzmann. In his excellent biography of Albert Einstein, *Subtle is the Lord*, Abraham Pais wrote that[8]:

From the point of view of physics in 1900, the logic of Planck's electromagnetic and thermodynamic steps was impeccable, but his statistical step was wild.

In 1911, Paul Ehrenfest demonstrated that Planck's statistical approach implied the existence of 'particles' of energy unlike any that had ever been invoked before. As we will see, Ehrenfest was right to be suspicious—Planck's particles of energy were not like just any other particles. Apart from this aspect of Planck's derivation, the remainder relied on standard methods of statistical thermodynamics, details of which can be found in Appendix 2. Planck submitted his derivation to the journal *Annalen der Physik* in January 1901, and in subsequent years, his energy elements of fixed size were referred to variously as 'energy particles', 'energy atoms', and 'energy quanta'. Thus it was that the term 'quantum' entered the language of the early-twentieth-century physics. The world would never be the same again.

Planck was a reluctant scientific revolutionary. Although his radiation formula could be derived from 'first principles' using Boltzmann's statistical approach, he did not really like

[8] Pais, Abraham (1982). *Subtle is the Lord: the science and life of Albert Einstein*. Oxford University Press, Oxford, p. 370.

the idea that energy could be exchanged between the oscillators and the radiation field only in fixed, discrete elements. Although, in his derivation, he had had to accept that the energy elements must be fixed as integer multiples of $h\nu$, he still tended to think of this as analogous to Boltzmann's approach of breaking up a continuously variable energy into 'buckets' as a means to introduce the mathematics of combinatorials. Classical physics said that energy was continuously variable, and the foundations of Planck's theory had been constructed entirely from Maxwell's classical equations of the electromagnetic field and classical thermodynamics, supplemented by Boltzmann's statistics.

Even if the principle of the quantization of energy were accepted, the question remained as to its origin, or what it was exactly that was supposed to be quantized. The experimental and theoretical evidence supporting the wave theory of electromagnetic radiation appeared irrefutable: surely nobody was now going to be so foolhardy as to propose that electromagnetic radiation was composed of discrete 'packets' (dare we even say 'corpuscles') of energy? In his entertaining and idiosyncratic telling of the early history of quantum theory, the physicist George Gamow neatly summarized Planck's position[9]:

Having let the spirit of quantum out of the bottle, Max Planck was himself scared to death of it and preferred to believe the packages of energy arise not from the properties of the light waves themselves but rather from the internal properties of atoms which can emit and absorb radiation only in certain discrete quantities.

Planck was 42 when he discovered the radiation formula and the fundamental constant that were to transform radically our understanding of the physical world and that today carry his name. A younger physicist with less of a reputation at stake might, perhaps, have been bolder. The decidedly non-classical concept of the quantization of energy had somehow worked its way into an otherwise entirely classical description, and it should not be surprising that those closest to its development did not yet recognize it (or want to recognize it) for what it was. In his definitive historical study of the emergence of quantum physics, Thomas Kuhn noted: 'If quantization is the subdivision of total energy into finite parts, then *Boltzmann* is its author'.[10] Nevertheless, quantization will forever be associated with Planck's name. He was awarded the 1918 Nobel prize in physics for letting the quantum cat well and truly out of the bag.

Quanta

Planck was not the only one to have mixed feelings about the interpretation of the radiation formula; most of the physics community was sceptical. While most physicists acknowledged the fact that Planck's radiation formula gave the correct results, some found it hard to believe that energy could be quantized. A few physicists initially believed that Planck's interpretation was so monstrous that the radiation formula itself (and hence also the experimental results) must be wrong.

However, the seeds of the quantum revolution had been sown. Planck's work was studied carefully by a young 'technical expert, third class' in the Swiss Patent Office in

---

[9] Gamov, George (1966). *Thirty years that shook physics*. Doubleday, New York (republished by Dover Publications, New York, in 1985), p. 22.

[10] Kuhn, Thomas S. (1978). *Black-body theory and the quantum discontinuity 1894–1912*. Oxford University Press, Oxford, p. 127. The italics are mine.

Bern. He was 26 and, with no academic reputation to protect, more than ready to think the unthinkable and say it. His name was Albert Einstein.

In 1905, Einstein expressed reservations about Planck's derivation, pointing out that Planck had been inconsistent in first assuming energy to be continuously variable and then assuming exactly the opposite, by setting $\varepsilon = h\nu$. But, unlike most other physicists, Einstein was prepared to accept the reality of quanta. His genius was to accept the 'impossible' and to use it to explain other puzzling phenomena, making predictions that could be tested experimentally. In a paper published in the same year, Einstein introduced his *light-quantum hypothesis*[11]:

Monochromatic radiation...behaves...as if it consists of mutually independent energy quanta of magnitude $[h\nu]$...this suggests an inquiry as to whether the laws of the generation and conversion of light are also constituted as if light were to consist of energy quanta of this kind.

In other words, Einstein was prepared not only to embrace the idea of light quanta (which were called *photons* by G. N. Lewis in 1926), but also to look seriously at its implications.

Science is a democratic activity. It is rare for a new theory to be adopted by the scientific community overnight. Rather, scientists need a good deal of persuading before they will invest belief in a new theory, especially if it provides an interpretation that runs counter to their intuition, built up after a long acquaintance with the old way of looking at things. The challenge often comes from younger scientists more ready, willing, and able to question the very foundations of the theoretical structures established by previous generations. The process of persuasion must, however, be backed up by hard experimental evidence, preferably from new experiments designed to test the predictions of the new theory. Only when a large cross-section of the scientific community believes in the new theory is it broadly accepted as 'true'.

So it was with quantum theory. Although Albert Einstein proposed his light-quantum hypothesis in 1905, it took about 20 years of hard work by both theoreticians and experimentalists before it was widely accepted. The resistance of many physicists to these new ideas is understandable: quantum theory was like nothing they had ever seen before.

The theory scored some notable early successes, largely through Einstein's inspired efforts. Einstein used his light-quantum hypothesis in 1905 to explain the photoelectric effect. This was another effect that had been puzzling physicists for some time. It was known that shining light on metal surfaces could lead to the ejection of electrons from these surfaces. However, contrary to the expectations of classical physics, the kinetic energies of individual emitted electrons show no dependence on the *intensity* of the radiation, but instead vary with the radiation *frequency*. This is a strange result if radiation is considered a purely wave phenomenon, because the energy contained in a classical wave depends on its amplitude (and hence its intensity), not its frequency.

Einstein solved this problem by suggesting that a light-quantum incident on the surface transfers all its energy to a single electron. That electron is ejected with a kinetic energy equal to the energy of the light-quantum less an amount expended by escaping to the surface and which is therefore characteristic of the metal (a property now known as the work function).

According to Planck, the energy of the light-quantum is given by $\varepsilon = h\nu$, and so the kinetic energy of the ejected electron is expected to increase with increasing frequency.

[11] Einstein, A. (1905). *Annalen der Physik*, **17**, 132.

Increasing the intensity of the radiation increases the number of light quanta incident on the surface, increasing the number of ejected electrons, but not their kinetic energies. Einstein's theory was very simple, and yet it made a number of important, testable predictions. These were confirmed in a series of experiments performed about 10 years later. Einstein's work on the photoelectric effect won him the 1921 Nobel prize for physics.

But Einstein's vision was broader than the light-quantum. He anticipated that quantization would likely be a much more universal phenomenon, and shortly after his work on the photoelectric effect, he directed his attention to an area that was ultimately to prove decisive for the new quantum theory as well as for the recognition of Einstein's efforts. Curiously, this was an area where there did not even appear to be a problem.

Specific heat capacity is a measurable property of materials in all forms—solids, liquids, and gases—and reflects the material's capacity to absorb and store energy within its internal structure. Towards the end of the nineteenth century, the specific heat capacities of solids had been studied extensively, and the experimental evidence pointed to a simple rule: for elements that form solids at normal temperatures, the specific heat is a constant, independent of the element. There were some badly behaved elements that did not quite conform to this rule, which was named for the French physicists Pierre Dulong and Alexis Petit, but these exceptions could be explained away with a little arm-waving.

Einstein reasoned that, just as Planck's radiation law was the result of quantum effects in vibrating electrons, so it should be possible to anticipate quantum effects in vibrating atoms and ions. The 'oscillators' were very different in each case, and the vibration frequencies would be very different, but the principles should be the same. Einstein simply took the expression for the internal energy of a single oscillator from Planck's derivation (Appendix 2), multiplied it by three (because, in a solid lattice, there are three directions or 'degrees of freedom' in which the atom or ion can vibrate), and used the result to develop an equation for the molecular heat capacity. Einstein's result showed that the heat capacities of different elements in solid form should indeed be the same at 'normal' temperatures. But the result also suggested that as the solids are cooled to very low temperatures, the heat capacities of all elements should tend towards zero. There was no precedent for this, and at the time, there were no experimental data to support it.

Einstein published these results in 1907 in the journal *Annalen der Physik*. Nobody seems to have taken the results very seriously at the time. The predicted low-temperature effects appeared to contradict everything that was then known and understood about the thermodynamics of solids, and, unlike black-body radiation, this was an area in which a considerable body of experimental and theoretical work had been built up over many years. There was simply no good reason to take Einstein seriously.

Things changed dramatically a few years later, as a result of new experiments and developments in the understanding of the thermodynamics of chemical reactions and chemical equilibrium. By 1909, the eminent scientist Walter Nernst had concluded that the evidence for a notable dependence of specific heats on temperature was much stronger than had been thought, and he began to wonder if the Dulong and Petit rule was the result of nothing more than coincidence. He became aware of Einstein's ideas on the subject and, in 1910, chose to pay Einstein a visit. Einstein by this time had moved to the Eidgenossische Technische Hochschule (ETH) in Zurich but was still a relative unknown. The visit by Nernst encouraged greater respect for Einstein and his work, with one Zurich colleague remarking: 'This Einstein must be a clever fellow, if the great Nernst comes so far from Berlin to Zurich to talk to him.'[12]

Nernst was instrumental in drawing attention to Einstein's approach and, unlike the situation for black-body radiation, there was already significant interest in and therefore a large audience in the scientific community for problems related to specific heat capacities. From early 1911, a growing number of scientists began to cite Einstein's papers and embrace quantum ideas.

This is wrong . . .

The concept of the atom, as an ultimate, indivisible quantity of matter, had survived as part of mankind's attempt to understand the world since the times of the ancient Greek philosopher Democritus. The reality of the chemical elements, their relative weights, and their rules of combination to give the enormous variety of chemical substances had been established through the work of John Dalton and others in the early nineteenth century, and provided strong evidence for the existence of atoms. Dalton was the first to assign different symbols to the different atoms of chemical elements as a means of describing the composition of chemical substances and to enable the easy calculation of their weights from the relative (atomic) weights of their constituent atoms (though it was Jöns Jacob Berzelius who gave us the notation—H_2O, CO_2, and so on—that we use today). J. J. Thomson's discovery of the negatively charged electron in 1897 implied that atoms, indivisible for more than 2000 years, now had to be recognized as having an *internal structure*.

Thomson was one in a long line of eminent directors of the Cavendish laboratory at Cambridge University in England, a line that had included James Clerk Maxwell and Lord Rayleigh. Thomson won the 1906 Nobel prize in physics for his discovery of the electron and immersed himself in the theory of atomic structure (his often disastrous relationship with experimental apparatus effectively ruling him out as a 'hands-on' experimentalist). He produced a theoretical model, now mentioned only in passing in school physics and chemistry courses and referred to as the Thomson 'plum-pudding' model—atoms consisting of uniform spherical sponges of positive charge in which the negatively charged electrons are embedded like currants. In the early years of the twentieth century, Cambridge was one of the leading centres of physics, and Thomson was universally admired not only for his contributions to the science but also for his irrepressible enthusiasm. It was therefore logical that a young Danish postdoctoral student, clutching a translation of his Ph.D. thesis and a stipend from the Carlsberg Foundation to support a period of study overseas, would want to work with Thomson.

The student was Niels Bohr. Unfortunately, Bohr's relationship with Thomson got off to a poor start from which it seems never to have recovered. As a young postdoctoral student, Bohr's grasp of the English language was not very good. Though always polite and courteous, Bohr's manner was often brusque and open to misinterpretation. So when, early in Bohr's stay in Cambridge, he entered Thomson's office with a copy of one of

---

[12] The comment is attributed to George Hevesy; see Kuhn, Thomas S. (1978). *Black-body theory and the quantum discontinuity 1894–1912*, Oxford University Press, Oxford, p. 215.

Thomson's books on atomic structure, pointed to a particular section, and said 'This is wrong...' it is hardly surprising that Thomson did not immediately warm to him.[13]

Bohr abandoned Cambridge after about a year and moved to Manchester to work with another giant of physics, the New Zealander Ernest Rutherford. Remarkably, Rutherford had been awarded the Nobel prize in 1908 for his work on radioactive substances, but the greatest discovery of his career was made *later*, in 1909. Together with his research collaborators Hans Geiger and Ernest Marsden, he had been conducting experiments in which energetic alpha particles, emitted with high velocity through the breakdown of certain radioactive elements, were fired through thin gold foil. One day, he simply asked Geiger and Marsden to see if they could obtain any effects from alpha particles scattered at oblique angles from the foil surface. Their eyebrows must have raised quite high. According to their understanding of the nature of the energetic alpha particles and the atomic structure of gold foil, this was rather like asking them to look to see if high-velocity machine-gun bullets might be deflected by tissue paper. To their astonishment, they found that about 1 in 8000 alpha particles were indeed deflected by the foil, sometimes by more than 90°. This was a result that simply could not be explained using Thomson's model of the atom and was subsequently interpreted by Rutherford to mean that most of the atom's mass is actually concentrated at its centre, the nucleus, with the much lighter electrons orbiting the nucleus much like the planets orbit the sun. As a visual image of the internal structure of the atom, this planetary model remains compelling to this day.

Compelling it might have been, but it was also impossible. Unlike the sun and planets, electrons and atomic nuclei are charged particles. Maxwell's equations made it quite clear that moving electrical charges produce diverging electromagnetic fields. The electrons in Rutherford's atomic model were therefore expected to behave like tiny radio antennae, but emitting radiation of a much higher frequency than their mechanical oscillations as they moved around a central nucleus. The electromagnetic waves would carry energy away from the electrons, leaving them at the mercy of the positively charged nucleus. The electrons would spiral down towards the nucleus as they lost their energy, and the atoms would collapse in on themselves within about one-hundred-millionth of a second. Something, somewhere, had gone horribly wrong.

Bohr's theory of the atom

Perhaps I have found out a little about the structure of atoms. Don't talk about it to anybody, for otherwise I couldn't write to you about it so soon. If it should be right it wouldn't be a suggestion of the nature of a possibility (i.e. an impossibility, as J. J. Thomson's theory) but perhaps a little bit of reality... You understand that I may yet be wrong; for it hasn't been worked out fully yet (but I don't think so); also, I do not believe that Rutherford thinks that it is completely wild...[14]

Niels Bohr wrote these words in a letter to his brother Harald in June 1912. He had become convinced that the inner electronic structure of the Rutherford atom was governed in some way by Planck's 'quantum of action'—a reference to the fact that the units of Planck's constant (energy multiplied by time) were those of the classical physical quantity

[13] Pais, Abraham (1991). *Niels Bohr's times*. Oxford University Press, Oxford, p. 120.

[14] Bohr, Niels, letter to Bohr, Harald, 19 June 1912, quoted in French, A. P. and Kennedy, P. J. (eds.) (1985). *Niels Bohr: a centenary volume*. Harvard University Press, Cambridge, MA, p. 76.

known as action. However, Bohr's was still a classical model on which quantum conditions were to be imposed and, as such, was rife with instabilities and contradictions. Bohr struggled to come to terms with his description through the rest of 1912 and into early 1913, when he was given a clue that would unlock the entire mystery. In February 1913, H. M. Hansen, a young professor of physics who had done experimental work on atomic spectroscopy at the University of Göttingen in Germany, drew his attention to the *Balmer formula*.

Spectroscopy is the study of the absorption and emission of electromagnetic radiation by atoms and molecules, a discipline that would not exist and a subject that would not merit more than half a page in any physics or chemistry textbook were it not for the fact that atomic and molecular energies are quantized. Instead of absorbing or emitting radiation over a continuous range of frequencies (thereby producing rather dull and predictable spectra), atoms and molecules display very discrete structures in their spectra which are related to their internal electronic structures or, in the case of molecules, to the collective internal motions of their constituent atoms. The simplest atoms exhibit the simplest spectra, and the hydrogen atom, consisting of a nucleus and just one electron, has a 'line spectrum' of discrete, narrowly defined frequencies. In 1885, Johann Jakob Balmer had measured one series of hydrogen emission lines and found them to follow a relatively simple pattern. He found that the frequencies of the lines are all proportional to the differences between the inverse squares of adjacent integer numbers (see Appendix 3). In itself, this was a purely empirical formula, and its origin in terms of the underlying physics was quite obscure. Bohr, however, immediately understood where the integer numbers had come from. They are *quantum numbers*.

Bohr still had to deal with the fact that a system of negatively charged electrons orbiting a positively charged nucleus is inherently unstable in classical terms. But by fusing this impossible classical picture with Planck's quantum concepts, he was able to argue that only certain electron orbits are 'allowed', and an electron moving from an outer, higher-energy orbit to a lower-energy inner orbit causes the release of energy as emitted radiation. The orbits are fixed, with energies that depend on integer numbers that can be counted outwards from the nucleus in a linear sequence (these integers are now known as the *principal quantum numbers*). The energy *differences* between the orbits are also therefore fixed, and Bohr was able to show that atomic emission can be observed only at those radiation frequencies corresponding to the energy differences. The differences, and hence the frequencies themselves, are therefore proportional to the inverse squares of the differences between the quantum numbers (see Appendix 3). Bohr was further able to show that the constant of proportionality, known as the Rydberg constant with a magnitude well known at the time from spectroscopic measurements, can be expressed as a collection of fundamental physical constants.

Bohr had tackled the problem by assuming that the electrons circulating around the nucleus are confined to elliptical orbits, exactly analogous to planetary orbits following Kepler's laws. This gave him some further discomforts, because in order to obtain the right result, he needed to assume that the mechanical frequencies of the electrons were no longer identical to the frequencies of the emitted radiation. In Planck's radiation theory, the frequency of the oscillators and the frequency of radiation were always identical and used interchangeably. Bohr tried to justify this change of approach but later abandoned this argument in favour of one originally proposed by J. W. Nicholson. Bohr's result follows from the fact that the orbital angular momentum of an electron moving around the nucleus in a *circular* orbit is a fixed quantity with a value of $h/2\pi$.

Bohr published his theory of the atom in 1913 in a series of three papers. Try to imagine the state of physics at the time, with physicists still uncertain about Planck's interpretation of his radiation law, with few caring overmuch for Einstein's light-quantum hypothesis and a great deal on confusion around, and you will begin to gain some idea of Bohr's breathtaking vision.

And there was more. It was apparent from Bohr's result that Balmer's formula is just a special case of a more general expression in which the quantum number of the lower-energy electron orbit has the value $n = 2$, and the quantum numbers for the higher-energy orbits have values $n = 3, 4, 5$, and so on. Bohr noted that setting the quantum number of the lower energy orbit to $n = 3$ gives another series of spectral lines named for Friedrich Paschen and, by setting the lower quantum number to $n = 1$ and $n = 4$ and 5, predicted further series in the ultraviolet and infrared that, at that time, had not been observed.

A further series of emission lines, known as the Pickering series, was thought by experimental spectroscopists also to belong to the hydrogen atom. However, at the time, the Pickering series was characterized by half-integer quantum numbers which are not possible in Bohr's theory. Instead, Bohr proposed that the formula be rewritten in terms of integer numbers, suggesting that the Pickering series belongs not to hydrogen atoms but to ionized helium atoms. An awkward mismatch between calculated and observed emission frequencies was later resolved by Bohr when he realized that he had neglected the effect of the motion of the heavy helium nucleus on the stable electron orbits of ionized helium. This correction gave a Rydberg constant for ionized helium some 4.00163 times greater than that for hydrogen (not 4 times greater, as Bohr had originally proposed). The experimentalists found this ratio to be 4.0016. When he heard about this result, Einstein described Bohr's theory as 'an enormous achievement'.

Bohr's idea of stable electron orbits had a further consequence. Transitions between the orbits had to occur in instantaneous 'jumps', because if the electron gradually moves from one orbit to another, it would again be expected to radiate energy continuously during the process. This is certainly not what is observed when an atom absorbs light. Thus, transitions between inherently non-classical stable orbits must themselves involve non-classical discontinuous 'quantum jumps'. Bohr wrote that[15]:

the dynamical equilibrium of the systems in the [stable orbits] can be discussed by help of the ordinary mechanics, while the passing of the systems between the different [stable orbits] cannot be treated on that basis.

Throughout all of this analysis, it was clear that the origin of quantization lay in the mechanics of the electrons moving in stable orbits around the nucleus, with the quantum numbers and Planck's constant making their appearance in the expressions for orbital energy and angular momentum. There was nothing in the analysis to suggest that radiation itself had to be quantized, and at this stage in his career, Bohr was not yet ready to believe in light quanta.

[15] Bohr, N. (1913). *Philosophical magazine*. Reproduced in French, A. P. and Kennedy, P. J. (eds.) (1985). *Niels Bohr: a centenary volume*. Harvard University Press, Cambridge, MA, p. 83.

Discontinuous physics

It is worthwhile noting that, from almost the very beginning, Einstein viewed the quantum interpretation as provisional, to be replaced eventually by a new, more complete theory that would explain quantum phenomena somewhat more rigorously. In 1916 and 1917, Einstein published his work on the spontaneous and stimulated emission of radiation by molecules, and incidentally laid the foundations of the theory of the laser. Einstein noted that the timing of a spontaneous transition, and the direction of the consequently emitted light-quantum, could not be predicted using quantum theory. In this sense, spontaneous emission is like radioactive decay. The theory allows the calculation of the *probability* that a spontaneous transition will take place, but leaves the exact details entirely to chance. Taken together with Bohr's quantum jumps, the theoretical evidence was building for a *discontinuous physics*, in which energy exchanges happened 'spontaneously', unpredictably, and with a complete lack of respect for the notion that there should be a direct connection between cause and effect.

Einstein was not at all comfortable with these ideas. Three years later, he wrote to Max Born on the subject of the absorption and emission of light, noting that he 'would be very unhappy to renounce complete causality'.[16] After pioneering quantum theory through one of its most testing early periods, Einstein was beginning to have grave doubts about the theory's implications. These doubts were eventually to turn Einstein into one of the theory's most determined critics.

We are today so used to the notion of a spontaneous transition that it is, perhaps, difficult to see what Einstein got so upset about. Let me propose the following (very imperfect) analogy. Suppose I lift an apple three metres off the ground and let go. This represents an unstable situation with respect to the state of the apple lying on the ground, and so I expect the force of gravity to act immediately on the apple, *causing* it to fall. Now imagine that the apple behaves like an excited electron in an atom. Instead of falling back as soon as I let go, the apple hovers above the ground, falling at some unpredictable moment that I can calculate only in terms of a probability. There may be a high probability that the apple will fall within a very short time, but there may also be a distinct, small probability that the apple will just hover above the ground for several days! An excited electron *will* fall to a more stable state; it is *caused* to do so by the quantum mechanics of the electromagnetic field. However, the exact moment of the transition appears to be left to chance.

By 1909, Einstein talked of radiation as though it were composed of 'pointlike quanta with energy $h\nu$', a clear reference to a particle description. However, one unambiguous way of demonstrating that something has a particle nature is to try to hit something else with it. The first 'something else' was an electron. In 1923, Arthur Compton and Peter Debye both used simple conservation of momentum arguments to show that 'bouncing' light quanta off electrons should change the frequencies of the quanta by readily calculable amounts. Compton compared his prediction with experiment and concluded that a light-quantum has a directed momentum, like a small projectile. The theory of light had come full circle. More than 200 years after Newton, light was once again thought to consist of particles.

But this was not a return to Newton's corpuscular theory. Experiments demonstrating the unambiguously wave-like properties of light, and their interpretation by Young in

[16] Einstein, Albert, letter to Born, Max, 27 January 1920.

terms of waves, were not invalidated by the Compton effect. Likewise, the electromagnetic theory created by the work of Faraday and Maxwell was not thrown out. What is now referred to as the 'old quantum theory' had grown out of the confusion of ideas that took the classical physics of Newton and Maxwell and imposed quantum conditions. These conditions could never arise from within the theory itself, and it was clear that something more fundamental was going on that likely required an abandonment of some or all parts of a cherished theoretical framework that had been familiar for two centuries.

Whatever was to replace classical physics had to confront the difficult task of somehow reconciling the wave-like and particle-like aspects of light in a single, coherent theory. That theory had to be based on some form of wave–particle *duality*.

2

Farewell to certainty

The 'old' quantum theory discussed in the last chapter was characterized by the entirely arbitrary imposition of quantum conditions on problems otherwise described in terms of the concepts and theories of classical physics. There was an element of magic about it. To get the exact shape of the curve of spectral density with frequency or wavelength of black-body radiation, one assumed $\varepsilon = h\nu$ without questioning the reasons too closely. To get the Balmer formula from an 'impossible' classical picture of a negatively charged electron in a circular orbit around the positively charged hydrogen atomic nucleus, one assumed the orbital angular momentum to be constrained to values $nh/2\pi$, where n is an integer. This was not very satisfactory. A more meaningful theory was needed in which the quantum conditions and their implications in terms of quantum numbers and discontinuous quantum jumps were the *result* of the theory and not input assumptions.

The answers came in a remarkable period of creativity in the 1920s and 1930s which led to not one, but two competing theories. The implications of both approaches were profound. If they were right, the subatomic world of the fundamental particles of matter and radiation was much stranger than anyone had ever imagined.

Wave–particle duality

The next substantial step forward was taken not by one of the great luminaries of early twentieth-century physics but by a prince, no less, who had studied medieval history at the Sorbonne but gained a passion for physics through his older brother and through his experiences serving in the French Army in field radio communications during the First World War. This was Louis de Broglie.[1] After the war, de Broglie joined a private physics laboratory, headed by his brother Maurice, which specialized in the study of X-rays. It was at this laboratory in 1923 that he 'suddenly' hit on the idea that, if electromagnetic waves possessed associated particle-like properties, it seemed reasonable to suppose that

[1] Pronounced 'de Broy'.

particles such as electrons might possess associated wave-like properties. His starting point was Planck's derivation of the radiation law and Einstein's special theory of relativity.

Einstein had introduced his special theory of relativity in 1905. He had struggled, and failed, to find a way of accommodating two general observations—the absence of an ether and an apparently universal speed of light, independent of the relative motion of the source—in any kind of Newtonian interpretation of space and time. Instead, he decided to accept these observations at face value and developed a new theory from the bare minimum of assumptions (or postulates) in which these observations would automatically result. He found that he needed only two.

He postulated that the laws of physics should be completely *objective*, that is, they should be identical for all observers. In particular, they should not depend in any way on how an observer is moving in uniform motion, relative to an observed object. In practical terms, this means that the laws of physics should appear to be identical in any so-called inertial frame of reference, and so all such frames of reference are equivalent. An observer stationary in one frame of reference should be able to draw the same conclusions from some set of physical measurements as another observer moving uniformly relative to the first (or stationary in their own moving frame of reference). Einstein also postulated that the speed of light should be regarded as a universal constant, representing an ultimate speed which cannot be exceeded. (The fact that this speed happens to be that of light is irrelevant—light happens to travel at the ultimate speed.)

From these simple postulates flowed a number of bizarre consequences. Out went any idea of an absolute frame of reference (and hence the idea of a stationary ether), together with absolute space, time, and simultaneity. In came all sorts of strange effects predicted for moving objects and clocks within a new four-dimensional space–time, all later confirmed by experiment. However, although the predictions of special relativity are rather strange, the theory is really one of classical physics (in the sense that it is not a quantum theory).

One of the consequences of special relativity was captured in a simple relation connecting energy to inertial mass, a relation now universally recognized and associated with Einstein's name. It is also a relation portending the nuclear age. The relation is, of course, $E = mc^2$. De Broglie later wrote[2]:

After long reflection in solitude and meditation, I suddenly had the idea, during the year 1923, that the discovery made by Einstein in 1905 should be generalized by extending it to all material particles and notably to electrons.

De Broglie noted that Planck's $\varepsilon = h\nu$ gave the energy of a light-quantum in terms of its frequency, a property unambiguously associated with wave motion. Special relativity identified mass and energy as equivalent and interchangeable, effectively reducing two laws of the conservation of mass and the conservation of energy to a single law for mass-energy. It implied that a light-quantum with energy $\varepsilon = h\nu$ could be thought to possess an inertial mass (a very particle-like property) related to its energy by $\varepsilon = mc^2$. Although de Broglie's derivation was a little more complex, his result is easily obtained by combining these two equations (see Appendix 4). The result was a new, 'tentative' theory of light quanta.

[2] De Broglie, Louis (1963). From the preface to his re-edited 1924 Ph.D. thesis, quoted in Pais, Abraham (1982). *Subtle is the Lord: the science and life of Albert Einstein*. Oxford University Press, Oxford, p. 436.

De Broglie went further: he suggested that his theory should hold for any moving particle and that moving particles should therefore exhibit corresponding wave-like properties characterized by a frequency or wavelength. In particular, he suggested that a beam of electrons could be diffracted. That this wave nature of particles is not apparent in 'everyday' macroscopic objects is due to the very small size of Planck's constant h. If Planck's constant were very much larger, the macroscopic world would be an even more peculiar place than it is (the physicist George Gamow has speculated on what it might be like to play quantum billiards). However, because Planck's constant is so small, the dual wave–particle nature of matter is apparent only in the microscopic world of fundamental particles. Of course, if Planck's constant were zero, there would be no duality, and the world would be entirely 'classical'.

De Broglie's work on radio waves and his interest in chamber music led him to think of an electron in an atomic orbit as a kind of musical instrument. Just as musical notes are produced by so-called *standing waves* in the strings or pipes of musical instruments, so de Broglie imagined the stable electron orbits to be the result of standing electron waves. Standing waves are familiar to every undergraduate scientist and are usually discussed in relation to pictures of standing waves generated in a string which is secured at both ends. A variety of standing wave patterns is possible, provided the patterns meet the requirement that they fit between the string's secured ends, that is, they must have zero amplitude at each secured end and therefore contain an integral number of half-wavelengths. Thus, the longest frequency standing wave is characterized by a wavelength equal to twice the length of the string (no *nodes*—points where the wave function passes through zero—between the ends). The next wave is characterized by a wavelength equal to the length of the string (one node between the ends), and so on. De Broglie reasoned that the 'whole number' requirement introduced by Bohr might then emerge naturally from this kind of picture through the simple requirement that, for standing waves to be produced in a circular orbit, the wavelengths must 'fit' in the circumference of the orbit. Put simply, by the time it has done one complete circuit around the orbit and returned to the starting point, the value of the wave amplitude and the phase of the wave (its position in its up–down cycle) must be the same as at the starting point. If this is not the case, the wave will not 'join up': it will interfere destructively, and no standing wave can be produced. To satisfy the requirement, the wavelength of the electron waves must be such that an *integral* number of wavelengths will fit into the circumference of the orbit. Bohr's quantum numbers could therefore be thought of as the number of wavelengths of the electron wave present in each atomic orbit.

De Broglie published his ideas in a series of three short papers in the *Comptes Rendus* of the Paris Academy in September and October 1923. He collected these papers together, extended his ideas, and presented them to his research supervisor, Paul Langevin, as a Ph.D. thesis. Langevin asked for a second copy to be sent to Einstein and requested his views on it. Einstein wrote back, saying that he found de Broglie's thesis 'quite interesting'.[3] Consequently, Langevin was happy to accept the thesis, which was eventually published in its entirety in the journal *Annales de Physique* in 1925. This work was subsequently to have an important influence on the Austrian physicist Erwin Schrödinger.

---

[3] Not everybody shared Einstein's views. Many other physicists were highly sceptical, and some dubbed the proposal 'la Comédie Française'.

Einstein and Bohr in conflict

Bohr and Einstein first met in 1920 and developed a strong friendship. However, in 1924, Bohr, in collaboration with Hendrik Kramers and John Slater, published a paper that contained proposals that alarmed Einstein, to the extent that Einstein regarded himself to be in *conflict* with Bohr. It was a conflict that was to have a profound impact on the further development of quantum theory and its interpretation.

Bohr did not like the idea of the light-quantum, and this dislike led him to develop a new approach to light absorption and emission by atoms. Bohr, Kramers, and Slater (BKS) proposed that the 'sudden leaps' (quantum jumps) associated with light absorption and emission meant that the ideas of energy and momentum conservation had to be abandoned. Einstein had thought of taking such a step himself about 10 years earlier but had finally decided against it. What alarmed Einstein most of all, however, was a further proposal that the idea of strict causality should also be abandoned. As I mentioned earlier, Einstein had already felt uneasy about the element of chance implied in spontaneous emission—that a light quantum could be ejected from an atom or molecule at some unpredictable moment determined by no apparent direct cause.

Although BKS suggested that there was no such thing as a truly spontaneous transition, their solution was to embrace the idea that probabilistic laws, involving so-called 'virtual' fields working in a non-causal manner, are responsible for inducing the transition. The BKS proposals immediately came under fire from all sides. They led to further experimental work on the Compton effect, which clearly demonstrated that energy and momentum are indeed conserved. When the accumulated evidence against the BKS theory was overwhelming, Bohr promised to give their 'revolutionary' efforts a decent burial and managed to overcome his resistance to the light-quantum. However, Bohr remained convinced that quantum theory still demanded a new interpretation. The stage was set for a debate on the meaning of quantum theory between Bohr and Einstein that was to be one of the most remarkable debates in the history of science.

Postscript: electron diffraction and interference

De Broglie suggested in 1923 that the wave-like nature of electrons could be demonstrated by the diffraction of an electron beam through a narrow aperture. Earlier, in 1912, Max von Laue had demonstrated the diffraction of X-rays by crystals, and this was quickly developed into a powerful analytical tool for determining crystal and molecular structures.

In 1925, Clinton Davisson and Lester Germer (accidentally!) obtained an electron diffraction pattern from large crystals of nickel. In the same year, G. P. Thomson and A. Reid demonstrated electron diffraction by passing beams of electrons through thin gold foils. Davisson and Thomson shared the 1937 Nobel prize for physics for their work on the wave properties of electrons. In a nice twist of history, G. P. Thomson won the Nobel prize for showing that the electron is a wave, whereas 31 years earlier, his father J. J. Thomson had been awarded the Nobel prize for showing that the electron is a particle! Today, electron diffraction is used routinely to determine the structures of molecules in the gas phase.

The wave-like nature of electrons should also give rise to interference effects analogous to those described for light by Thomas Young. Double-slit interference of a beam of electrons has long been discussed by physicists but was demonstrated in the laboratory for the first time only in 1989. The interference patterns obtained are shown in Fig. 2.1. In this sequence of photographs, each white spot registers the arrival of an electron that has passed through a double-slit apparatus. With a few electrons, it is impossible to pick out any pattern in the spots—they seem to appear randomly. But as their number is increased, a clear interference pattern, consisting of 'bright' and 'dark' fringes, becomes discernible.

The appearance of distinct spots suggests that each individual electron has a particle-like property (each spot says an electron struck here), and yet the interference pattern is obviously wave-like. In anticipation of some fun to come in subsequent chapters, you might like to imagine what happens to an individual electron as it passes through the double-slit apparatus.

Wave mechanics

On 23 November 1925, Erwin Schrödinger gave a presentation on de Broglie's thesis at a seminar organized by physicists from the University of Zurich and the ETH. In the discussion that followed, Peter Debye commented that he thought this approach to wave–particle duality to be somewhat 'childish'. After all, said Debye, 'to deal properly with waves one had to have a wave equation...'[4]

A few days before Christmas, Schrödinger left Zurich for a vacation in the Swiss Alps, leaving his wife behind but taking an old girlfriend (Schrödinger was noted for his womanizing) and his notes on de Broglie's hypothesis. We do not know who the girlfriend was or what influence she might have had on him, but when he returned on the 9 January 1926, he had discovered wave mechanics.

It is impossible to provide a physically or mathematically rigorous derivation of the quantum mechanical Schrödinger equation starting from classical physics. In many textbooks on quantum theory, the equation is simply given and then justified through its successful application to systems of interest to chemists and physicists. However, the equation had to come from somewhere, and it is indeed possible to 'derive' the Schrödinger equation using somewhat less rigorous methods.

Schrödinger's first wave equation was actually a relativistic one, although when he finally published his work he chose to present his derivation of the non-relativistic version. It is possible to follow his reasoning from notebooks he kept at the time. His starting point was the well-known equation of classical wave motion, which interrelates the space and time dependences of the waves. Although his published derivation was somewhat more obscure, in fact all he had done was take the classical wave equation and substitute for the wavelength using the relation presented by de Broglie which connected wavelength with linear momentum (see Appendix 5). This introduced particle-like properties into an otherwise purely wave description.

Schrödinger presented his wave mechanics to the world in a paper he submitted to the journal *Annalen der Physik* towards the end of January 1926, barely 3 weeks after he had made his initial discovery. In this paper, he not only offered his rather obscure derivation

---

[4] Quotation from Bloch, Felix (1976). *Physics Today*, **29**, 23.

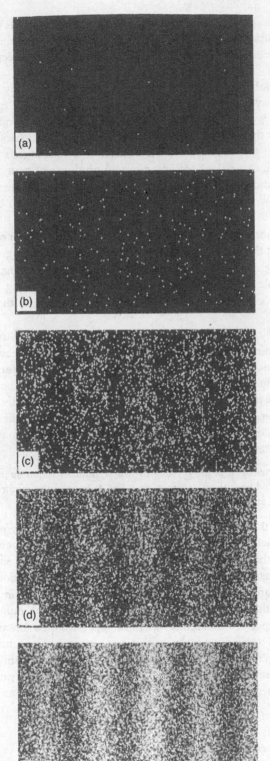

Fig. 2.1 Build-up of an electron-interference pattern. In (a), the passage of 10 electrons through a double-slit apparatus has been recorded. In (b)–(e), the numbers recorded are 100, 3000, 20,000, and 70,000, respectively. (Reprinted with permission from Tonomura, A., Endo, J., Matsuda, T., Kawasaki, T., and Ezawa, H. (1989). *American Journal of Physics*, **57**, 117–120.)

of the wave equation but also applied the new theory to the hydrogen atom. It was this first application of wave mechanics that caught the attention of the physics community. Had he simply presented the wave equation, perhaps few physicists would have been convinced of its significance.

The earlier Rutherford–Bohr model of the hydrogen atom is essentially a planetary model, consisting of a massive central nucleus, the proton, orbited by a much lighter electron. The potential energy of the nucleus is spherically symmetric, and so a more logical coordinate system for the problem is one of spherical polar coordinates rather than traditional Cartesian (x, y, z) coordinates. Transformation of Schrödinger's wave equation for the hydrogen atom into a polar coordinate system produces quite a complicated differential equation, and although Schrödinger was an accomplished mathematician, he needed help to solve it. However, assistance was at hand in the form of a colleague at Zurich, Hermann Weyl.

Schrödinger's aim was to show that the quantum numbers introduced in a rather *ad hoc* fashion by Bohr emerged 'in the same natural way as the integers specifying the number of nodes in a vibrating string'.[5] This is exactly analogous to de Broglie's suggestion, but the problem is more difficult for the hydrogen atom since we are dealing with three-dimensional standing waves confined by a spherical potential. However, the principles are the same.

In order to obtain 'sensible' solutions of the wave equation for the hydrogen atom, it is necessary to restrict the range of functions that can be admitted as acceptable. In particular, the acceptable functions must be single valued (only one value for a given set of coordinates), finite (no infinities), and continuous (no sudden 'breaks' in the functions). The last requirement must be met because the wave equation is a second-order differential equation (see Appendix 5), and a discontinuous function has no second differential.

Imposing these conditions on the wave functions is all that is necessary to produce the quantum numbers. Schrödinger wrote:

What seems to me to be important is that the mysterious 'whole number requirement' no longer appears, but is, so to speak, traced back to an earlier stage: it has its basis in the requirement that a certain spatial function be finite and single-valued.

Thus, the integer numbers that appeared as if by magic in Bohr's theory of the atom are generated naturally in Schrödinger's. These integer numbers, the quantum numbers, are an intrinsic part of the acceptable solutions of Schrödinger's wave equation and hence, also, of the energies associated with these functions. The quantization of energy therefore follows from the standing-wave condition applied to the electron in an atom.

We might add here that the differential equations that arise in Schrödinger's wave mechanics have a special property: a differential operator[6] operates on a function to yield the same function multiplied by some quantity (in this case, energy). The functions satisfying such equations are given the special name *eigenfunctions*, and the quantities produced as a result of the operation are called *eigenvalues*. Thus, when Schrödinger published his first paper on his new wave mechanics in 1926, its title was 'Quantization as an eigenvalue problem'.

---

[5] Schrödinger, E. (1926). *Annalen der Physik*, 79, 361.

[6] An operator is simply an instruction to do something to a mathematical function, such as multiply it, differentiate it, and so on.

The results that Schrödinger obtained for the wave functions of the hydrogen atom are familiar to every undergraduate scientist who has taken an introductory course in quantum mechanics. They are the electron orbitals and their three-dimensional shapes alone—which depend on the 'azimuthal' quantum number and the 'magnetic' quantum number—explain a great deal of chemistry. Their energies depend only on the principal quantum number and are given by the same expression that had been deduced earlier by Bohr (see Appendix 3).

Interpreting the wave functions

Schrödinger's application of his new wave mechanics to the hydrogen atom was hailed as a triumph. However, although the new theory explained the rules of quantization, it had merely shifted the burden of explanation from the rules to the physical meaning of the wave functions themselves. A real understanding of the behaviour of subatomic particles, encompassing the full details of the relationship between the mechanics and the underlying physical reality, could only come through an interpretation of the wave functions. What were they?

In his first few papers on wave mechanics, Schrödinger referred to the wave function as a 'mechanical field scalar', a suitably obscure title for a function whose meaning was far from clear. Schrödinger was in fact convinced that the underlying physical reality was one of wave motion—that quantum theory was essentially a wave theory. He initially interpreted the wave function as representing a vibration in an electromagnetic field, 'to which we can ascribe more than today's doubtful reality of the electronic orbits'.[7]

Schrödinger supposed that the transitions between standing waves representing the stationary quantum states of an atom are smooth and continuous. He was hopeful that he could explain the apparent non-classical properties of atoms with essentially classical concepts, and thereby recover some of the cherished notions of determinism and causality that quantum theory seemed to wish to abandon.

He therefore viewed an atomic electron not as a particle but as a collection of wave disturbances in an electromagnetic field. He proposed that the electron's particle-like properties are really manifestations of their purely wave nature. When a collection of waves with different amplitudes, phases, and frequencies are superimposed, it is possible that they may add up to give a large resultant wave located in a specific region of space (see Fig. 2.2). Such a superposition of waves is commonly called a wave 'packet'. Schrödinger argued that, since the square of the amplitude of the resultant wave is related to the strength of the electromagnetic field as a function of position, the movement of a wave packet through space might, therefore, look to all intents and purposes like the movement of a particle. This is in many ways analogous to the relationship between geometrical optics (or ray optics) and wave optics. According to this view, the dual wave–particle nature of subatomic particles is replaced by a purely wave interpretation, with the wave functions representing the amplitudes of a field.

This explanation was not entirely satisfactory, as Hendrik Lorentz pointed out to Schrödinger in a letter dated May 1926. A wave packet can persist for an appreciable

[7] Schrödinger, E. (1926). *Annalen der Physik*, **79**, 361.

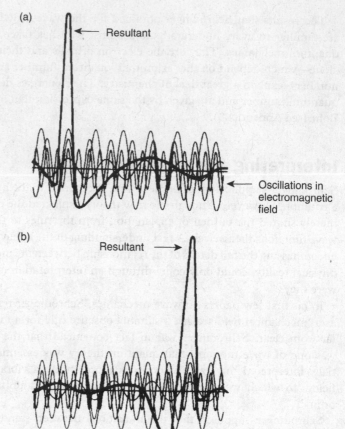

(a)

Resultant

Oscillations in
electromagnetic
field

(b)

Resultant

Fig. 2.2 Motion of a wave
packet. (a) The amplitudes,
phases, and frequencies of a
collection of waves combine and
constructively interfere to form a
resultant wave packet with a
large amplitude confined to a
specific region of space. (b) As
all the individual waves move, so
too does the region of
constructive interference.

time only if its dimensions are large compared with the wavelength. When confined to
move in small region of space, tightly grouped superpositions of waves in a wave packet
are expected to spread out rapidly, dispersing into a more uniform amplitude distribution.
This is obviously not what happens to subatomic particles like electrons.

Schrödinger had other problems too. He did not like the fact that his wave functions
could be complex (i.e. that they contain 'imaginary' numbers based on the quantity i,
representing the square root of -1). He believed that any description of a microphysical reality worth its salt ought to involve only 'real' functions. In addition, the wave
functions for complicated systems containing two or more particles are functions not
just in three spatial coordinates but of many coordinates. In fact, the wave function of
a system containing N particles depends on $3N$ position coordinates and is a function
in a $3N$-dimensional 'configuration space'. This might be all very well for a mathematician, but Schrödinger was searching for a reality that was supposed to lie beneath his
wave mechanics, and it is difficult to visualize a reality in an abstract, multidimensional
space.

To a certain extent, these problems of interpretation were clarified by Max Born, to be discussed below. But Born's interpretation of the wave functions was not to everyone's taste. Schrödinger and Einstein in particular did not like it one bit.

Max Born wrote a short paper about the quantum mechanics of collisions between particles, which was published in 1926 at about the same time as Schrödinger's fourth paper in the series 'Quantization as an eigenvalue problem'. Born rejected Schrödinger's wave approach. He had been influenced by a suggestion made by Einstein that, for photons, the wave field acts as a strange kind of 'phantom' field, 'guiding' the photon particles on paths which could therefore be determined by wave interference effects. Thus, reasoned Born, the square of the amplitude of the wave function in some specific region of configuration space is related to the *probability* of finding the associated quantum particle in that region of configuration space.

At first sight, Born's interpretation seems unremarkable. After all, we know that the square of the amplitude of a light wave in a specific region of space is related to its intensity, and, from the photoelectric effect, we know that the intensity is in turn related to the number of photons present in the same region of space. However, Born's way of thinking represented a marked break with classical physics. Unlike Schrödinger, who wanted to invest an element of physical reality in the wave functions, Born argued that they are actually much more abstract. He envisaged the wave function as some kind of 'probability wave', with the modulus square[8] of the wave function representing the measurable probability for particles in specific states. In other words, he believed that the wave function represents our *knowledge* of the state of a physical object.

Born's interpretation solved many of the problems raised by Schrödinger's wave mechanics. According to Born, the wave functions are not 'real' (in the same sense that water waves are real disturbances in a real substance), and so it does not matter that they may sometimes be imaginary. The probabilities derived from the wave functions *must* be real, since they refer to measurable properties of quantum particles. Likewise, the tendency for Schrödinger's wave packet to spread out and disperse is a problem only if the wave functions are assumed to be physically real. No such problem arises if the wave functions represent the evolution of our state of knowledge of the quantum system.

Born's interpretation was also consistent in some respects with the view being developed by Bohr and (as we will see) the German physicist Werner Heisenberg. Born argued that wave mechanics tells us nothing about the state of two quantum particles (such as electrons) following a collision: we can use the theory only to obtain probabilities for the various possible states resulting from the collision. Just as Einstein had discovered for spontaneous transitions, quantum theory appeared to have no respect for the link between cause and effect.

In a later paper, Born showed that his interpretation allowed the calculation of the probability of a quantum transition between stable states, such as an atomic absorption or emission between electron orbitals. The model of a classical electron moving between

---

[8] We use the modulus square, or the square of the absolute value of the wave function, or the wave function multiplied by its complex conjugate, because the probability depends only on the absolute value of the amplitude and not on its sign (positive or negative) or whether it is real or imaginary (i.e. contains the square root of -1).

stable orbits fails because such an electron would be expected to radiate energy during the transition. Born argued that Schrödinger's interpretation in terms of vibrations in an electromagnetic field did not help, since it could not be used to explain how an electron removed completely from a stable orbit (i.e. ionized) could produce a discrete track in an ionization chamber. He therefore combined wave mechanics with the idea of quantum jumps implied by Bohr's theory of the atom. Born admitted that such a quantum jump[9]:

can hardly be described within the conceptual framework of Bohr's theory, nay, probably in no language which lends itself to visualizability.

Cause and effect was once again threatened by instantaneous quantum jumps.

This, of course, was exactly what Schrödinger had been hoping to avoid. The purpose of his wave mechanics was to restore a classical interpretation for the mechanics of the atom, albeit one of waves rather than particles. To add quantum jumps to this picture simply added insult to injury. In a heated debate between Schrödinger, Bohr, and Heisenberg on the interpretation of quantum theory, an exasperated Schrödinger pleaded with an unyielding Bohr[10]:

You surely must understand, Bohr, that the whole idea of quantum jumps necessarily leads to nonsense . . . If we are still going to have to put up with these damn quantum jumps, I am sorry that I ever had anything to do with quantum theory.

The relation between cause and effect has become deeply ingrained as a result of our everyday experience, and is not something to be given up lightly. Many physicists found Born's interpretation unpalatable. Ironically, Born claimed that he had been influenced by Einstein, and yet in December 1926, Einstein wrote a letter to Born which contains the phrase that has since become symbolic of Einstein's lasting dislike of the element of chance implied by quantum theory[11]:

Quantum mechanics is very impressive. But an inner voice tells me that it is not yet the real thing. The theory produces a good deal but hardly brings us closer to the secret of the Old One. I am at all events convinced that *He* does not play dice.

Matrix mechanics

Before leaving Göttingen to join Niels Bohr in Copenhagen, Werner Heisenberg developed a completely novel approach to quantum theory which became known as *matrix mechanics*. In this theory, physical quantities are represented not by their values as in classical physics, but by arrays of time-dependent complex numbers that represent *relations* between physical quantities. Thus, instead of dealing with the energies of atomic orbitals, matrix mechanics deals with the radiation frequencies and relative intensities of transitions *between* orbitals with different energies. Unlike Schrödinger, Heisenberg was not particularly concerned to find some underlying physical reality—he was simply after a framework through which connections could be made between physical quantities in ways that would

---

[9] Born, M. (1926). *Zeitschrift für Physik*, **40**, 167.

[10] Quotation from Heisenberg, Werner. (1969). *The part and the whole*. Verlag, Munich, reproduced in Moore, Walter (1989). *Schrödinger, life and thought*. Cambridge University Press, Cambridge, p. 227.

[11] Einstein, Albert, letter to Born, Max, 4 December 1926.

fit the known facts. Heisenberg's theory was essentially a mathematical algorithm—plug in the right numbers in the right way, and you got the right answer.

Heisenberg was working in Göttingen alongside Max Born and Born's assistant Pascual Jordan. His starting point was to express the position dependence of an electric field generated by an electron as a mathematical series known as a *Fourier series* and—following Born and Jordan—to connect the strength of transitions between different stationary states of the electron to the squares of the amplitudes of different components of the series. To obtain physically meaningful results from this description, Heisenberg had to make some assumptions about how the amplitudes of different Fourier series should be multiplied together. Born recognized that Heisenberg's rules were actually those of *matrix multiplication*. A matrix is an array of numbers which can take the form of a column, row, rectangle or square. Just as there are rules for combining ordinary numbers in addition, subtraction, multiplication, and division, so there are rules for combining matrices. One very important consequence of the rule for matrix multiplication is that the result can sometimes depend on the order in which two matrices are multiplied together. In other words, the product of two matrices **A** times **B** is not necessarily equal to the product of **B** times **A**. Matrices with this property are said to be *non-commuting*. Obviously, for ordinary numbers A times B is always equal to B times A, and so ordinary numbers always commute.

Some years later, Heisenberg described the discovery as follows[12]:

I replaced the positional coordinates, therefore, with a table of amplitudes that was meant to correspond to the classical Fourier series, and wrote out the classical equation of motion for it, making use as I did so... of the multiplication of amplitude-series that had proved itself in dispersion theory. Only much later did I learn from Born that it was simply a matter here of multiplying matrices, a branch of mathematics that had hitherto remained unknown to me.

Born, Jordan and Heisenberg reformulated Heisenberg's original theory as matrix mechanics. They showed that the matrix versions of position **q** and momentum **p** (called conjugate variables in classical mechanics) were required by the imposition of quantum conditions to be non-commuting matrices, with the difference **pq** − **qp** equal to $ih/2\pi$, where i represents the square root of -1. This approach was very successful—it too explained many of the otherwise inexplicable features of quantum phenomena. In 1926, Wolfgang Pauli showed how the theory could be used to explain the hydrogen-atom emission spectrum.

But now physicists had yet another problem to confront: matrix mechanics and wave mechanics were formulated and presented at about the same time (late 1925 to early 1926), and although the predictions of the theories were the same, the theories themselves were quite different. As George Gamow later wrote[13]:

they started from entirely different physical assumptions, used entirely different mathematical methods, and seemed to have nothing to do with each other.

Which was right?

To a certain extent, the answer to this question was provided by Schrödinger in a paper published in 1926. He demonstrated that matrix mechanics and wave mechanics give

[12] Heisenberg, Werner, manuscript of a lecture intended for delivery in Göttingen, May 1975, subsequently published in 1983, *Encounters with Einstein*. Princeton University Press, Princeton, NJ, p. 45.

[13] Gamov, George (1966). *Thirty years that shook physics*. Doubleday, New York (republished by Dover Publications, New York, 1985), p. 98.

the same results because the two theories are mathematically equivalent: they represent two different ways of addressing the same problem. Of course, Schrödinger argued, wave mechanics was to be preferred because it offered a conceptual basis for understanding the behaviour of quantum particles, which, as a mathematical algorithm, matrix mechanics could never provide. Many physicists tended to agree, although some dissented. For example, although Lorentz expressed a preference for wave mechanics for the description of systems in simple Cartesian coordinates, he struggled with the physical interpretation of quantum waves in systems with a higher number of degrees of freedom, and in these circumstances preferred to use matrix mechanics. However, all physicists were a little uneasy in the knowledge that a theory as important as quantum theory could be expressed in two such radically different ways.

The real connection between matrix mechanics and wave mechanics was made clear by the mathematician John von Neumann in the early 1930s. He showed that wave mechanics could be expressed in an algebra based on *mathematical operators*. Where matrix mechanics depends on the properties of non-commuting matrices, wave mechanics can be derived from the properties of non-commuting operators.

Heisenberg's uncertainty principle

Any introductory course on quantum mechanics will contain a discussion of Heisenberg's famous uncertainty principle. This principle is often presented to students in a matter-of-fact way. Students are told what it is, how it fits into the structure of quantum theory, and how it applies to physical systems. It all seems very neat, but in fact, the uncertainty principle was formulated in the midst of heated arguments about the interpretation of quantum theory. Despite the fact that today we know (or think we know) quite a bit more about the theory and its applications, arguments about the meaning of the uncertainty principle are no less heated and confusing now than they were in 1926.

Towards the end of that year, it had become clear that Schrödinger's views were winning out. Many physicists who expressed a preference opted for wave mechanics because it appeared to offer the best prospects for further interpretation. Bohr and Heisenberg tried hard to persuade Schrödinger of the importance of the idea of quantum jumps but failed. Bohr and Heisenberg did not give up, however. They became more determined than ever to resolve the difficulties of interpretation by taking a radical new approach.

The problem with matrix mechanics lay in its abstract nature. Whereas Born's probabilistic interpretation of Schrödinger's wave function seemed to be at least consistent with the idea of an electron path or trajectory, no such trajectory is defined in matrix mechanics, and Heisenberg firmly resisted any attempt to connect his theory with a specific description. But then, Schrödinger's own interpretation of his wave mechanics was self-contradictory: the motion of a wave packet could not be used to describe the path of an electron because of the tendency of the wave packet to disperse. Anyone who had looked at the track left by an electron in a cloud chamber could be convinced of the reality of the electron's particle-like properties, and yet this was something that Schrödinger's interpretation seemed unable to rationalize in terms of the prevailing understanding of physics.

The situation was very confusing. It was at this point that Bohr and Heisenberg decided to go right back to the drawing board. They began to ask themselves some fairly searching, fundamental questions, such as: What do we actually *mean* when we speak about the *position* of an electron? The track caused by the passage of an electron through a cloud chamber seems real enough—surely it provides an unambiguous measure of the electron's position? But wait: the track is made visible by the condensation of water droplets around atoms that have been ionized by the electron as it passes through the chamber. This process of ionization is a quantum process and therefore subject to the rules, and open to the probabilistic interpretation, of quantum mechanics. According to this interpretation, it is the large number of probabilistic (and hence individually indeterminate) ionizations which allows what seems to be a classical, deterministic path to be made visible.

In 1927, Heisenberg decided that to talk about the position and momentum of any object requires an operational definition in terms of some experiment designed to measure these quantities. To illustrate his reasoning, Heisenberg developed a 'thought' experiment involving a hypothetical gamma-ray microscope. Supposing we wished to measure the path of an electron—its position and velocity (or momentum) as it travels through space. The most direct way of doing this would be to follow the electron's motion using a microscope. Now the resolving power of an optical microscope increases with increasing frequency of radiation (shorter wavelengths), and so a gamma-ray microscope would be necessary to give the spatial resolution required to 'see' an electron. The gamma-ray photons bounce off the electron, and some are collected and used to produce a magnified image.

But now, according to Heisenberg, we have a problem. Gamma rays consist of high-energy photons (remember $\varepsilon = h\nu$), and, as we know from the Compton effect, each time a gamma-ray photon bounces off an electron, the electron is given a severe jolt. This jolt means that the direction of motion and the momentum of the electron are changed in ways that are generally unpredictable. According to Born's interpretation of the wave function of the electron, only the probability for scattering in certain directions and certain momenta can be calculated using quantum mechanics. Although we might be able to obtain a fix on the electron's instantaneous position, the sizeable interaction of the electron with the device we are using to measure its position means that we can say nothing at all about the electron's momentum.

We could use much lower-energy photons in an attempt to avoid this problem and so measure the electron's momentum, but the use of lower-energy (longer wavelength) photons would mean that we lose resolution and must then give up hope of determining the electron's position. Heisenberg reasoned that the exact position and momentum of a quantum particle could not be measured simultaneously. To determine these quantities requires two quite different kinds of measuring apparatus, and the measurement of one property with certainty excludes the simultaneous measurement of the other.

Heisenberg used Born's probabilistic interpretation of the wave function to derive an expression for the 'uncertainties' (actually root-mean-square deviations) of the position and linear momentum of a particle confined to move in one dimension. He found that the product of the uncertainties in position and momentum has a lower limit of $h/4\pi$. Fixing the position of an electron exactly (no uncertainty) implies an infinite uncertainty in the electron's momentum, and vice versa. Extending the same arguments to the measurement of energy and time,[14] Heisenberg found a lower limit for the product of the uncertainties

---

[14] Strictly speaking, this is energy and *lifetime* (see below). Actually the status of the energy-time uncertainty relation in the structure of the theory is somewhat problematic, as the relation can be traced back to the properties of the mathematical operators used to derive measureable quantities from the wave function (see Chapter 4), and there is no operator for time in quantum theory.

of these quantities to be again $h/4\pi$. This is often presented as an energy–time uncertainty relation, but is actually a reworking of the position–momentum uncertainty relation in the context of the time-dependent Schrödinger equation (see Appendix 5). The energy–time relation is usually interpreted in a practical sense to signify that the lifetime of an emission (the amount of time taken for the intensity of the light emitted to decay to some specific proportion of its initial intensity) will be uncertain by an amount related to the uncertainty in its energy. The uncertainty in the lifetime can be translated into an uncertainty in the exact moment of emission of a quantum particle. In other words, the more sharply we can measure (in time) the lifetime and hence the moment of creation of a quantum particle, or follow its passage through an apparatus, the more uncertain will be its energy, and *vice versa*.

Some physicists have argued that the uncertainty principle represents the starting point from which the whole of quantum mechanics can be deduced. It is apparent that Heisenberg himself thought something along these lines. He did not believe that it was necessary to use terms like 'wave' or 'particle' when talking about quantum phenomena and preferred, instead, to continue with the supposition that the theory merely provided a 'consistent mathematical scheme [that] tells us everything which can be observed'. This is a purely 'instrumentalist' approach—the theory (and, in particular, the uncertainty principle) tells us that there are limits on what is *measurable*, and it is impossible to do anything other than speculate on what is not measurable.

Bohr vehemently disagreed with Heisenberg on this point. For him, it was wave–particle duality that lay at the heart of quantum mechanics. All the rest—including the uncertainty principle—were the physical and mathematical consequences of using two diametrically opposed classical concepts, waves and particles, to describe something that was fundamentally non-classical. According to Bohr, quantum theory tells us not what is measurable but what is *knowable*.

There were several aspects of Heisenberg's treatment that Bohr found objectionable. For one thing, the reasoning that Heisenberg had used in his hypothetical gamma-ray microscope experiment was fatally flawed. Heisenberg had almost failed to secure his doctorate at the University of Munich because he had been unable to derive expressions for the resolving power of a microscope, incurring the wrath of one of his examiners, Wilhelm Wien, who had covered all the required background in his lectures.[15] Heisenberg had been mortified by this experience, as had his doctoral supervisor, Arnold Sommerfeld. He was now struggling with the theory all over again. Bohr pointed out that it was not necessary to invoke the Compton effect as an explanation of the origin of uncertainty; it was sufficient that light waves were scattered by the electron into the *finite* aperture of the microscope—a fundamental limitation on the resolving power of all such instruments. But this required that a wave interpretation be adopted for the gamma rays, with the connection between wavelength and momentum being provided by de Broglie's relationship. The uncertainty relation for position and momentum followed immediately from these simple considerations, but Heisenberg was determined to avoid contaminating his argument with the assumption of one description or the other. And besides, waves were identified with his rival, Schrödinger, and he wanted to stress both the corpuscular and

[15] Cassidy, David C. (1992). *Uncertainty*. W. H. Freeman, New York, p. 152.

essentially discontinuous and unpredictable nature of quantum phenomena. Bohr was equally adamant that Heisenberg's paper—by now in press—should be withdrawn.

Bohr put Heisenberg under intolerable pressure—so much so that harsh words were exchanged on all sides, and at one point Heisenberg was reduced to tears. They finally managed to reach a compromise, and in the paper in which Heisenberg presented his uncertainty principle for the first time, he added a footnote containing the statement[16]:

Above all, the uncertainty in our observation does not arise exclusively from the occurrence of discontinuities, but is tied directly to the demand that we ascribe equal validity to the quite different experiments which show up in the corpuscular theory on one hand, and in the wave theory on the other hand.

He also commented on the issue with the gamma-ray microscope experiment and in subsequent presentations based his derivation of the uncertainty relation on arguments based on the resolving power of the microscope combined with de Broglie's relationship.[17] He acknowledged his discussions with Bohr and the importance of the viewpoint that Bohr had developed. This was to become Bohr's notion of *complementarity*.

Heisenberg summarized this period of intense debate as follows:

I remember discussions with Bohr which went through many hours till very late at night and ended almost in despair; and when at the end of the discussion I went alone for a walk in the neighbouring park I repeated to myself again and again the question: Can nature possibly be as absurd as it seemed . . . ?

[16] Heisenberg, Werner. (1927). *Zeitschrift für Physik*, **43**, 172, republished in English translation in Wheeler, J. A. and Zurek, W. H. (eds.) (1983). *Quantum theory and measurement*. Princeton University Press, Princeton, NJ, p. 62.

[17] See, for example, Heisenberg, Werner. (1930). *The physical principles of the quantum theory*. University of Chicago Press, Chicago (republished in 1949 by Dover, New York), p. 21.

3

An absolute wonder

Bohr's theory of the atom did a fine job of explaining the absorption and emission spectra of one-electron atoms in terms of the quantum rules. Schrödinger's wave mechanics did better in the sense that the rules of quantization were given a firm mathematical basis. However, problems still abounded. Experimental spectra revealed quite a number of problems for which wave mechanics (and matrix mechanics) appeared to have no solutions. In particular, some atomic emission lines predicted by the theory were seen to be split in the presence of a magnetic field into several quite distinct lines (a phenomenon known as the Zeeman effect). Neither wave mechanics nor matrix mechanics could explain this phenomenon without some further, largely *ad hoc*, assumptions. Worse still, when the spectra of atoms with two or more electrons were studied, certain predicted lines were found to be *missing*.

It did not take too long to work out that even with the refinements of wave or matrix mechanics, something fundamental was still missing from the quantum description of electrons in atoms. There appeared to be nothing in the description to prevent all the electrons in a multi-electron atom simply collapsing down into the lowest-energy orbit. For neutral atoms, increasing the number of electrons implies an increase in the total positive charge of the nucleus. The greater the nuclear charge, the smaller the radius of the innermost orbit, as the electrons are pulled closer to the nucleus. Now we would expect that the repulsion between increasingly closely packed electrons would tend to resist collapse of the atom into ever-smaller orbits and hence ever-smaller volumes, but it is relatively straightforward to show that electron–electron repulsion cannot prevent heavier atoms from shrinking dramatically in size as the charge of the central nucleus is increased. The repulsion between neighbouring electrons simply is not strong enough to overcome the force of attraction as the charge of the central nucleus is increased. Atomic volumes, easily calculated from atomic weights and densities of the elements, follow a complex variation with atomic charge but they do *not* systematically shrink with increasing charge.

Pauli's exclusion principle and the self-rotating electron

There was no alternative but to introduce an argument that said that each electron orbit can contain only two electrons. As soon as an orbit was 'full' in this sense, the argument went, further electrons had to be inserted into higher-energy orbits. This 'exclusion principle', first stated by the physicist Wolfgang Pauli, was all that was needed to explain the entire periodic table of the elements and the basis for their chemical reactivity. This was a fantastic achievement. But, still, why only *two* electrons per orbit?

Perhaps, argued a young physicist named Ralph de Laer Kronig, this was because the electron exhibited a basic 'two-valuedness' associated with *self-rotation*. In January 1925, Kronig had travelled from the United States, where he had recently completed a Ph.D. at Columbia University, to work with Alfred Landé in Europe. Landé was an expert on the atomic spectroscopy of rare earth elements and deeply engaged in the study of the Zeeman effect. Landé showed him a letter from Pauli describing the problem, and Kronig saw immediately how it might be resolved if the electron was imagined to rotate about its own axis, in much the same way that the earth rotates on its axis as it orbits the sun. A spinning electric charge moving in an electromagnetic field generates a small, local magnetic field and behaves like a tiny bar magnet. This behaviour can be characterized by a *magnetic moment* that can become aligned with, or aligned against, the lines of force of the applied magnetic field, giving two states of different energy. In the absence of this splitting, there is only one state and hence only one line in the atomic emission spectrum. Kronig saw that it was possible to explain the results of the spectroscopic measurements if the angular momentum of self-rotation was assumed to be $\hbar/2$, where $\hbar$ is a shorthand for Planck's constant h divided by 2π. In addition, he required that the ratio of the magnetic moment and the angular momentum due to self-rotation, a characteristic factor known as the 'g-factor' for the electron, g_e, had to have the value 2. This latter requirement was a little curious, as the ratio of the electron orbital magnetic moment to the orbital angular momentum is 1, as predicted by classical mechanics. Pauli, noted both for his talents as a physicist and for his biting wit (Paul Ehrenfest had once called him 'God's whip', a term he later used of himself) completely dismissed Kronig's suggestion, declaring it to be: 'very clever but of course [it] has nothing to do with reality'. Kronig talked to Bohr, Kramers, and Heisenberg, who were similarly dismissive. Kronig subsequently dropped the idea.

Later that year, Samuel Goudsmit and George Uhlenbeck, based in Leiden in the Netherlands, independently reached the same conclusion. Uhlenbeck understood that the three quantum numbers described in the prevailing atomic theory represented quantization within three 'degrees of freedom' of the electron in an atom. A fourth quantum number therefore implied a fourth degree of freedom, which Goudsmit and Uhlenbeck ascribed to electron rotation. They summarized their conclusion in a paper they submitted to the journal *Naturwissenschaften*. After submitting their paper, they talked to Hendrik Lorentz about their proposal, and he advised them that this was simply impossible in classical electron theory. Fearing that they had made a significant error, they scrambled to withdraw their paper before it could be published, but it was too late.[1]

---

[1] Tomonaga, Sin-Itiro (1997). *The story of spin*. Chicago University Press, Chicago, p. 36.

There were other problems with Goudsmit and Uhlenbeck's paper, but it nevertheless caught the attention of the physics community. Despite his initial reservations, Bohr became a strong advocate and may have been the first to use the term 'spin' to describe electron self-rotation. This is a term that has stuck, despite the fact that its meaning in quantum theory is considerably far removed from its classical interpretation. Pauli declared it 'a new Copenhagen heresy', but eventually began to warm to the idea and regretted his earlier dismissal of Kronig's proposal, a dismissal that had arguably prevented Kronig from developing the idea further and from becoming the 'discoverer' of electron spin.[2] Of course, a classical spinning object would not be constrained to only two positions of alignment of the magnetic moment. It was reasoned that this restriction must be somehow due to the quantum nature of the electron. It was just not clear how.

Electron spin

There was, at least, a bonus. Two puzzles, one involving the splitting of an atomic transition into two and the other concerning missing lines in the emission spectra of multi-electron atoms, appeared to be one and the same puzzle. As the English physicist Paul Dirac subsequently proposed, if the electron could be considered to possess two possible spin orientations, this perhaps explained why each atomic orbit could accommodate only two electrons. The two electrons had to be of opposite spin to 'fit' in the same orbit. If electron spin was assumed to be governed by a spin quantum number with only two possible values, Pauli's 'exclusion principle' was the same as saying that no two electrons could possess the same set of four—principal, azimuthal, magnetic, and spin—quantum numbers. An orbit could hold a maximum of two electrons provided their spins were *paired*.

Whilst the concept of a spinning electron was very picturesque, its problems were apparent even in Bohr's old planetary model of the atom. The equator of a tiny, spinning bit of charged matter would necessarily be required to spin around at speeds faster than light—a possibility forbidden by Einstein's special theory of relativity. The problems worsened as soon as Bohr's old model was abandoned for Schrödinger's wave mechanics. Where was electron spin in Schrödinger's wave mechanics of the hydrogen atom? Furthermore, the restriction imposed by quantization was unlike any restriction encountered so far. The restriction on orbital energy and orbital angular momentum expressed through the principal and azimuthal quantum numbers was by now very familiar, but placed no upper bounds on the values of these numbers. In contrast, the electron spin angular momentum was clearly restricted to just two possible values, associated with two values of the spin quantum number. Why this limitation?

A half-way solution was provided by Pauli in 1927. He introduced a new kind of variable into the time-dependent Schrödinger wave equation that produced two coupled differential equations. The variables were called *Pauli spin matrices* and were actually a collection of two-by-two square matrices (one matrix for each Cartesian coordinate), which adequately accounted for two possible orientations for the electron spin. In Pauli's theory, the wave

---

[2] A verse penned some time later summarized the situation: '*Der Kronig hätt' den Spin entdeckt, hätt' Pauli ihn nicht abgeschreckt.*' (Kronig would have discovered the spin if Pauli had not discouraged him.) See Enz, Charles P. (2002). *No time to be brief: a scientific biography of Wolfgang Pauli.* Oxford University Press, Oxford, p. 117.

function was now no longer a simple function but a column matrix with two components, each a wave function representing different possible spin orientations. Pauli was never really satisfied with this approach. He was conscious of the fact that the introduction of spin was still all rather *ad hoc*. Furthermore, the theory still did not meet the requirements of special relativity. It was becoming increasingly apparent that electron spin was somehow closely associated with relativistic quantum effects.

Dirac's theory

At the end of 1926, Heisenberg and Dirac agreed a bet on how soon spin could be understood within the framework of quantum theory. Heisenberg suggested 3 years, Dirac 3 months. Neither was exactly on the mark, but Dirac was closer. A little over a year later, on 2 January 1928, Dirac himself submitted a paper to the *Proceedings of the Royal Society* which set out the correct relativistic quantum theory of the electron. Electron spin emerged naturally from Dirac's theory.

Einstein's special theory of relativity is in many ways all about the correct treatment of time as a kind of fourth dimension, on an equal footing with the three conventional spatial dimensions, x, y, and z. Schrödinger's original time-dependent wave equation is 'unbalanced' in this regard, being a second-order differential equation in the three spatial coordinates but only a first-order differential equation in time (see Appendix 5).

A number of attempts had been made since the appearance of Schrödinger's original papers to develop a proper relativistic version of wave mechanics. In fact, Schrödinger himself had started out with a relativistic version, but it did not produce very satisfactory agreement with experiment, and Schrödinger therefore decided to continue with a non-relativistic alternative, which did. Schrödinger's relativistic equation was independently rediscovered in the spring of 1926 by Oscar Klein and, with some refinements by Walter Gordon, became known as the Klein–Gordon equation. As a second-order differential in time, this achieved the required balance but still suffered from the problems that had led Schrödinger to abandon it in the first place. And, once again, it was impossible to extract electron spin from the Klein–Gordon equation.

Dirac's breakthrough came from the realization that the relativistic equation had to remain a *first-order* differential equation in time. There was by now too much at stake, not least Born's probabilistic interpretation of the wave function, which could continue to apply only if the equation remained first-order in time. The challenge, then, was to restructure the equation in such a way as to make both its spatial and time components first-order differentials. This could easily be done but led to some rather ugly-looking square-root operators which implied a differential equation of infinite order. Dirac needed to find a way to 'linearize' the square-root operators. The answer, when it came, was very similar to the Pauli spin matrices with one exception. Instead of two-by-two matrices, Dirac found that he needed four-by-four matrices and suddenly had twice as many solutions on his hands than he thought he had wanted (see Appendix 6).

The use of matrices with the same form as Pauli's spin matrices meant that the property of electron spin had automatically 'dropped out' of Dirac's relativistic wave equation. The introduction of a four-dimensional space–time into the equations of quantum theory had resulted in a fourth 'degree of freedom' for the electron which, in turn, *demanded* a fourth quantum number, as Goudsmit and Uhlenbeck had reasoned. But whatever it was, the property of electron spin did not correspond in any way to the notion of an electron spinning on its axis. It was a purely relativistic quantum property with no counterpart in classical physics. To see why, it is useful to look at how classical properties can be derived from quantum properties in the limit that Planck's constant h tends to zero. The orbital angular momentum of an electron in an atom is related to the azimuthal quantum number multiplied by h, and there is no restriction or upper bound on the value that the quantum number can take. So, as we reduce h to zero, we can compensate by increasing the quantum number to infinity, with the result that the orbital angular momentum does not disappear but instead tends towards its classical value. However, the same is not true of electron spin. The spin quantum number is constrained to one value for the electron and cannot be increased to infinity as h tends to zero. The property of electron spin therefore disappears for classical objects.

Although its interpretation is obscure, we do know that electron spin produces effects that give rise to a small electron magnetic moment. This moment can become aligned in the direction of an applied magnetic field or against that direction. We have learned to think of these as possibilities as 'spin-up' and 'spin-down'. In a magnetic field, the two possible orientations of the electron's magnetic moment give rise to two energy levels that are characterized by magnetic spin quantum numbers corresponding to the spin-up and spin-down states, and the two levels give rise to two lines in an atomic emission spectrum.

Reworking Dirac's equation to describe an electron moving in an electromagnetic field produced results that reproduced the Zeeman effect, in which the spin magnetic moment is twice the magnitude that would be anticipated on the basis of classical mechanics. As stated above, the magnetic moment associated with the electron's orbital angular momentum is in accordance with classical preconceptions. The difference in this spin versus orbital behaviour was sometimes referred to as the 'magnetic anomaly of the spin' and rationalized on the basis that electron spin has no counterpart in classical mechanics (and should not, therefore, be expected to conform to classical preconceptions). Dirac's theory predicted that the spin magnetic moment is related to the spin quantum number of the electron multiplied by the 'g-factor' for the electron, g_e. In classical mechanics, g_e would be exactly 1. As in Kronig's original proposal, so in Dirac's theory g_e is exactly 2.

The two possible orientations of the electron spin angular momentum account for half of the solutions available from the Dirac equation. The remainder correspond to electron states of negative energy and are an inevitable consequence of using the correct relativistic expression for the total energy of a freely moving particle, in which energy appears as E^2 (see Appendix 6). The temptation of most physicists when faced with negative energy solutions is to dismiss them as 'unphysical' and continue only with the positive-energy solutions. This was fine when considering problems in classical physics, in which energy changes are continuous and systems that start with positive energy cannot suddenly jump to negative energy states (with energy less than or equal to minus mc^2). However, quantum mechanics allowed exactly this kind of sudden, discontinuous jump, and it was perfectly feasible for a positive energy electron to jump to a negative energy state. Dirac had no choice but to take the negative-energy solutions seriously.

The negative-energy solutions had some bizarre properties. Where the 'ordinary', positive-energy electrons would accelerate under an applied force, the particles described by the negative energy solutions actually slowed down the harder they were pushed. Dirac's struggle to rationalize these solutions led him to the realization that negative-energy, negatively charged particles might behave, to all intents and purposes, like positive-energy, positively charged particles. Suppose, he said, that the universe is filled with a 'sea' of negative-energy states all occupied by spin-paired electrons. We would have no way of knowing of the existence of such a sea because, when filled, it would not interact with anything and would merely serve as some kind of backdrop against which positive energy increments would be measured. Indeed, such negative energy states are not really characterized by negative absolute values of energy—they are negative only in relation to an arbitrary energy zero. However, if an electron were to be promoted out of the sea (to become an observable, positive-energy electron) it would leave behind a 'hole'. The negative-energy hole would behave exactly like a positive-energy, positively charged particle.

For reasons of his own, Dirac proposed that the positively charged particle created by the hole in the negative-energy sea was, in fact, a *proton*. The neutron had yet to be discovered, atoms were thought to consist of protons and electrons, and making protons out of 'electron holes' had an nice symmetry about it that fed Dirac's desire to seek a unitary description of the fundamental constituents of matter. The proposal was roundly criticized on all sides because of, among other things, the huge difference between the masses of the proton and the electron. By 1931, Dirac had revised his position and claimed that the positive-energy, positively charged particle created by the hole in the electron sea was a hypothetical *anti-electron*, a positively charged version of the electron with otherwise equivalent properties. When asked many years later by the physicist Murray Gell-Mann why he had not immediately predicted the anti-electron, Dirac replied: 'Pure cowardice'.[3]

The particle did not remain hypothetical for very long. Carl Anderson identified a light, positively charged particle in cosmic-ray experiments conducted in 1932, although he did not recognize it for what it was until 1933, when he threw caution to the winds and claimed that he had found a positive electron, or *positron*, as it came to be called. Though Anderson knew about Dirac's hypothetical anti-electron, he did not immediately identify this as his positron. This was left to the British and Italian physicists Patrick Blackett and Guiseppe Occhialini.

News of Dirac's breakthrough spread quickly. He wrote a letter to Max Born in Göttingen ahead of publication setting out his approach. Léon Rosenfeld described its reception:[4]

It was immediately seen as *the* solution. It was regarded really as an absolute wonder.

Dirac shared the 1933 Nobel prize in physics with Schrödinger and Heisenberg.

Quantum electrodynamics

The burst of creativity through the 10-year interval from 1923 to 1933 had taken quantum theory from an *ad hoc* imposition of quantum rules on to classical models and turned it into a self-consistent structure in which quantum rules arose naturally as the solutions of suitably adapted relativistic wave or matrix equations. John von Neumann completed the task of providing a formal foundation for quantum theory in terms of a series of axioms or postulates, which declared the fundamental nature of the wave functions and the means of extracting from them the values of 'observable' quantities such as position, momentum, and energy. These are described in the next chapter.

[3] Gell-Mann, Murray (1994). *The quark and the jaguar*. Little Brown, London, p. 179.
[4] Kragh, Helge (1990). *Dirac: a scientific biography*. Cambridge University Press, Cambridge, p. 62.

However, despite appearances, all was far from satisfactory. Dirac's breakthrough had created a relativistic wave equation for the electron which predicted the property of electron spin and the existence of the anti-electron. This was indeed a triumph but also quickly became a dead end. Physicists had by now acknowledged that a quantum theory of *fields* was required, starting with a quantum version of Maxwell's electromagnetic field. This was not simply a return to Schrödinger's wave concepts but a recognition that (non-classical) fields were more fundamental than particles and that a proper quantum field description should yield particles as the *quanta of the fields* themselves. It seemed clear that the photon was the quantum of the electromagnetic field, created and destroyed when charged particles interacted. This is known as *second quantization*. The recognition of the wave nature of particles (and the particle nature of waves) is referred to as first quantization, and second quantization is the ability to create and destroy quanta in various types of interaction. The quantum theory of 1933 appeared to provide an adequate description (or, at least, a rationalization) of first quantization but could not yet accommodate second quantization.

The problem was that as soon as interactions were introduced into Dirac's equation, divergent integrals broke out like a rash, creating awkward infinities. Then, new data from cosmic-ray experiments appeared to demonstrate that the theory broke down at high energies.[5] In the midst of all this, there was the long-standing problem of radioactive beta decay, involving the expulsion of a high-velocity electron from a radioactive nucleus and transformation of a neutron into a proton. To ensure energy conservation in this process, Pauli had proposed the physicists' perfect escape clause—a massless, chargeless particle that interacted with virtually nothing—later called a *neutrino*.

Dirac himself became quite dissatisfied with the versions of quantum electrodynamics that he and others were wrestling with, to the point where he rejected his own relativistic quantum theory of the electron for which he had been awarded the Nobel prize. He seized on some (as it turned out, incorrect) experimental results to argue for an abandonment of strict energy conservation and a return to the BKS theory of 1924. Dirac's ideas were in turn rejected as nonsense, and Bohr in particular argued strongly against a return to his old theory. The problems, which possessed an entire generation of quantum physicists in the 1930s, were summarized by the American J. Robert Oppenheimer in a letter to his brother Frank[6]:

As you undoubtedly know, theoretical physics—what with the haunting ghosts of neutrinos, the Copenhagen conviction, against all evidence, that cosmic rays are protons, Born's absolutely unquantizable field theory, the divergence difficulties with the positron, and the utter impossibility of making a rigorous calculation of anything at all—is in a hell of a way.

Quantum field theory was declared to be in crisis. Some promising theorists left the field altogether. Others, such as Einstein and Schrödinger, who had become sceptical of the path that quantum theory was following, were not entirely unhappy that the theory had got itself in a mess. And then, in 1939, a crisis in human affairs overshadowed the lives of all the world's leading physicists.

---

[5] Although it was actually the experiments themselves that were being misinterpreted—the discovery in 1937 of the 'heavy electron', or muon, helped resolve some of the difficulties.

[6] Oppenheimer, J. Robert (1934). Letter to Oppenheimer, Frank. Quoted in Kragh, Helge (1990). *Dirac: a scientific biography*. Cambridge University Press, Cambridge, p. 165.

Shelter Island

The challenge was taken up again in earnest after the end of the Second World War. New experimental data, which seemed at first to make things worse, actually pointed the way towards a solution. The g-factor of the electron was found not to be exactly 2, as required by Dirac's relativistic theory, but to have a slightly larger value, more like 2.00236. This was a small difference, perhaps, but such differences were now well within experimental uncertainty and were providing increasingly critical tests of the theory. And, in a moment of historical repetition, in April 1947, American physicist Willis Lamb announced the results of new, more detailed spectroscopic studies of the hydrogen atom. Where Dirac's theory said that there should be just one line in the microwave spectrum of hydrogen, there were, in fact, two. The hydrogen atomic energy levels $2S_{1/2}$ and $2P_{1/2}$ have the same value of the principal quantum number but different values of the azimuthal quantum number and were predicted in Dirac's theory to have the same energy (such levels are said to be *degenerate*). Because they were predicted to have the same energy, they were expected to produce indistinguishable lines in absorption or emission spectra. However, experiment was now saying these levels are *not* degenerate. One level is shifted relative to the other (called the *Lamb shift*), and two lines can be distinguished in the spectrum.

Against this background of despondency, a small elite group of theoretical physicists convened on 2 June 1947 at a conference on the fundamental problems of quantum mechanics, at Long Island's Shelter Island near New York City. Although the older generation of physicists that had grappled with the theory in its earliest years was not represented, Oppenheimer and von Neumann were present, and there were several from the youngest generation—including David Bohm (who was to have a profound impact on quantum theory in the years to come) and two New York rivals, Julian Schwinger and Richard Feynman. They heard the full details of Lamb's latest experimental results directly from Lamb himself on the first day of the conference. Now here was a challenge. The problems with quantum electrodynamics were, in fact, nothing to do with zeros or infinities. The problems were to do with the very finite g-factor for the electron and the very finite differences between energy levels of the hydrogen atom.

The solution, when it came, was a quantum electrodynamics that appeared stranger than anything that had gone before.

Sum-over-histories

In his excellent lectures on quantum electrodynamics, published in book form with the title *QED: the strange theory of light and matter*, Richard Feynman argued that all the phenomena of light and electrons arise from three deceptively simple actions[7]:

—Action #1: A photon goes from place to place.
—Action #2: An electron goes from place to place.
—Action #3: An electron emits or absorbs a photon.

Though simply stated, even the first of these actions has something of an air of mystery about it, and understanding how (and with what probability) a photon goes from place

---

[7] Feynman, Richard P. (1985). *QED: the strange theory of light and matter*. Penguin, London, p. 85.

to place can take us straight to some of the most fundamental principles of quantum electrodynamics.

Most physics students will have spent some time in the teaching laboratory investigating the properties of light using small projection lamps, slits, mirrors, lenses, and prisms. We learn from geometrical, or ray, optics all about reflection and refraction. We observe that light rays travel in straight lines, and we discover that light does this because the straight lines represent the least amount of time required for light to travel from its source to its destination, a principle first enunciated by Pierre de Fermat (of 'last theorem' fame) in 1657. The principle is so old and well known that, as young science students, we may never have thought to ask how light is supposed to 'know' in advance what the path of least time *is*. We certainly never see light setting off in the wrong direction, realizing this in mid-course and turning around to get back on track.

The use of phrases such as 'setting off' and 'turning around' implies a particulate view of light—bundles of light energy shooting out from the source like tiny projectiles. Here, though, it is the wave nature of light that provides a solution to this particular mystery. Light does not need to know in advance which is the path of least time because it takes *all paths* from its source to its destination. We can picture what happens by reference to the plane electromagnetic wave illustrated in Fig. 1.2 on p. 8. The wave oscillates up and down as it moves forward through space. Clearly, a second wave with the same wavelength, starting out from the same place but heading for a destination just slightly displaced from the destination of the first wave, will look very similar in terms of the time and space dependences of the amplitude. However, as we increase the displacement of the destination of the wave further and further, the time and space dependences of the waves become increasingly 'misaligned': at specific points in time and space, peak no longer lines up with peak, and trough no longer lines up with trough. The result is destructive interference and a loss of *coherence* of the light. Clearly, we get constructive interference and maximum coherence for light paths that do not differ significantly in terms of distance and therefore time. The mystery is now resolved. When light travels through a single medium (such as air), the light paths that do not differ significantly in terms of distance and time are all clustered around the shortest, straight-line path from source to destination, which is also the path of least time.

The principle is one of least time and not least distance. If the light passes from one medium into another at an oblique angle (for example, from air into water), the path of least distance remains the straight line drawn from source to destination. But light slows down in water, and this path of least distance does not minimize the amount of water that light has to travel through to get to the destination. The more water the light has to travel through, the longer it takes. We can identify a 'path of least water' which enters the water vertically and goes straight down to the destination, but this would involve a very roundabout path, lengthening the distance, and hence the time travelled, through the air. The preferred path—that of least time—represents a compromise. It is not the path of least distance (which is why a stick half immersed in water appears bent), nor is it the path of least water.

Feynman elevated these relatively simple physical principles into an alternative formulation of non-relativistic quantum theory equivalent to wave and matrix mechanics. He represented the passage of a quantum particle from one place to another in terms of an integral over all the possible paths the particle could take or, alternatively, in terms of the sum over all possible 'histories' of the particle's motion. This is a determinedly particle representation—the wave aspect of quantum entities is captured in terms of the *phases* of these entities, whatever their origin. Combine the phases and you get interference effects, which look like the effects of wave interference.

Feynman published his 'sum-over-histories' formulation of quantum theory in the journal *Reviews of Modern Physics* in 1948. He had struggled to publish the paper elsewhere, and, as it offered no new results and appeared to offer no new insights, it seems to have been largely ignored at the time.

Feynman diagrams

As 1948 wore on, Julian Schwinger began to make headway with his formulation of quantum electrodynamics, but it was tough going. There was the problem of dealing with all the infinities. The solution was a mathematical trick called *renormalization*, which involved what at first sight seems an absurdity. The infinities in the theory were removed simply by *subtracting* them. If we consider the interactions involved in bringing a free electron into a specific atomic orbit, quantum electrodynamics provides a value for the 'observable' mass of the free electron, which includes a finite mass contribution from the 'bare' electron combined with an infinite *self-mass* associated with the electron interacting with its own electromagnetic field. When the electron is brought into a bound state in an atomic orbit, the theory similarly predicts an observable mass for the bound electron consisting of an infinite self-mass. In mass renormalization, the infinities are subtracted in a way that satisfies the requirements of special relativity, and the result is the *finite* difference between the masses of the free and bound electrons, this difference reflecting a contribution to the binding energy of the electron from the electron mass according to $E = mc^2$. Astonishingly, subtracting the infinities leads not to nonsense but to a finite result that can be compared with experiment.

Schwinger's approach was very formal, somewhat laboured, and lacking any real physical insight, but when theoretical physicists reconvened at a follow-up to the Shelter Island conference, this was accepted to be their best hope. They met on 30 March 1948 in the Pocono mountains in Pennsylvania. This time, both Bohr and Dirac were present. Schwinger's approach was well received; Feynman's more intuitive approach was not. Feynman was only momentarily daunted. With help from a young theoretical physicist called Freeman Dyson, he began to set out his vision for quantum electrodynamics and made a concerted effort to win over the opinions of his colleagues.

The 'sum-over-histories' approach could be used to describe how photons and electrons went from place to place, the first two actions that Feynman believed were necessary for a complete description of light and electrons. Now there was the question of how best to deal with the interactions between electrons and photons, the quanta of the electromagnetic field. These interactions could be introduced into quantum theory as specific terms in the equations. These terms could themselves be assumed to be expansions in a series, the first few terms of which represented the most significant contributions to the interactions, with subsequent terms providing ever smaller corrections. Feynman developed a very singular, appealing, and intuitive diagrammatic way of describing and keeping track of these corrections. These became known as *Feynman diagrams*.

Feynman diagrams have only two axes, a vertical axis for time and a horizontal axis for 'space', effectively reducing any three-dimensional view of a quantum interaction to a single dimension. We can use the diagrams to visualize the paths of quantum entities

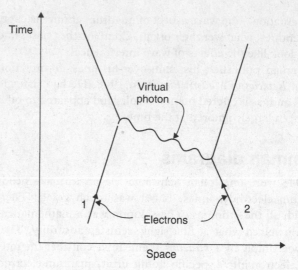

Fig. 3.1 A Feynman diagram illustrating the interaction between two electrons involving exchange of a virtual photon. Reproduced from Feynman, Richard P. (1985). *QED: the strange theory of light and matter*, with the permission of Penguin Books Ltd. and Princeton University Press.

such as electrons and photons in both space and time. Figure 3.1 shows the interaction between two electrons. The electron paths are pictured here as continuous lines. As the electrons move closer to each other, they 'feel' a repulsive electromagnetic force and move apart. This force is carried by a 'virtual photon' as the quantum of the electromagnetic field ('virtual' because we never see it) and is represented in the diagram as a wavy line connecting the electron lines at their point of closest approach. Reading this diagram logically, we might conclude that electron 2 emits a photon which is absorbed some short time later by electron 1. However, the symmetry of the interaction is such that there is in principle nothing to prevent us from concluding that the photon is emitted by electron 1 and travels *backwards in time* to be absorbed by electron 2.

Whilst the diagrams certainly serve as a useful aid to the visualization of the process, they serve primarily as a book-keeping device for all the different ways that quantum entities can interact in going from some initial state to a final state. For each diagram, there is a term in the series expansion containing an amplitude function, which, when its modulus is squared, gives a probability for the process as a contribution to the total probability of conversion from initial to final state. By analogy with his 'sum-over-histories' approach, Feynman reasoned that to calculate correctly the energy associated with the transition from initial to final state, it is necessary to consider *all* the possible routes that the quantum system can take, no matter how seemingly absurd. The most important contribution to the interaction would be the 'direct' transition from initial state to final state, but all the 'indirect' routes represented increasingly smaller corrections to the interaction term and had to be included.

This is clearly seen in the first few Feynman diagrams required for evaluation of the g-factor of the electron. Figure 3.2(a) depicts the interaction of an electron with a photon from a magnet. This is the simplest, most direct interaction involving an electron in some initial state at time/position 1 to some other final state at time/position 2. Evaluating the electron magnetic moment from the interaction term gives the Dirac result, a g-factor of exactly 2, but there are other ways in which this process can happen. Figure 3.2(b) shows the same interaction, but this time, a virtual photon is emitted and absorbed *by the same*

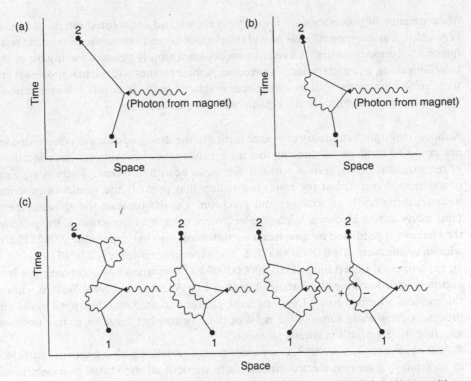

Fig. 3.2 (a) Feynman diagram for the interaction of an electron with a photon from a magnet. Considering this interaction alone would lead to the prediction of a g-factor for the electron of exactly 2, as given by Dirac's theory. (b) Same process described in (a), but now with electron self-interaction depicted as the emission and reabsorption of a virtual photon. Including this process leads to a slight increase in the g-factor for the electron. (c) Further 'higher-order' processes involving emission and reabsorption of two virtual photons. The right-hand side of the diagram shows the spontaneous creation and annihilation of an electron–positron pair. Reproduced from Feynman, Richard P. (1985). *QED: the strange theory of light and matter*, with the permission of Penguin Books Ltd. and Princeton University Press.

electron, viewed either forwards or backwards in time. This represents an electron interacting with its own electromagnetic field, and although the probability of this occurring is small, it is not zero. Including this term in the series expansion makes the g-factor of the electron slightly larger than 2. The first three diagrams in Fig. 3.2(c) represent different possibilities for self-interaction via two virtual photons. The last diagram in Fig. 3.2(c) shows a single virtual photon, as depicted in Fig. 3.2(b), but this time instead of simply being emitted and reabsorbed, the virtual photon first creates an electron–positron pair, which then mutually annihilate to form another photon, which is then absorbed.

Processes such as this last example seemed to imply that the rules of energy conservation had again been thrown out of the window. But it quickly became apparent that creation and destruction of virtual particles were entirely allowed within the constraints of Heisenberg's energy–time uncertainty relation. The energy required to create a photon or an electron–positron pair literally out of nothing could be 'borrowed' so long as it was 'given back' within the time frame dictated by the uncertainty relation. The probabilities of occurrence of processes with evermore complex and convoluted interactions may be very small, but they nonetheless provide important corrections to the interaction terms in quantum electrodynamics. A free electron does not simply persist as a tiny bit of matter travelling along a predetermined Newtonian path; it seethes with virtual processes arising from self-interactions with its own electromagnetic field. It is this self-interaction that causes divergent integrals in the equations leading to infinities.

Although this quick summary does little justice to the theory of quantum electrodynamics, one could be forgiven for thinking that the involvement of virtual processes and the trick of renormalization suggests a solution that is, at best, half-baked. There is an element of madness about it, but the result is a theory that predicts the results of experiments to astonishing levels of accuracy and precision. On the basis of the above corrections (and many others involving 'higher-order', more complex interactions), the g-factor for the electron is predicted by quantum electrodynamics to have the value 2.00231930476, with an uncertainty of ±0.00000000052. The experimental value is 2.00231930482, with an experimental uncertainty of ±0.00000000040.[8] Feynman has compared this level of accuracy with knowing the distance between Los Angeles and New York to within the thickness of a human hair.[9] In essence, the photons created and destroyed in the virtual processes carry away some of the mass of the electron but leave its charge unchanged, affecting the electron's magnetic moment.

And what of the Lamb shift? A very crude way of visualizing what happens to an electron in an atom in quantum electrodynamics is to think of all the virtual processes resulting in a 'wobble' in the electron motion in addition to its orbital and spin motions. This wobble effectively smears the electron further over a small region of space and is most noticeable when the electron occupies an orbital that keeps it (in probabilistic terms) close to the nucleus. The end result is a small energy increase for the $2S_{1/2}$ level in which the electron occupies the spherically symmetric s-orbital, relative to the (otherwise degenerate) $2P_{1/2}$ level, in which the electron occupies the dumb-bell shaped p-orbital.

Feynman published a string of papers through 1948 and 1949. In his biography of Feynman, James Gleick has written[10]:

No aspiring physicist could read these papers without thinking about what space was, what time was, what energy was. Feynman was helping physics live up to the special promise it made to its devotees: that this most fundamental of disciplines would bring them face to face with the primeval questions. Above all, however, to young physicists the diagrams spoke loudest.

Within a few years, Schwinger's algebra had given way to Feynman's diagrams as the preferred approach to quantum electrodynamics. Both approaches could be quickly demonstrated to yield the same results. Schwinger's approach was more formal; Feynman's more intuitive. Perhaps inevitably, Schwinger did not like Feynman's approach because it was much less formal and lacking in rigour. Having stayed for a time at Schwinger's home in

[8] These quantities are subject to constant refinement, both experimental and theoretical. The values quoted here are taken from Dodd, J. E. (1984). *The ideas of particle physics*. Cambridge University Press, Cambridge, p. 32. The value 2.002319304386(20), where the numbers in parentheses represent the uncertainty in the last two digits, is recommended by the CODATA Task Group—see Mills, I. M. *et al.* (1988). *Quantities, units and symbols in physical chemistry*. Blackwell Scientific, Oxford, p. 81.

[9] Feynman, Richard P. (1985). *QED: the strange theory of light and matter*. Penguin, London, p. 7.

[10] Gleick, James (1992). *Genius. Richard Feynman and modern physics*. Little Brown, London, p. 272.

Cambridge, Massachusetts, Murray Gell-Mann liked to report subsequently that he had searched everywhere for Feynman diagrams, but could find none. However, one room had been locked.

Feynman, Schwinger, and Japanese physicist Sin-itiro Tomonaga shared the 1965 Nobel prize for physics for their work on quantum electrodynamics.

Quarks and the standard model

The task of extending quantum theory to other fields and other particles was overtaken in the early 1960s by the discovery that neutrons and protons are not in themselves fundamental particles but are composed of *quarks*. A proton consists of two 'up' quarks, each with a positive charge of $\frac{2}{3}$, and a 'down' quark with a negative charge of $\frac{1}{3}$.[11] The appearance of fractional charges is somewhat disconcerting until you realize that this combination predicts a charge of +1 for the proton, and the quarks and their fractional charges can never be seen in isolation—they are predicted to be permanently trapped (or 'confined') within particles which themselves possess whole number charges. A neutron consists of two 'down' quarks and an 'up' quark and hence has no net charge. The properties 'up' and 'down' (and a third property introduced around this time called 'strange') are actually further quantum numbers and are referred to as *flavours*. The use of a name such as flavour to describe the properties of quarks reflects the fact that there are simply no classical counterparts to these properties and no meaningful analogies from classical dynamics. One name is therefore as good as any other.

Quarks, like electrons, have a spin quantum number of $\frac{1}{2}$. Such particles are collectively called *fermions*. The various possible combinations of quark flavours initially seemed to violate Pauli's exclusion principle, which states that no two fermions can share the same set of quantum numbers. To circumvent this problem, theoreticians proposed that quarks have a further quantum number, which they called *colour*, with values 'red', 'green', and 'blue', and combination rules such that protons and neutrons are 'white'; that is, they consist of one quark of each colour, much like white light can be created by mixing red, green, and blue light in equal proportions.

The quarks are confined inside protons and neutrons by a very strong force, carried by *gluons*. Just as photons are the quanta of the electromagnetic field and are responsible for carrying the electromagnetic forces of attraction or repulsion, so gluons are the quanta of the 'colour field'. There are big differences, however. The photon is massless and chargeless, and has a spin quantum number of 1. Particles with integral spin quantum numbers are collectively called *bosons*. Gluons are also massless bosons with spin quantum numbers of 1 but, because there are three possible quark colours, the gluons are required to carry 'colour charge', and eight different types of gluons are required to account for all the possible interactions between quarks with different colour quantum numbers. This means that the 'colour force' does not disappear over long distances and is the principal reason why quarks and gluons remain confined inside larger particles. The strong nuclear force, responsible for holding neutrons and protons together inside atomic nuclei, is believed to be literally a byproduct of the colour force binding quarks and gluons inside the nucleons

---

[11] This does not necessarily mean that there are only three quarks inside a proton—it may mean that there is a *net surplus* of two quarks with 'up' quantum numbers and one with a 'down' quantum number. See Treiman, Sam (1999). *The odd quantum*. Princeton University Press, Princeton, NJ, p. 226.

themselves.[12] The quantum field theory of coloured quarks and gluons is called *quantum chromodynamics.*

If the colour field is indirectly responsible for the strong nuclear force, what of the forces resulting from quark flavours? These are weaker forces and involve not only quarks but a family of particles collectively called *leptons*—including the electron, the electron neutrino (yes, it does appear to exist![13]), the heavier brothers of the electron, called the muon and the tau, and their neutrinos. All are fermions with flavour (but not colour). Through the late 1950s and early 1960s, Sheldon Glashow, Steven Weinberg, and Abdus Salam developed a quantum field theory, which unified the electromagnetic force and the weak nuclear force responsible for nuclear beta decay. This is sometimes referred to as *quantum flavourdynamics.*

This unified field theory produced many more predictions. The mechanism of beta decay was proposed to involve the transition of a 'down' quark in a neutron into an 'up' quark (thereby turning a neutron into a proton) with the creation of a W particle, the purported carrier of the 'electroweak' force. The W particle then quickly decays into a high-velocity electron (a beta particle) and an anti-neutrino. Unlike the photon, which carries the electromagnetic force over reasonable distances, the range of the W particle is much smaller implying (from the energy–time uncertainty relation) that the particle has considerable mass. It is also required to carry electric charge and can come in two forms, W^+ and W^-. A further, massive neutral particle, called Z^0, was also predicted, as was a fourth quark flavour, called 'charm'. Such was the confidence of the Nobel prize committee in the veracity of the unified theory that Glashow, Weinberg, and Salam were awarded the 1979 prize *before* any of these particles were identified experimentally. Evidence for the W^+, W^-, and Z^0 particles and quark charm has been subsequently obtained.

But it was not quite over yet. The detection of the tau lepton in the mid-1970s could be explained only if there were two further possible flavours for the quarks. These were termed 'top' and 'bottom'. Evidence for the bottom quark was quickly forthcoming, but the top quark remained elusive for a further 20 years before evidence for it was eventually found.

The theory of quarks, leptons, and the field quanta responsible for carrying the forces between them as described by quantum chromodynamics and quantum flavourdynamics is known as the 'standard model'. As it stands, the model has been enormously successful, but much remains to be done. The forces that the different quantum field theories describe are very similar but have defied further unification. The theories appear to require a very large number of particles. Six quark flavours multiplied by three possible colour values gives 18 different types of quark. Add the leptons—the electron, muon, and tau—and their neutrinos, and we have 24 fermions. Then, there are the antiparticles of all these, making 48, to which we need to add the field quanta, the photon, eight different types of gluon and the W^+, W^-, and Z^0 particles, making 60 in total. To this, we should add the Higgs boson—a new, hypothetical particle with a spin quantum number of zero introduced into

[12] Readers familiar with the weak van der Waals forces responsible for some types of interaction between molecules might like to know that the strong nuclear force is, in effect, a van der Waals-type force resulting from the colour force that binds quarks inside nucleons.

[13] The neutrinos may not actually be massless. See Treiman, Sam (1999). *The odd quantum.* Princeton University Press, Princeton, NJ, p. 217.

the standard model to provide fermions and the quanta of the weak interaction with non-zero masses in a way that does not lead to irritating infinities. The Higgs boson makes 61 fundamental particles. This hardly seems the stuff of a fundamental theory. Then, there is the significant number of arbitrary constants, many required to pin down the values of the particle masses. And then, still stuck out in the cold and defying all attempts at unification, there is the force of gravitation, thought to be carried by 'gravitons', the hypothetical quanta of the gravitational field.

Is a more fundamental theory possible? The answer is yes[14]:

The third possibility is that a simple theory underlies the elementary particle system, according to which the number of such particles can be regarded as infinite, with only a finite number accessible to experimental detection at available energies. Superstring theory falls into this category of explanation.

Heterotic superstring theory attempts to describe particles as vibrations in tiny loops of 'string', as an alternative to point masses, thereby avoiding the problems with infinities that plague the traditional quantum theory. Instead of interactions being described by space–time lines in Feynman diagrams, the interactions between loops of string are described by space–time *tubes*, and the sharp changes resulting from interactions between point masses are smoothed out in superstring theory. Needless to say, the loops are much too small to see, but they offer a potential way of accommodating a system of particles and their interactions, which is less fraught with mathematical difficulties.

This breathless romp through about 50 years of development in theoretical physics and experiments with particle accelerators brings us reasonably up to date. But it should also leave us with the clear impression that the fundamental principles of quantum theory are those discovered in the 1920s and 1930s. These have persisted largely unchanged. The structure rests on the concepts of wave–particle duality, quantum jumps—with their implications for the connection between cause and effect—the wave functions and their interpretation, the uncertainty principle and its interpretation, and Pauli's exclusion principle. Although considerable progress was made in the second half of the twentieth century, this has been largely a matter of properly accommodating interactions and second quantization into the theoretical framework. This progress has certainly sharpened the predictive power of the theory, but it has done little to sharpen our understanding of it. All the fundamental problems of interpretation remain.

[14] Gell-Mann, Murray. (1994). *The quark and the jaguar*. Little Brown, London, p. 198.

Part II

Formalism

4

Quantum rules

The development of quantum theory was driven by a consistently repeated failure of classical physics to accommodate the results of ever more sophisticated experiments on the microphysical world of radiation, elementary particles, atoms, and molecules. Each breakthrough in the development of the theory's conceptual structure was followed (sometimes very rapidly) by new experimental data that just did not fit, pointing to the need for yet further refinements. As is typical in such a radical change in the approach to acquiring scientific understanding, old and often unsuitable concepts were forced into service to meet the needs of the new description. Confusion reigned. The major figures in the theory's early development took up radically different positions on the theory's interpretation, and, as we will see, the debate became polarized in ways that remain relevant today.

Some believed that in order to make progress, it was necessary to throw out much of the conceptual baggage that early quantum theory carried around with it and re-establish the theory on a much firmer mathematical foundation. It was at this critical stage, perhaps, that the search for deeper insights into the underlying physical reality was set aside in favour of mathematical expediency. The pragmatism and instrumentalism typical of the younger generation of theoreticians involved in the theory's early development, such as Heisenberg, Dirac, and von Neumann, called for a coherent mathematical framework which *worked*. To these physicists, it did not matter too much that the deeper meaning of the theory's concepts appeared to become increasingly disconnected from the reality that the theory was trying to describe. And yet, reality itself, it could be argued, had taken a profoundly bizarre turn.

Schrödinger's wave mechanics and Heisenberg, Born, and Jordan's matrix mechanics pointed to a structure with non-commutation as one of its central concepts, arising from either mathematical operators or matrices. That it tends to be wave mechanics rather than matrix mechanics that predominates in today's university teaching programmes is entirely due to our relative familiarity with ordinary functions and their mathematical operators, our relative unfamiliarity with the methods of matrix algebra, and the problems associated with handling matrices of infinite dimensions. Schrödinger himself demonstrated the equivalence of the two theories and noted that the 'observable' quantities that entered the equations of classical physics directly had been replaced in quantum physics by mathematical operations required to extract the values of observables from his mysterious wave function.

The heated arguments about the nature and interpretation of the quantum wave functions and the need to leave behind the misconceptions of classical wave motion hinted at the need for a new structure that did not promote this tendency. In his *Principles of quantum mechanics*, first published in 1930, Dirac set out a powerful new theoretical structure designed deliberately to look nothing like its intellectual predecessors. But this was criticized by the mathematician John von Neumann for its lack of strict mathematical rigour. It was von Neumann who, in his seminal work *Mathematical foundations of quantum mechanics*, first published in Berlin in 1932, went on to provide a mathematically unassailable foundation for the new quantum physics.

In 1926, von Neumann had attended a lecture on matrix mechanics given by Heisenberg in Göttingen. The great mathematician David Hilbert had also been in the audience and had expressed personal difficulties in comprehending the significance of matrix mechanics and its relationships with Schrödinger's wave mechanics. Von Neumann volunteered an explanation in a language that Hilbert could understand. Whilst the matrix and wave-mechanical formalisms could be connected by a set of rules within a general 'transformation theory', von Neumann believed that this equivalence belied a deeper, more fundamental mathematical framework. The problem was that the two quantum descriptions were so very different in their approaches. Matrix mechanics was a theory in a 'space' of index values identifying the locations of elements within the theory's abstract 'tables of numbers'. Wave mechanics was a theory in an abstract multidimensional 'configuration space'. The theories could not be brought together easily. Von Neumann wrote[1]:

That this cannot be achieved without some violence to the formalism and to mathematics is not surprising. The spaces . . . are in reality very different, and every attempt to relate the two must run into great difficulties.

However, von Neumann argued, the spaces themselves were not so important. It is the relation between functions in the two spaces that is essential for the physical interpretation of the theory's predictions. Von Neumann showed that these functions are *isomorphic*—it is possible to establish a direct, one-to-one correspondence between them:

they are identical in their intrinsic structure (they realise the same abstract properties in different mathematical forms)—and since they . . . are the real analytical substrate of the matrix and wave theories, this isomorphism means that the two theories must always yield the same numerical results.

The matrices act in much the same way as vectors in classical physics, except that where classical vectors operate in 'everyday', three-dimensional Euclidean space, the system of functions in matrix mechanics operate in *Hilbert space*, named for David Hilbert. As Dirac had done in his *Principles*, so Von Neumann identified 'abstract Hilbert space' to be the common denominator—the space in which the functions and operators of both matrix and wave mechanics belong—and used the mathematical structure of Hilbert space itself as the foundation for a unique formalism for quantum theory. The paper that von Neumann

[1] Von Neumann, John (1955). *Mathematical foundations of quantum mechanics*, Princeton University Press, Princeton, NJ, p. 28.

drafted in 1926 to explain the relationships between matrix and wave mechanics to Hilbert was subsequently to become the *Mathematical foundations of quantum mechanics*.

Much of the material covered in this and the next chapter is concerned with quantum theory's mathematical structure. It is included principally to set us up for our detailed examination of recent experimental explorations of the theory's integrity and interpretation in Part IV. However, even here, it is not possible to talk about the theory without contaminating the discussion with problems of interpretation. That this theory is the best description of the microphysical world is not in dispute. It is what the theory means— or does not mean—in terms of the relationship between its concepts and an underlying physical reality that continues to cause so much argument and debate. In the account that follows, the concepts of the theory are presented within the 'traditional' interpretation usually given by design or default to modern undergraduate students of physical science. This interpretation is not necessarily accepted by every physical scientist.

The axiomatization of physics

A second key ingredient of the formalism can be traced back to the influence of David Hilbert. In a lecture delivered to the International Congress of Mathematicians in Paris in 1900, Hilbert outlined a long list of key problems that he believed would occupy the 'leading mathematical spirits of coming generations'. The list has become known as *Hilbert's problems*.[2] The sixth of these concerns the mathematical treatment of the axioms of physics. Drawing on the foundations of geometry as an example, Hilbert stated that an important goal for future mathematicians would be 'to treat in the same manner, by means of axioms, those physical sciences in which mathematics plays an important part...' He went on to say:

If geometry is to serve as a model for the treatment of physical axioms, we shall try first by a small number of axioms to include as large a class as possible of physical phenomena, and then by adjoining new axioms to arrive gradually at the more special theories... The mathematician will have also to take account not only of those theories coming near to reality, but also, as in geometry, of all logically possible theories. He must be always alert to obtain a complete survey of all conclusions derivable from the system of axioms assumed.

Axioms are self-evident truths that are not in themselves in need of further proof, and represent the foundation stones of the mathematical structure derived from them. The introduction of the axiomatic method by Hilbert and his disciples represented an almost pathological drive to eliminate any form of intuitive reasoning from mathematics, arguing that the subject was far too important for its truths to be anything other than 'hard-wired' from its axioms through to its theorems. The drive for mathematical rigour and formality inevitably resulted in a disconcerting increase in obscure symbolism and mathematical abstraction contributing, according to the philosopher Roland Omnès, to the 'fracture': the near impossibility for anyone of average intelligence but without formal training in mathematics or logic to fully comprehend aspects of modern fundamental science.[3]

A similar axiomatization of quantum theory created a number of 'quantum rules' on which the formalism was to be constructed. However, these rules are hardly self-evident

---

[2] Hilbert, David (1902). *Bulletin of the American Mathematical Society*, **8**, pp. 437–479 (English translation by Newson, Mary Winton).

[3] Omnès, Roland (1999). *Quantum philosophy*. Princeton University Press, Princeton, NJ, p. 81.

truths. We can accept them as axioms if we adopt a modern interpretation of the term, as propositions belonging to a formal language[4] which are assumed to be true by hypothesis. They are also often referred to as *postulates*—statements of the rules that must be accepted at face value and without proof (and often without question) and which are subsequently justified through agreement between prediction and experiment. We will use the term postulate here.

Vector spaces

By any set of measures, John von Neumann was an extraordinary individual. He mastered calculus at the age of 8, and was reading professional mathematics books at age 12.[5] Discouraged from pursuing a career in mathematics by his father (who felt that it was not a subject from which a decent living could be made), he studied chemical engineering at the University of Budapest, receiving a diploma in 1926. This was all a diversionary tactic, however. He would leave to study mathematics in Berlin at the beginning of each term, returning at the end to take examinations. Whilst in Berlin, he would often journey to Göttingen to study mathematics under David Hilbert. From Hilbert, he learned about the axiomatic method, which he applied to set theory in his doctoral dissertation. Following periods at the Universities of Berlin and Hamburg, where he worked mainly on quantum physics, he moved to Princeton in 1930 and joined Einstein as a professor at the newly formed Institute for Advanced Study in 1933. During the Second World War, he worked on the Manhattan project, contributing crucial insights on the implosion mechanism as a means of creating a critical mass of fissionable material, thereby triggering a nuclear chain reaction and an atomic explosion. He became a post-war 'hawk', arguing stridently for the development of the hydrogen bomb and, purportedly, serving as one of the models for Stanley Kubrick's 1963 film, *Dr Strangelove*, memorably portrayed by Peter Sellers.[6] A vocal conservative, he never really got along with Einstein, whom he tended to regard as politically naive. With Oskar Morgenstern, he created game theory in 1944, with the publication of *Theory of games and economic behaviour*, and went on to make many important contributions to the theory of computers.

Towards the end of the 1920s, the mathematical principles of Schrödinger's wave mechanics were reasonably well established. A quantum system had to be described in terms of a wave function which somehow 'contained' all the information about its physical state—position, momentum, energy, etc. This information could be extracted from the wave function by applying an appropriate mathematical operator. According to Max Born, the wave functions represent probability 'amplitudes', and the modulus-square of the wave

---

[4] A formal language is a set of symbols and a set of rules for combining the symbols to form proposition statements.

[5] Nasar, Sylvia (1998). *A beautiful mind*. Faber & Faber, London. Though a biography of the mathematician John Nash, chapter 7 is an excellent biographical sketch of von Neumann.

[6] In 1955, von Neumann was diagnosed to have bone cancer and was confined permanently to a wheelchair in January 1956. Although there are many parallels between the character of Dr Strangelove and von Neumann, in truth the character is likely to be a composite of several individuals. Werner von Braun and Edward Teller have also been suggested, and Peter Sellers said that he had based his interpretation of the character's idiosyncrasies on Henry Kissinger. See Poundstone, William (1992). *Prisoner's dilemma: John von Neumann, game theory and the puzzle of the bomb*. Random House, New York, p. 190.

function at some point in space or at some energy therefore gives a probability 'density' for the quantum particle in a particular physical state.

Schrödinger had tried hard to hang on to some semblance of classical continuity and determinism by claiming that the wave functions described an essentially wave-like underlying reality, with particle-like properties appearing as the result of quantum wave-packet states. But by now, it had become apparent that the wave functions themselves were functions with many dimensions that far exceeded those of ordinary space. A system consisting of a single particle could be described in terms of three spatial coordinates but a two-particle system demanded a six-dimensional configuration space. Aside from the problem of having to deal with an abstract, multi-dimensional space, there was the further issue of dealing with the 'dimensions' associated with electron spin, described by wave functions that had now become column matrices rather than simple algebraic functions and which have no classical counterpart. Add to this the fact that the wave functions are complex, that is, they are functions containing the square root of -1, and it was clear that the formalism was going to be an abstract one.

Von Neumann recognized that many of the mathematical properties of the functions in both matrix and wave mechanics were reproduced in the theory of *vector spaces*, and this is the structure taught today to more advanced students.[7] This theory frees us from the mental prison of 'ordinary' three-dimensional space and allows us to define a space with whatever number and nature of dimensions we need. By definition, these dimensions are not constrained to be visualizable in a Cartesian frame. A vector space is defined by the laws chosen for combining vectors and for multiplying them by scalar (magnitude-only) quantities.[8] The combining laws chosen to create a quantum theory mirror those of the wave functions of standard wave mechanics but cannot be derived from within the earlier theoretical structures: rather, they are assumed and later proved by agreement between prediction and experiment. The vector space can be complex, yet it provides a very simple, straightforward, and intuitive connection with quantum probabilities. All the manipulations required to yield the values of observable quantities in matrix and wave mechanics can be reproduced in a quantum theory based on vector spaces.

Despite the apparent flexibility achieved through the introduction of these concepts, the requirements of quantum theory and the connection with real quantities measurable in the real world must still be satisfied. This is particularly important in situations where an infinite number of dimensions is required. The results of calculations involving an infinite series of terms had to *converge*. Vector spaces with the right kinds of convergence properties are singled out for special attention and are referred to as Hilbert spaces.

Whilst it might be difficult to grasp the significance of complex vectors in a peculiar, non-Euclidean space, it is relatively easy to understand their basic properties by looking back at the classical vectors of Newtonian mechanics. Classical vectors are special cases of a more general structure in the sense that their magnitudes are always real, and they are constrained to directions in Euclidean space. We think of a vector, **v**, with a specific magnitude (which we represent as the length of its 'arrow') and direction in three-dimensional space.

[7] In fact, the formal structures and concepts of wave mechanics, matrix mechanics, Dirac's theory, and vector spaces tend to be used somewhat interchangeably, emphasizing that, even in rigorous application, there is still considerable freedom of choice. More sophisticated treatments use yet another framework, called C^* algebra.

[8] See Isham, Chris J. (1995). *Lectures on quantum theory*. Imperial College Press, London, p. 22.

We know we can resolve this vector into its components along the x, y, and z coordinates. These components can be written in terms of the 'projection' of $\mathbf{v}$ onto the x-axis (say), v_x, which depends on the angle between $\mathbf{v}$ and the x-axis, multiplied by the unit vector along the x-axis, traditionally given the symbol $\mathbf{i}$. Because Euclidean space is three-dimensional, we need only specify three unit vectors, one for each direction, to provide a complete *representation* of any classical vector in terms of its components.

The properties of the classical unit vectors are linked to simple geometrical considerations, and yet, as we will see below, they provide a powerful analogy with the mathematical requirements of the quantum description. The relative simplicity and elegance of the formalism are, however, bought at a price.

Quantum states

In this representation of quantum theory, the wave functions of wave mechanics are replaced by *state vectors*. We will reach immediately for the extremely elegant shorthand notation devised by Dirac in 1939[9] which not only simplifies the description but also allows significant additional mathematical insight. The inner (or 'dot', or scalar) product of two classical vectors $\mathbf{m}$ and $\mathbf{n}$ would normally be written as $(\mathbf{m}, \mathbf{n})$ or $(\mathbf{m} \cdot \mathbf{n})$. In the Dirac notation, the equivalent inner product or 'overlap' function for two state vectors is written as a *bracket*—$\langle m|n \rangle$—where the symbols m and n denote state vectors with different sets of quantum numbers. Dirac split this symbol up and treated the two halves as independent entities, defining the state vector as a *ket*—$|n\rangle$, which has all the properties and significance we have so far invested in the concept of a wave function. The ket is combined with a *bra*, $\langle m|$, to create a bracket.[10] This brings us to the first of four postulates of quantum theory:

Postulate 1.   The state of a quantum mechanical system is *completely* defined by the state vector $|n\rangle$.

This is tantamount to saying that anything of relevance or significance in the description of the physical state of a quantum system is assumed to be contained in the mathematical structure of its state vector. There is one important proviso. Quantum theory is probabilistic in nature, and much of the information contained in the state vector is available only in the form of a probability density, derived from the modulus-square of the state vector, $\||n\rangle\|^2$. We use the modulus-square rather than the square of the state vector as the structure of the vector may itself be complex, and yet the probability measure must be a real quantity if it is to represent a prediction for the magnitude of an observable property of the system. For example, if we wished to know the probability of finding a quantum particle at a specific location in space, it is necessary to multiply the probability density

---

[9] Dirac introduced this notation in 1939, but he used it himself only some years later, and it did not receive wider acknowledgement until the publication of the third edition of Dirac's *Principles of quantum mechanics* in 1947.

[10] The bra can often be written as the complex conjugate of the ket but should not be misinterpreted as equivalent to the complex conjugate of a simple wave function. In fact, the bracket in Dirac's notation replaces the wave-mechanical *integral* over all dimensions of the configuration space of a wave function multiplied by its complex conjugate. Such integrals appear frequently in wave mechanics, and the direct analogy with the inner or dot products of vectors demonstrates the flexibility of choice of mathematical description. For a strict definition of a bra, see Isham, Chris J. (1995). *Lectures on quantum theory*. Imperial College Press, London, p. 35.

calculated from the state vector by the volume element to which we refer (much as we would calculate the mass of some material by multiplying its mass density by its volume). Moreover, the probabilities themselves refer to individual quantum particles, and it is therefore important that these sum to unity (there is only one quantum particle, and it must be somewhere). A condition of *normalization* is therefore also usually imposed such that the inner product of the state vector $|n\rangle$ with itself, $\langle n|n\rangle$, is equal to 1.

It is very important to note that most practising physicists seek to preserve their sanity by thinking of the above only in terms of *repeated measurements on a series of identically prepared quantum systems*, meaning quantum systems in the same *state*. Such a multiplicity of identically prepared quantum particles will tend to exhibit a spread of probabilities. Some particles will have a high probability of being found 'here', and others may have a high probability of being found 'over there'. After many measurements, the map of measured probability density over spatial coordinates therefore reflects the variation of $\||n\rangle|^2$ through space and can be thought of as a map of *relative frequencies* with which the particles are found in specific locations. For these physicists, observable quantities have meaning *only* in the context of a series of identically prepared systems (sometimes called an *ensemble*), and it is, for them, meaningless to talk about the observables of *individual* quantum states. However, this view imposes an interpretation on the theory that is certainly not an inherent feature of the theory's structure. We will be returning often to this point.

Operators and observables

We can see from Appendix 5 that the Schrödinger wave equation can be 'derived' from the equations of classical wave motion using the de Broglie relationship. The same result can be obtained by replacing the momentum, p, in the classical equation for the total (kinetic plus potential) energy of a system with its quantum-mechanical operator equivalent. In quantum theory, therefore, physical quantities no longer appear directly in the theory. Rather, the theory is a recipe for extracting a probability that a system will yield a certain value of a physical quantity when the system undergoes certain manipulations that can be expressed theoretically in terms of the manipulations of an appropriate mathematical operator. This places special emphasis on the role of *measurement* in quantum theory, and to reflect this point, physical quantities are often referred to as observable quantities, or just observables. This brings us to the second postulate:

Postulate 2.   Observables are represented in quantum theory by mathematical operators that act on the relevant Hilbert space.

It would seem that every observable in quantum theory should therefore have a corresponding operator, and conversely, every operator should correspond to an observable. There are, however, some qualifications. If we regard the mass of a quantum particle as an observable, it should be admitted that there is no operator for mass in the conventional quantum theory taught to most undergraduate students of physical science. Mass is therefore sometimes referred to as a *structural* quantity, but in truth, more sophisticated quantum treatments do include an operator for mass, which on application yields the masses of the elementary particles.

Taking the converse, it may well be that all operators that can be admitted in the theory do correspond to observables of some form or another, but in practice, for many of the operators, it is difficult to conceive the apparatus that would be required actually to perform such measurements. Most calculations of interest therefore use the operators required to yield 'standard' observables such as position, momentum (both linear and angular), and energy.

Once again, the need to connect the theory with real observables in the real world places constraints on the structure of the operators themselves. The operators must be linear, they must conform to certain rules of combination, and they must be of a type known as *self-adjoint* or *hermitian* (see Appendix 7). This last requirement is of fundamental importance as the eigenvalues of a self-adjoint operator are exclusively real numbers, and only real numbers can represent the values of observables, which brings us neatly to the third postulate:

Postulate 3.   The mean value of an observable is given by the expectation value of its corresponding operator.

The expectation value has a functional form that mirrors an equivalent quantity in probability calculus and emphasizes once again the probabilistic nature of quantum theory. Basically, it implies that it is pointless to assume that values of observable quantities can be somehow assigned to, or are inherent in, a specific quantum state. Instead, we tend to interpret the results of making repeated measurements on a series of identically prepared quantum particles as yielding a spread of values concentrated around the mean value of the observable. And, as the expectation value in probability calculus provides a true mean value only when we make an infinite number of measurements, we can expect that the experimental value will only approximate the mean value predicted by the theory.

The theory is not all about averages, however. If the state vector $|n\rangle$ is an *eigenstate* of the operator in question, the expectation value of the operator is *exactly* the corresponding eigenvalue (see Appendix 7). In these circumstances, it is possible to be very specific about the value of the observable.

Complementary observables

Suppose I throw an apple into the air, and you photograph it as it falls to the ground using a camera fitted with a rapid autowind facility. You take a sequence of photographs at fixed time intervals as the apple falls. Each photograph is a record of the position of the apple (its height above the ground, x) at a particular time in its motion. We could analyse the sequence of photographs to find the values of the apple's position and velocity as a function of time. A separate measurement of the apple's mass allows us to calculate the linear momentum, p_x (mass times velocity) of the apple in the x-direction.

Further suppose that, for some obscure reason, we need to calculate the product of the apple's position and momentum. I choose to determine the product by multiplying position by momentum. You choose to multiply momentum by position. No matter. We know that for ordinary numbers and their units xp_x is equal to $p_x x$, and so the results should be the same. In other words, the classical physical quantities x and p_x *commute*, and xp_x minus $p_x x$ is equal to zero.

This might seem trivial, but as the discoverers of quantum theory found in the 1920s, experiment demands that, for quantum particles, it *does* matter what order you multiply these quantities together, and xp_x is *not* equal to $p_x x$. Clearly, these quantities can no longer be represented by ordinary numbers. Either they are matrices (matrix mechanics), or they are operators (wave mechanics or quantum theory based on vector spaces). In the former case, the rules of matrix multiplication are such that the order of multiplication is important. In the latter case, it is the order in which operations are carried out on a function that can be important. We achieve correspondence between the predictions of quantum theory and the results of experiments if xp_x minus $p_x x$ is assumed to be equal to $i\hbar$ where i represents the square root of -1 and $\hbar$ is Planck's constant divided by 2π. This is known as the quantum-mechanical position-momentum *commutation relation*. Because Planck's constant h is very small, measurements made on macroscopic objects such as apples will never reveal behaviour that contradicts our 'common-sense' classical presumption that xp_x and $p_x x$ commute. However, for microscopic objects like electrons, the magnitude of Planck's constant becomes extremely significant. The above considerations demonstrate, once again, that classical mechanics can be recovered from quantum mechanics in the limit that h is approximated to zero.

Observables whose operators do not commute are said to be *complementary*, and there is a direct link between the property of non-commutation and the uncertainty principle. We can measure one or the other complementary observable with arbitrarily high precision but not both simultaneously (see Appendix 8). As Heisenberg discovered, position and linear momentum are complementary observables. However, the uncertainty relation for position and momentum should not be taken to imply that we cannot measure either observable with arbitrary precision. The relation says that we cannot measure both observables *simultaneously* with arbitrary precision. If we set up an experiment to measure *exactly* the position of a quantum particle, the uncertainty relation implies that the momentum will have an infinite uncertainty, and we can therefore say nothing meaningful about the particle's momentum. Conversely, exact measurement of momentum implies an infinite uncertainty in position.

We need to be a little careful here over our use of the word 'simultaneous'. If we adopt an interpretation based on repeated measurements on a series of identically prepared quantum systems, the uncertainty relation refers to the root-mean-square deviations of the measured values around the mean and does not say anything about our ability to measure these quantities simultaneously. Heisenberg's original interpretation was based on something more fundamental than this: that there is a limit to what we can measure because of the 'clumsiness' of our macroscopic experimental apparatus in relation to individual microscopic quantum particles. Bohr argued instead that there is a fundamental limitation on what is *knowable*, which arises through our use of 'complementary' classical descriptions—waves and particles—and that to measure quantities such as position and momentum, two completely different types of apparatus are required. Measurement of one complementary observable *precludes* simultaneous measurement of the other.

The time evolution of state vectors

The last piece of foundation architecture in the formalism of quantum theory concerns the time evolution of quantum states, summarized in the fourth postulate:

Postulate 4. In a 'closed' system with no external influences, the state vector will evolve in time according to the time-dependent Schrödinger equation.

Recall that the time-dependent Schrödinger equation can be loosely derived by imposing quantum conditions (notably, $E = h\nu$) on the classical equations of wave motion (see

Appendix 5). There is nothing in this procedure that makes the result any less continuous or deterministic than the classical equations themselves. When not subject to an external influence, the time evolution of a quantum state is entirely continuous and deterministic. Once in a state described by some generalized state vector, which we denote as $|\psi\rangle$, the system will evolve continuously in time. The time-dependent form of Schrödinger's wave equation cannot describe the instantaneous, discontinuous, indeterministic transition from one state to another that we call a quantum 'jump'.

In order to describe transitions between quantum states, Max Born had been forced to *combine* the deterministic equations of Schrödinger's wave mechanics with the indeterministic quantum jumps demanded by Bohr's atomic theory. The latter *cannot* be derived from the former. More on this later.

The expansion theorem

Although the language of a quantum theory based on vector spaces no longer contains explicit references to the classical concepts of waves or particles, much of the mathematical apparatus required to make the theory work effectively has obvious parallels in the classical structures. This should not be surprising. Modern quantum theory was created from the fusion of classical wave and particle concepts. It may have been recast in a more appropriate formalism to avoid difficulties of interpretation arising from the use of classical concepts in an inherently non-classical domain, but that formalism nevertheless retains many characteristics inherited from its classical progenitors.

This means that the formalism carries over one of the most powerful features of functional analysis from the world of classical wave forms. This is *Fourier analysis*, named for the French mathematician Joseph Fourier, who was the first to state in 1807 that a completely arbitrary function can be described over a given interval by an infinite sum of simple sine and cosine functions. Before Fourier, functional analysis had focused on the properties of infinite power series—polynomials containing terms of increasing power but whose ever-diminishing coefficients could force the series to convergence. Fourier's approach had immediate application in the analysis of musical notes in terms of a series of harmonics, a hint that this was an approach that was to have very important applications in physics generally.

We can take any arbitrary wave function (such as a 'saw-tooth' function, for example) and decompose it into an infinite series of sine or cosine functions, each function possessing a higher frequency with coefficients (or amplitudes) that describe how much of each simple function we need to mix together to recreate the arbitrary function. This procedure works even in situations where the arbitrary function has 'sharp edges' or discontinuities, although we pay a penalty here as it is obviously impractical to calculate an infinite sum, and we are therefore often forced to *truncate* the series at some point. The consequence of this procedure is a result called the *Gibbs phenomenon*—the calculated function tends to deviate from the function we are trying to recreate both before and after the discontinuity.

The simple sine and cosine functions used in Fourier analysis are examples of *basis functions*, and the series required to recreate any arbitrary wave function is called the *basis set*. The simple functions are inherently orthogonal.[11] Another way of looking at this kind of analysis is to say that the wave function or state vector we are seeking to model is

---

[11] This means that if $|m\rangle$ and $|n\rangle$ are basis functions, the product $\langle m|n\rangle$ is equal to zero. Functions that are both orthogonal and normalised are said to be orthonormal.

a function in a particular space. The simple sine and cosine functions serve as suitable basis functions in this space and can therefore be used to recreate the state vector by computing the coefficients or amplitudes of each term in the series. This is a theorem of classical vibrations and waves but one that can be extended to form a powerful theorem of quantum state vectors. In quantum theory, the expansion theorem states that:

The expansion theorem: Any arbitrary state vector $|\psi\rangle$ can be expanded as a linear combination (sometimes called a superposition) of basis vectors. The basis vectors are formed from the complete set of eigenstates of any self-adjoint operator operating in the relevant Hilbert space.

Instead of simple sine and cosine functions, the basis set becomes the set of eigenstates of a specific operator and can be chosen to recreate the arbitrary state vector $|\psi\rangle$ depending on what kind of information we are trying to extract from it. The arbitrary state vector must be 'well behaved' in the sense that it should closely resemble the eigenstates to be used to model it and should conform to the same set of boundary conditions. The basis set should be 'complete' in the sense that all of the eigenstates of the operator are included. Care is required when dealing with operators in an infinite-dimensional vector space (with an infinite number of eigenstates) to ensure that the series has the appropriate convergence properties. As we have already stated, these convergence requirements form the definition of the particular vector space as a Hilbert space.[12] Appendix 9 provides an example of the use of the expansion theorem in the very simple case where only two eigenstates are required.

The expansion theorem is extremely important. Almost any problem in quantum mechanics for which the functional form of the state vector cannot be easily deduced can be solved in principle by expanding the state vector as a linear combination of eigenstates of the operator corresponding to the property in which we are interested (often the total energy).

We are completely free to choose whatever set of eigenstates we like, but it makes sense to choose ones that bear some resemblance to the problem we are trying to solve. For example, the state vectors of electrons in molecules can be modelled using a basis of atomic states, and the atomic states themselves can be modelled using simple Gaussian functions. Unfortunately, because we need all of the eigenstates to form a complete set, including so-called continuum states associated with ionization, it is very difficult to reproduce the state vector of interest exactly. However, if we are happy to accept a small truncation error, a judicious choice of basis will mean that we can get away with a much smaller number of basis vectors.

One of the charges levelled at modern quantum theory concerns its general lack of visualizability. It is certainly true that in the hands of the more mathematically inclined physicists, quantum theory evolved a language that appears to have become somewhat disconnected from the objects of study in the physical world. However, some of the abstract mathematics, particularly where the language of wave functions is applied, can be brought into focus through quantum-mechanical phenomena that *can* be visualized not only in pictures but in computer-generated animations.[13] The charge of lack of visualizability is poorly founded in this sense and no more than a distraction from the real issues which, as we will see, concern quantum measurement and interpretation.

---

[12] See Isham, Chris J. (1995). *Lectures on quantum theory*. Imperial College Press, London, p. 41.
[13] See, for example, Thaller, Bernd (2000). *Visual quantum mechanics*. Springer-Verlag, New York.

Projection amplitudes

The contribution that each eigenstate makes to an arbitrary state vector $|\psi\rangle$ is represented by the *expansion coefficient* for that eigenstate. Through some very simple manipulations, it is possible to show that these expansion coefficients are the *projections* of the arbitrary state vector onto the individual basis vectors. Thus, if the eigenstate $|n\rangle$ represents a suitable basis vector for the state $|\psi\rangle$, the corresponding coefficient in the linear expansion of $|\psi\rangle$ is given by $\langle n|\psi\rangle$ (see Appendix 9). Consequently, these coefficients are sometimes referred to as *projection amplitudes*.

The use of the term *projection* is very evocative. It cements the relationship between the ideas of vectors in classical physics defined in three-dimensional Euclidean space and quantum state vectors defined in a multi-dimensional Hilbert space. Imagine a classical vector **v** pointing in some arbitrary direction in Euclidean space. Such a vector might represent the instantaneous motion of an apple thrown high into the air. The apple is heading in a specific direction (up into the air, say) with a certain velocity. We draw an arrow to represent the direction of the vector, and the length of the arrow represents its magnitude. We define this length to be unity in some arbitrary unit system. Now, suppose we wish to 'map out' the vector in terms of its components along defined coordinates (say x and y). We 'resolve' the vector into two orthogonal components. Each component is also a vector which we represent as a coefficient multiplied by the unit vector corresponding to the coordinate. In other words, the vector **v** is equal to the sum of $v_x\mathbf{i}$ and $v_y\mathbf{j}$, where v_x and v_y are the (scalar) coefficients, and **i** and **j** are the unit vectors along the x and y coordinates, respectively (see Fig. 4.1).

The coefficients v_x and v_y are the projections of the vector **v** onto the x and y coordinates. They can be calculated as the inner or dot products of **v** and the unit vectors: v_x is equal to $(\mathbf{i}\cdot\mathbf{v})$ and v_y is equal to $(\mathbf{j}\cdot\mathbf{v})$. The contribution that each unit vector makes to the arbitrary vector **v** is therefore determined by the magnitude of the projection of **v** on that unit vector's corresponding coordinate. This projection can be calculated from the inner product of the unit vector and the arbitrary vector. In the quantum theory based on vector spaces, an arbitrary state vector $|\psi\rangle$ is decomposed into orthogonal components formed from the set of basis vectors, which are eigenstates of a specific operator in the relevant Hilbert space. The contribution that each basis vector makes is determined by the magnitude of the projection of $|\psi\rangle$ along the 'direction' corresponding to the basis vector. This projection is calculated as the inner product or projection amplitude of the

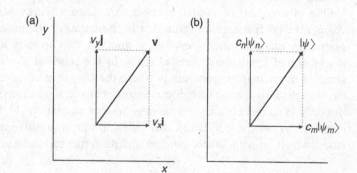

Fig. 4.1 Comparison of (a) unit vectors in Euclidean space and (b) basis vectors in Hilbert space.

basis vector and the arbitrary state vector $|\psi\rangle$. The analogy is complete: *the basis vectors are the unit vectors of Hilbert space.*

Euclidean space is three-dimensional, and so only three unit vectors, usually symbolized by $\mathbf{i}$, $\mathbf{j}$, and $\mathbf{k}$, are needed to specify completely an arbitrary classical vector. In contrast, Hilbert space has as many dimensions as we need to specify completely an arbitrary quantum state vector. The parallels between state vectors and classical unit vectors are drawn out in Appendix 10.

The idea of the state vector thus brings with it many of the mathematical properties we associate with vectors in classical physics. To some extent, this is very helpful. Because we are familiar with the concepts of classical vectors and can visualize what they are and how they combine, we are provided with a ready-made interpretation of state vectors that is intrinsically appealing. However, we should be under no illusions. The state vectors have properties that classical vectors can never have. The state vectors belong in a mathematically defined space, and they can show interference effects. Whereas we can 'measure' vectors in classical physics, we cannot measure state vectors directly: only the modulus-squares of the state vectors are accessible to experiment. The analogy between state vectors and classical unit vectors is a *mathematical* one: it offers us little help in understanding what a state vector *is*.

Indistinguishable particles

The quantum formalism is constructed on a set of four postulates, together with the position–momentum commutation relation, the convergence properties of Hilbert space, and the expansion theorem. The last remaining ingredient to consider is also one of the most puzzling. It is the mathematical treatment of indistinguishability.

If I were to acquire two apples that had exactly the same shape, size, and colouration, and I were to place them side by side on my desk, we might, perhaps, agree that these apples are indistinguishable. But would this be strictly true? After all, I can use a metre rule to measure off the distances to each apple from the front and left-hand edges of my desk (the x and y coordinates) and note that apple 1 has coordinates x_1, y_1 and apple 2 has coordinates x_2, y_2. These two sets of coordinates must be different, otherwise the apples would occupy the same space (they would be the same apple). Thus, the apples are distinguishable because they occupy measurably different regions of space.

However, electrons are quantum 'wave-particles' represented in quantum theory by the appropriate state vectors. We saw in Chapter 2 why we must abandon the idea that we can keep track of an electron as it orbits the nucleus of an atom. Instead, we tend to think of electrons in terms of delocalized probability densities, and the three-dimensional shapes of their density 'maps' correspond to our familiar pictures of atomic orbitals. If two electrons occupy the same atomic orbital, how can we distinguish between them? We cannot now measure the coordinates of the two electrons in the same way as we can measure the coordinates of apples on my desk. The fact is that the electrons, like all quantum wave-particles, are indistinguishable.

The statistics of counting distinguishable particles are completely different from those for counting indistinguishable particles. Recall that Planck had decided to use Boltzmann's statistical approach in deriving his radiation law, but instead of assuming his energy elements to be distinguishable (as Boltzmann had always assumed when applying his methods to atoms and molecules), Planck purposefully made them indistinguishable. Paul Ehrenfest pointed out in 1911 that in doing this, Planck had given his quanta properties that were simply impossible for classical particles. Of course, photons and electrons are not classical particles. They possess wave-like properties too, and these properties lead to behaviour that is completely counter-intuitive if we try to think of photons and electrons as tiny, self-contained particles.

What does this mean for state vectors in Hilbert space? For a two-particle state (a state consisting of two electrons, for example), we would normally calculate the resulting state vector as the *product* of the state vectors of the two particles. Thus, if particle 1 is in a state described by the state vector $|m\rangle$, and particle 2 is in a state described by the state vector $|n\rangle$, the appropriate product state can be written $|m\rangle_1|n\rangle_2$, where the subscripts indicate which particle is in which quantum state.

But these particles are supposed to be indistinguishable. We have labelled the particles as 1 and 2, but if they are indeed indistinguishable, we have no way of telling them apart. We can certainly distinguish between the possible quantum states $|m\rangle$ and $|n\rangle$, since they can correspond to states with different quantum numbers, energies, angular momenta, etc., but we cannot tell experimentally which particle is in which state. Thus, the product $|n\rangle_1|m\rangle_2$ is just as acceptable as $|m\rangle_1|n\rangle_2$. (Note that the order in which we write the functions down is irrelevant; $|m\rangle_1|n\rangle_2$ is equal to $|n\rangle_2|m\rangle_1$.)

Because both of these product state vectors are equally 'correct', we have to assume that the total two-particle state vector can be written as a suitably normalized linear combination of both possibilities $|m\rangle_1|n\rangle_2$ *and* $|n\rangle_1|m\rangle_2$. This mixture must contain equal proportions of each product state (because they must be equally possible), and so it follows that the modulus-squares of the corresponding expansion coefficients must be equal. This means that the expansion coefficients must have the same magnitude, but they can have different *signs*. Although the only quantities accessible to us through experiment are the modulus-squares of the state vectors (the probability densities), the signs of these expansion coefficients can certainly have important, measureable effects, as we will see.

Fermions and bosons

There are obviously two ways in which the expansion coefficients can have equal modulus. They can have an equal magnitude and be of opposite sign, or they can be of equal magnitude and be of the same sign. Both possibilities are considered in Appendix 11. These simple considerations lead to some powerful conclusions. The two possibilities for the signs of the expansion coefficients lead to two possibilities for the structure of the two-particle state vector.

In the first possibility, the two-particle state vector is *antisymmetric* with respect to the exchange of the particles, which means that the state vector changes sign on exchange of the particles. This can have no direct consequence in any practical measurement, as only the modulus-square of the state vector is accessible to experiment. Nevertheless, it is relatively straightforward to show that particles whose two-particle state vectors are antisymmetric with respect to exchange are *forbidden* from occupying the same quantum state (see Appendix 11). If this sounds familiar, that is because it is. This is the Pauli exclusion principle. The exclusion principle was developed for electrons, but when couched in terms of pairwise exchange, it becomes a general rule which applies to all quantum particles with antisymmetric multi-particle state vectors. Such particles have half-integral spin quantum numbers and are collectively called fermions (see Chapter 3), named for the Italian physicist Enrico Fermi. Examples include electrons, protons, neutrons, and some atomic nuclei.

In the second possibility, the coefficients are equal in magnitude and sign, and the two-particle state vector is consequently symmetric (it does not change sign) on exchange of the particles (see Appendix 11). Particles whose two-particle state vectors possess this property are called bosons (see Chapter 3), named for the Indian physicist Satyendra Bose, and have zero or integral spin quantum numbers. Examples include photons and some atomic nuclei. They are *not* prevented from occupying the same quantum state, and in the phenomenon known as *bose condensation*, systems are prepared experimentally in which many bosons are present in the same quantum state. Bose had been struggling to get a paper on the quantum theory of light published in a reputable journal, and in 1924, he decided to send the paper to Einstein with a request for help to get it published in a German-language physics journal. Einstein read the paper with interest, translated it into German himself, and sent it to a journal with a personal note of recommendation. It was Einstein who first speculated that it might be possible for bosons to condense into the lowest-energy quantum state at sufficiently low temperatures.

One of the best examples of bose condensation is provided by lasers. A laser can be fashioned from a suitable medium (gas, liquid, or solid) in which a *population inversion* can be created, with many more of the medium's constituent atoms or molecules driven into some excited state than remain in a lower-energy state. Spontaneous emission is then reflected back into the medium by mirrors that form an optical cavity. This stimulates further emission and amplifies the circulating radiation, which, after a very short time interval, is established as a standing wave in the cavity. The very large number of photons circulating in such a cavity can be thought of as being in a single quantum state. Useful output can be extracted from the laser by making one of the mirrors partially transmitting.

A further example is provided by the phenomenon of superfluidity. A ^{4}He atom consists of two protons, two neutrons, and two electrons, and therefore an even number of fermions. It therefore acts as a boson, and large numbers of ^{4}He atoms can therefore condense into a single quantum state. The result is a 'superfluid' state: at around 2.2 K (the so-called 'lambda point'), the viscosity of liquid ^{4}He falls by about six orders of magnitude. Cooperative motions in this bose condensed state produce a number of rather peculiar properties, including frictionless 'creeping' of the fluid as a thin film up, over, and around surfaces.

A final example is provided by superconductors. When cooled below a certain critical temperature, a number of metals and metal alloys cease to exhibit any resistance to the flow of electric current, and they become superconductors. The explanation for this behaviour was obtained in the 1950s. Electrons passing through a superconductor cooled below its critical temperature no longer repel one another. In fact, they display a weak attraction. In essence, what happens is that an electron passing close to a positively charged ion in the metal lattice pulls the ion out of position, distorting the lattice slightly. The electron moves on, but the distorted lattice continues to vibrate. This vibration produces a region of excess positive charge which attracts a second electron, with the result that two electrons move through the lattice cooperatively, their attraction mediated by lattice vibrations. The two electrons (both fermions) collectively make a boson, and so the pair of electrons (called Cooper pairs, after the American physicist Leon Cooper) can condense into the same quantum state.

In its general form, the Pauli principle applies to all quantum particles. This principle states that particles with half-integral spin quantum numbers—fermions—must have two-particle (or, in general, many-particle) state vectors that are antisymmetric with respect to the pairwise interchange of particles. Particles with integral spin quantum numbers—bosons—must have symmetric many-particle state vectors. The Pauli exclusion principle is an expression of the more general Pauli principle as applied to electrons: the requirement for an antisymmetric state vector for electrons means that electrons are excluded from occupying the same quantum state or possessing the same set of quantum numbers. These symmetry requirements arise naturally when the effects of special relativity are introduced in the quantum mechanics of many-particle quantum states, as was shown by Pauli himself.

We could be forgiven for thinking that the Pauli principle provides yet another layer of mysterious formalism for quantum systems containing many particles on top of an already quite impenetrable formalism for single particles. There are, in fact, not that many mysteries. At the heart of the Pauli principle lies the indistinguishability of all quantum particles, with fermions differing from bosons in the symmetry properties of their many-particle state vectors. As we have seen, the assumption of indistinguishable energy elements was a necessary part of Planck's 'act of desperation', which led ultimately to the development of quantum theory itself.

Indistinguishability is a property of quantum particles that is intrinsically linked to their wave-particle nature, as is the position–momentum commutation relation and Heisenberg's uncertainty principle. All these problems are one problem.

5

Quantum measurement

Measurement lies at the core of the quantum formalism. This may seem a statement of the intuitively obvious for any theoretical structure with a pretension to predict the outcomes of observations made or experiments performed on objects in the real world. But classical mechanics is a theory of the physical properties of objects and their behaviour when exposed to different kinds of forces, and there is no particularly significant role for measurement other than a passive role involving the recording of those properties or behaviour. In contrast, measurement in quantum theory takes on a far deeper, almost ominous, significance.

In his debate with Heisenberg over the interpretation of the uncertainty principle, Bohr had argued that the use of diametrically opposed classical concepts—waves and particles— prevents us from knowing what is really happening to quantum particles until they are exposed to some form of measuring device. This device operates in a macroscopic world and, Bohr contended, operates according to classical principles. Our choice of device then dictates the kind of behaviour we could expect to see. If we choose to interrogate a quantum system using an apparatus designed to reveal particle positions, we get particle-like behaviour: the particles are 'here' and 'there'. If we choose to interrogate the quantum system using an apparatus designed to reveal interference effects, we get wave-like behaviour: the particles are neither 'here' nor 'there', and instead, we see interference fringes. To quote the character Kerner in Tom Stoppard's play *Hapgood*: 'The act of observing determines the reality'.[1]

It is at this stage that we sense our grasp on reality starting to slip away. The postulates of quantum theory described in the last chapter provide a powerful and robust foundation, with a consistency established by its own internal rules, whether these are written in the language of wave mechanics, matrix mechanics, or vector spaces. But, without measurement, the theory is an empty framework. And yet *all* the conceptual and philosophical problems of the theory are to be found here.

---

[1] Stoppard, Tom (1988). *Hapgood.* Faber & Faber, London, p. 12.

Quantum probabilities

Max Born declared that quantum theory is inherently probabilistic in nature and can furnish us only with information about the probabilities of the outcomes of measurements, not certainties. It is therefore important that we fully appreciate the nature of probability in quantum theory.

It will help to run through a simple example in which we deduce and use statistical probabilities from an 'everyday' classical perspective and then see how these compare with their equivalents in quantum theory. Imagine that I toss a coin into the air, and it falls to the ground. When the coin comes to rest flat on the ground, I take note of the outcome—heads or tails—and enter this in my laboratory notebook. I denote the result 'heads' as R_h and the result 'tails' as R_t. I repeat this 'measurement' process N times, where N is a large number.

At the end of this experiment, I add up the total number of times the result R_h was obtained and denote this as N_h. Similarly, N_t denotes the number of times R_t was obtained. As there are only two possible outcomes for each measurement,[2] I know that the sum of N_h and N_t will be equal to N. I now define the *relative frequency* with which the result R_h was obtained as the ratio N_h/N. Similarly, the relative frequency with which the result R_t was obtained is N_t/N, and the relative frequencies for both results sum to 1 (as there are only two possible outcomes).

In principle, the outcome of any one measurement is determined by a number of variables, including the force and torque exerted on the coin as I toss it into the air, the interactions between the spinning coin and fluctuations in air currents, and the angle and force of impact as the coin hits the ground. These variables could be controlled, for example, by using a computer-operated mechanical hand to toss the coin and by performing measurements in a vacuum. Alternatively, if we knew these variables precisely, we could, in principle, use this information to calculate the exact trajectory of the coin. It is therefore not impossible to imagine that, by carefully controlling the conditions or by acquiring sufficiently detailed information, the outcome of a particular individual measurement could be traced right back to the initial conditions—the coin placed heads up or tails up on the hand—and therefore predicted with certainty.

In the absence of such control or knowledge of the variables, we assume that our measurements on the coin serve to 'calibrate' the system and allow us to make predictions about its behaviour in experiments yet to be performed. For example, if we discover that N_h/N is equal to $\frac{1}{2}$, we would conclude that the *probabilities* P_h and P_t of obtaining the results R_h and R_t, respectively, for any subsequent measurement would also be equal to $\frac{1}{2}$. We would further conclude that the coin is 'neutral' with regard to the measurement; that is, both possible outcomes are obtained with equal probability. The coin does not have to be neutral: it could have been biased in favour of one of the results, and this would have been reflected in the measured frequencies. We should note that the definition of probability that we are using here is an intuitive one. In practice, coin tossing is subject to chance fluctuations that can often lead to some completely unexpected sequences of results. However, for our present purposes, it is sufficient to propose that our coin and method of tossing are unbiased and that we can make N so large that the effects of chance fluctuations become negligible. Having established that P_h is equal to P_t, which in turn is

---

[2] I am assuming that the coin falls flat each time it is tossed.

equal to $\frac{1}{2}$, we expect to obtain the result R_h or the result R_t with equal probability in all future measurements.

In this example, we know what the variables are, but we do not control them. We may therefore have no difficulty in accepting that the outcome of a particular measurement is *predetermined* the moment the coin is launched into the air. Just as for Boltzmann's statistical mechanics, it is our lack of knowledge of the many individual variables at work which forces us to resort to statistical probabilities.

Now consider a quantum system, described by the state vector $|\psi\rangle$, on which a measurement also has two possible outcomes. An example of such a measurement could be the determination of the direction of an electron spin vector in some arbitrary laboratory frame.[3] There are two possible outcomes—spin up and spin down. If we assume that we make repeated measurements on a 'beam' of identically prepared quantum particles, then by analogy with the classical example given above, we can record the number of times we achieve the spin-up result, denoted R_u, and the number of times we achieve the spin-down result, R_d. We should not be surprised that after a large number, N, of such measurements, we find that N_u/N is equal to N_d/N, which in turn is equal to $\frac{1}{2}$. We come to the conclusion that the probabilities of obtaining spin-up and spin-down results are equal, and the analogy with tossing a coin seems complete.

But despite appearances, the probabilities we are dealing with here are very different. In the coin-tossing example, the dual property of heads and tails is obviously inherent in the coin before, during, and after the measurement process. Also, we tend to make repeated measurements on the same coin (although there is nothing in principle preventing us from making repeated measurements on a large bag of identical coins, with one measurement on each coin). In the case of the quantum measurement of spin vectors, we might assume initially that each individual quantum particle is present *either* in a spin-up *or* a spin-down configuration, and in the beam of identically prepared particles, there is an equal number in both possible states. The measurement then tells us which individual particle is in which of the two possible states. However, we would be wrong on several counts.

Firstly, in its most common interpretation, probability in quantum theory does *not* attach any meaning to the outcomes of measurements on individual particles, but rather applies to the distribution of results of repeated measurements on a collection of identically prepared systems. Secondly, if we force the issue and think at the level of individual particles, it is generally wrong to suppose that the particles possess the properties we are measuring *prior to their measurement*. This is not pedantry and not a statement of the simple fact that we can know nothing for certain until we have observed or measured it. It is a statement of the much more complex fact that assuming quantum particles possess properties *prior* to their measurement yields predictions that *conflict* with the results of actual measurements. Proof of this comes in experiments to be described in Part IV.

In a sense, quantum measurement is itself responsible for *creating the outcomes*. Quantum probabilities do not reflect our ignorance of the intricate details of some underlying physical reality, as do classical, statistical probabilities. They are rather an expression of the likelihood that interaction between the quantum system and the measuring device will create specific measurement outcomes.

---

[3] This experiment can be conducted using a beam of silver atoms (rather than electrons) passed between the poles of a magnet in what is known as a Stern–Gerlach apparatus.

Linear polarization

It will help in what follows in this and subsequent chapters to focus on some actual quantum properties of actual particles and use these to explore the theoretical approach to quantum probability and measurement. The polarization properties of light provide just such a reference and have the added advantage that these properties have also provided fertile ground for rigorous testing of the theory. We will begin with a discussion of the polarization properties of light that is based almost entirely on classical concepts.

In the seventeenth century, Isaac Newton noticed that there appeared to be two different 'types' of light, but it was Christiaan Huygens and, later, Thomas Young who produced an explanation. In terms of Maxwell's theory of electromagnetism, the electric vectors of transverse light waves confined to oscillate in one dimension only (plane waves) can take up two possible orientations that are mutually orthogonal and also perpendicular to the direction of propagation. If we assume that the direction of propagation of some plane light wave is along the z-axis, the electric vector of vertically polarized light is confined to oscillate only in the x-direction, and for horizontally polarized light, the electric vector is confined to oscillate in the y-direction (see Fig. 5.1).

Most readers will be familiar with polarizing filters—pieces of plastic film (often called 'Polaroid' film after the manufacturer's trademark) consisting of an array of polymer molecules which shows a preference for absorption of light along one specific axis. Imagine that we take two pieces of Polaroid film placed one on top of the other, and we arrange for them to be illuminated from behind by a suitable light source. We arrange them so that the maximum amount of light is transmitted through both filters. Now, we slowly rotate one of the filters through 90° and note how the intensity of transmitted light falls until, when the axes of maximum transmission of the two filters are at right angles, no light is transmitted at all (the filters are said to be 'crossed').

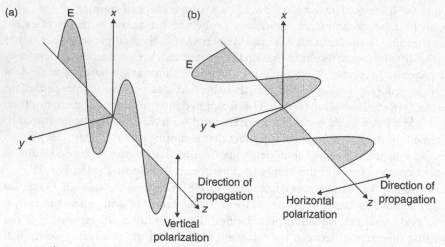

Fig. 5.1 (a) Vertically and (b) horizontally polarized plane electromagnetic waves. Only the electric vector of the waves is shown: the magnetic vector oscillates at right angles to both the electric vector and the direction of propagation.

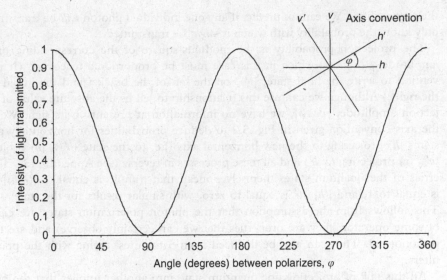

Fig. 5.2 The intensity of light transmitted through two polarizing filters varies as $\cos^2 \varphi$, where φ is the angle between the filters' axes of maximum transmission. When φ is equal to 90° or 270°, no light is transmitted, and the filters are said to be 'crossed'.

The eye is a powerful but non-quantitative light detector. If we were to measure the amount of light transmitted through both filters using a device such as a photomultiplier or a photodiode, we would discover that the intensity falls off according to the cosine-squared of the angle between the transmission axes of the filters (see Fig. 5.2). This is Malus's law.

We can readily interpret this law using classical concepts. We can suppose that the first filter transmits light polarized predominantly along its axis of maximum transmission—the filter provides a source of (in this case, we assume) vertically polarized light. As the angle, φ, between the two filters is changed, the second filter transmits only the component of the electric vector of the vertically polarized light that lies along its axis of maximum transmission. This component is the projection of the electric vector of the vertically (v) polarized light onto the new vertical (v') axis and therefore depends on the cosine of the angle between them. Since the intensity of light is proportional to the square of the modulus of the electric vector, the intensity of light transmitted through both filters varies as $\cos^2 \varphi$.

Photon-polarization states

How should Malus's law be interpreted in terms of photons? According to quantum theory, we can assign the photons transmitted through the first filter to a state of vertical polarization. We denote such a state by the state vector $|v\rangle$. As the second filter is rotated, the photons are projected into a new state, $|v'\rangle$, with a probability equal to $\cos^2 \varphi$ (see the axis convention given in Fig. 5.2). The intensity of light transmitted through both filters depends on the number of photons detected. This number is determined by the probability that the photons described by $|v\rangle$ are projected into the state $|v'\rangle$ and so transmitted by the second filter. We cannot predict if any one individual photon *will* be transmitted: we only know the probability with which it *might* be transmitted.

The projection probability is the modulus-square of the corresponding projection amplitude—$|\langle v'|v\rangle|^2$—which in this case must be proportional to $\cos^2\varphi$. (It is a convention to write the final state, $|v'\rangle$, on the left of the bracket and the initial state on the right.) Although we can use this relationship to tell us the absolute value of the projection amplitude, $|\cos\varphi|$, we have no information at present on its sign. We can use the axis convention given in Fig. 5.2 to deduce probabilities for horizontally polarized states $|h\rangle$ projecting to the new horizontal axis (i.e. to the state $|h'\rangle$), $|h\rangle$ projecting to $|v'\rangle$, $|v\rangle$ projecting to $|h'\rangle$ and all these processes in reverse (see Appendix 12). The properties of the quantum states themselves mean that $|\langle v|v\rangle|^2$ is equal to $|\langle h|h\rangle|^2$, which is equal to 1, and $|\langle h|v\rangle|^2$ is equal to zero, with similar results for the states $|v'\rangle$, $|h'\rangle$. This follows from the assumption that the photon polarization states are eigenstates of some operator (they are properties that we can certainly observe) and are therefore orthonormal. They can also be deduced from 5 minutes' toying with the polarization filters.

All this talk of projecting one quantum state into another implies that the polarizing filters are taking a very active role in the process rather than passively transmitting a proportion of the light *without* changing its polarization properties. Any doubt about this active role can be quickly dispelled by performing the following simple experiment. Take two polarizing filters and orient them so that they are crossed. No light is transmitted because, according to the above arguments, the projection probability $|\langle h|v\rangle|^2$ is given by $\cos^2 90° = 0$ (see Fig. 5.2).

Now insert a third filter, oriented at any angle φ greater than 0° and less than 90°, *between* the two crossed filters, as illustrated in Fig. 5.3. If we assume that the middle filter simply transmits a proportion of the vertically polarized light without changing it in any way, this light will surely be blocked by the final, crossed filter, and we would therefore expect that no light would be transmitted through all three filters. But this is not so. Light *is* transmitted through all three filters with an intensity we can deduce from the projection probability for the first and middle filters, $|\langle v'|v\rangle|^2$, which, from our considerations above, we know is proportional to $\cos^2\varphi$, multiplied by the projection probability for the middle and final filters, $|\langle h|v'\rangle|^2$, which is proportional to $\cos^2(90-\varphi)$, or $\sin^2\varphi$. Thus, if $\varphi = 45°$, we would expect 25 per cent of the incident light to be transmitted.

Photon spin

Photons are bosons. They are quantum particles with spin quantum numbers s equal to 1. Like the electron, a photon can have only one value of s, and there are different ways that the photon spin can be 'aligned' in a magnetic field, corresponding to the different values of the magnetic spin quantum number, denoted m_s. For electrons, s is equal to $\frac{1}{2}$, and so we have two possibilities: $m_s = +\frac{1}{2}$ (spin up) and $m_s = -\frac{1}{2}$ (spin down). As a general rule, quantum theory predicts the existence of states with values of m_s in the series $s, (s-1), (s-2), \ldots, 0, \ldots, -(s-2), -(s-1), -s$. This rule would lead us to predict that photons should have three possibilities for m_s, corresponding to $m_s = +1, 0$ and -1. However, relativistic quantum theory forbids an $m_s = 0$ component for particles travelling at the speed of light. This leaves us with just two possibilities, and, by convention, we associate the $m_s = +1$ component with left-circularly polarized light and the $m_s = -1$

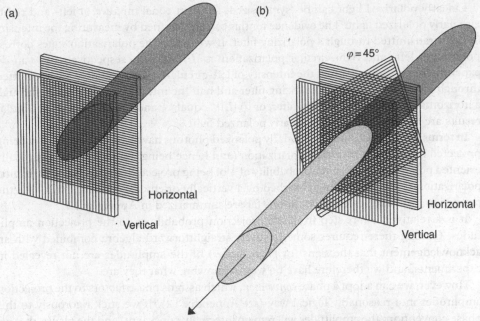

Fig. 5.3 (a) With two polarizing filters orientated at right angles, they block both vertical and horizontal components of the light, and no light is transmitted. (b) Inserting a third polarization filter at an angle $\varphi = 45°$ between the two crossed filters allows 25 per cent of the incident light to pass through the combination.

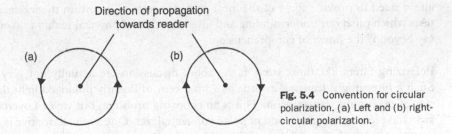

Fig. 5.4 Convention for circular polarization. (a) Left and (b) right-circular polarization.

component with right-circularly polarized light. This makes sense if we remember that the spin property of a quantum particle is manifested as an intrinsic angular momentum. We define left circular as a counterclockwise rotation of the electric vector of light viewed along the direction of propagation and travelling towards the observer (see Fig. 5.4).

Although the spin property of a quantum particle should never be interpreted as if the particle were literally spinning on its axis, it is nevertheless manifested as an intrinsic angular momentum. Thus, a beam containing a large number of circularly polarized photons (such as in a laser beam) will impart a measurable torque to a target. However, this angular momentum is not only a collective phenomenon: in the absorption of an individual photon resulting in electron excitation in an atom or molecule, the angular momentum intrinsic to the photon is transferred to the excited electron, and total angular momentum is conserved. That transfer has important, measurable effects on the absorption spectrum.

Linearly polarized light can be 'synthesized' from an equal mixture of left- and right-circularly polarized light. The evidence for this can be obtained by measuring the intensity of light transmitted through a polarizing filter. If we define the polarization states corresponding to left- and right-circular polarization as $|L\rangle$ and $|R\rangle$, respectively, we can do experiments to show that half the intensity of left-circularly polarized light is transmitted through a vertically oriented polarizing filter and half the intensity is transmitted through a horizontally oriented polarizing filter, or $|\langle v|L\rangle|^2$ equals $\frac{1}{2}$ and $|\langle h|L\rangle|^2$ equals $\frac{1}{2}$. Similar results are obtained for right-circularly polarized light.

In terms of photons, the left-circularly polarized photons have a probability of $\frac{1}{2}$ of being projected into a state of vertical polarization (and hence being transmitted by a vertically oriented polarizing filter) and a probability of $\frac{1}{2}$ of being projected into a state of horizontal polarization (and hence being absorbed by a vertically oriented polarizing filter). All the projection probabilities considered thus far are summarized in Appendix 12.

It is a relatively short step from the projection probabilities to the projection amplitudes. Getting there requires some relatively straightforward algebra combined with an acknowledgement that the signs (or *phase factors*) of the amplitudes are not revealed in experiments, and we therefore have no way of knowing what they are.

However, we can adopt a phase *convention*, which assigns phase factors to the projection amplitudes in a reasonably logical way (see Appendix 12). If we stick rigorously to this phase convention, the amplitudes will remain internally consistent, and the results that we would predict by using them are consistent with the experimental projection probabilities. The projection amplitudes for all of the possible combinations of vertical, horizontal, v', h' and left- and right-circularly polarized photons are summarized in Appendix 12. We can use these amplitudes to express any photon polarization state in any basis (e.g. $|v\rangle$ in terms of $|v'\rangle$ and $|h'\rangle$ or $|L\rangle$ and $|R\rangle$). The information contained in Appendix 12 is in fact all we need to provide some of the most stringent tests of quantum theory ever devised, tests which push our understanding and interpretation of physical reality to (some would say beyond) the limits of comprehension.

Polarizing filters like those used in the above discussion are actually not very efficient. Such a filter might transmit as little as 70 per cent of linearly polarized light through its axis of maximum transmission. This is an annoying problem, but we can overcome it by switching to an alternative kind of polarization analyzer. One such alternative is a piece of calcite, a naturally occurring crystalline form of calcium carbonate.

Calcite is naturally *birefringent*; it has a crystal structure which has different refractive indices along two distinct crystal planes. One offers an axis of maximum transmission for vertically polarized light, and the other offers an axis of maximum transmission for horizontally polarized light. The vertical and horizontal components of light, which is a mixture of polarizations, are therefore physically separated by passage through the crystal, and their intensities can be measured separately. With careful machining, a calcite crystal can transmit virtually all of the light incident on it.

There are a number of ways of obtaining a source of left circularly polarized light. These vary from a standard (i.e. unpolarized) light source passed through an optical device known as a quarter-wave plate to an atomic source that relies on the quantum mechanics of photon emission. An example of the latter is a beam of atoms that are excited to some electronically excited state from which emission occurs. If angular momentum is to be conserved in the process, the emitted photon must carry away any excess angular momentum lost by the excited electron as it returns to a more stable quantum state.

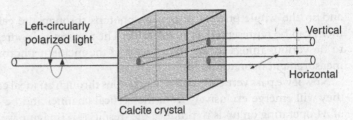

Fig. 5.5 A calcite crystal splits left circularly polarized light into two equal vertical and horizontal components.

An appropriate choice of states between which the transition occurs can give rise to the emission of photons exclusively with $m_s = +1$. We will meet this kind of source again in Part IV.

A beam of left-circularly polarized light entering a calcite crystal will split into two beams, one of vertical and one of horizontal polarization (see Fig. 5.5). We can use detectors (such as photomultipliers), placed in the paths of the emergent beams, to confirm that each has half the intensity of the initial beam. This is consistent with the photon projection amplitudes and probabilities given in Appendix 12.

Von Neumann's theory of measurement

John von Neumann's *Mathematical foundations of quantum mechanics* provided not only an unassailable mathematical foundation for the theory but also an approach to the interpretation of quantum measurement that shaped virtually all subsequent thinking on the subject. Indeed, the language used in the above discussion, with its reference to projection amplitudes and projection probabilities, is derived almost entirely from von Neumann's original approach.

When a quantum system interacts with a measuring device, according to von Neumann this interaction is subject to, and described by, the laws of quantum mechanics. Whereas Bohr took great pains to distinguish between events occurring at the quantum level and macroscopic measuring devices understood at the level of the classical physics of waves or particles, von Neumann deliberately drew no boundary between a 'quantum world' and a 'classical world'. He believed that the only legitimate language of physics is the language of quantum physics and that we can define the measuring device any way we like (including human observers, as necessary) without in any way invalidating the application of quantum mechanical principles.

Adherence to this view leads us to the following logical consequence. If the observables of a particular quantum system are eigenvalues of some operator, then the results of measurements we make in the laboratory are similarly eigenvalues of some 'measurement operator', which describes the measuring device as a quantum mechanical system. Of course, we cannot access the eigenvalues of a quantum system directly: we have access only to the eigenvalues of the system following interaction with the measuring device, and we *interpret* the results in terms of the eigenvalues of the former.

In our example developed above, we have put together a device to decompose incident light into vertical and horizontal polarization components, which are then detected. According to von Neumann, the measuring device can be thought of as a quantum system, and so the whole process of passing photons through the calcite crystal and detecting them can be represented as an operator. The apparatus is, after all, a set of 'instructions' to do various things to the state vector of the incident photons. We can denote such a measurement operator as $\hat{M}$.

Now if we pass vertically polarized photons through an ideal calcite crystal, we know that they will emerge exclusively from the vertical channel and be detected. Thus, the effect of $\hat{M}$ operating on $|v\rangle$ is to produce the 'result' $|v\rangle$. Imagine we have the apparatus rigged so that a red light comes on if photons are detected through the vertical channel. We may conscientiously enter this result in our laboratory notebook, perhaps representing it by writing R_v. We conclude that $|v\rangle$ is an eigenstate of the operator $\hat{M}$ with eigenvalue R_v. This is merely the same as saying that the apparatus is set up to measure vertical polarization.

If we set up the apparatus so that a blue light comes on if photons are detected through the horizontal channel, we can use similar arguments to show that $|h\rangle$ is an eigenstate of $\hat{M}$ with eigenvalue R_h. Note that it is unnecessary for us to figure out the exact mathematical form of $\hat{M}$: its properties and effects on the state vectors are defined by the way we have the apparatus set up.

The state vector of a collection of identically prepared left-circularly polarized photons can be expressed as a linear superposition of $|v\rangle$ and $|h\rangle$ (see Appendix 12). When we pass left circularly polarized photons through our measuring apparatus, the measurement operator $\hat{M}$ operates on both the $|v\rangle$ and $|h\rangle$ components of $|L\rangle$. If we assume $|L\rangle$ to be normalized, the expectation value of the measurement operator is given by $\frac{1}{2}(R_v + R_h)$ (see Appendix 13), that is, we expect to see the red light and the blue light come on with equal probability. This does not mean that both lights are 'half on'. It means that, on average, the red light comes on for half of the photons detected, and the blue light comes on for the other half. The theory does not allow us to predict with certainty which light will come on for a given incident photon.

Another way of looking at the measurement process is to say that, in order to obtain the result corresponding to the eigenvalue R_v, the initial photon state $|L\rangle$ must be projected into the state $|v\rangle$. The subsequent effect of $\hat{M}$ on the state vector is to yield the eigenvalue R_v. From Appendix 12, we note that the corresponding projection probability, $|\langle v|L\rangle|^2$, is equal to $\frac{1}{2}$.

The 'collapse of the wavefunction'

But let us now reduce the intensity of the incident left-circularly polarized light so that, on average, only one photon passes through the crystal at a time. What happens to the photon? It cannot split into two, one half following one path and the other half following the other path, because the photon is a 'fundamental' particle. Besides, if we really could split a photon in half, we would necessarily halve the energy (and, from $\varepsilon = hv$, the frequency). A simple experiment to measure the frequency of each transmitted photon confirms that it has the same value as the incident photon. If the photon follows only one path through the crystal, it must emerge *either* from the vertical 'channel' *or* from the horizontal 'channel'.

An incident left circularly polarized photon *must* be detected in either the vertical or the horizontal channel of the calcite crystal. Prior to measurement, the quantum state of the photon can be described as a linear superposition of the two possible measurement eigenstates, $|v\rangle$ and $|h\rangle$. After measurement, the photon is inferred to have been in one, and only one, of the two possible measurement eigenstates. Somewhere along the way, the state vector has changed from one consisting of two measurement *possibilities* $|v\rangle$ *and* $|h\rangle$ to one *actuality*, $|v\rangle$ or $|h\rangle$.

Von Neumann recognized that measurement involved two very different types of 'intervention'. The first is like a quantum jump: an instantaneous, discontinuous, indeterministic transition of the state of the quantum system before measurement to the state after measurement. This process is now commonly known as the 'collapse of the wavefunction' or the 'reduction of the state vector'.[4] Its introduction into the quantum formalism is generally known as von Neumann's *projection postulate*. The second type of intervention is the continuous and entirely deterministic evolution of the measuring system in response to the collapse, which is described by the time-dependent Schrödinger equation. As already stated in the discussion of postulate 4 in Chapter 4, the time-dependent form of the Schrödinger equation cannot describe the discontinuous transition characteristic of a quantum jump. Both types of intervention are therefore required in order to interpret the quantum measurement process. Von Neumann wrote[5]:

The two interventions are ... fundamentally different from each other ... the development of a state according to [the discontinuous process] is statistical, while according to [the continuous process] it is causal.

The projection postulate goes directly to the heart of the meaning of quantum theory. What does the superposition of the states $|v\rangle$ and $|h\rangle$ actually represent when used to refer to an individual quantum particle? Might it not merely reflect the fact that left-circularly polarized light is, aside from a phase factor, really a 50:50 mixture of vertically and horizontally polarized photons, and we use it because, prior to measurement, we are ignorant of the actual polarization state of any one individual photon? If this is the case, each individual photon is present in a predetermined $|v\rangle$ or $|h\rangle$ state; each follows a predetermined path through the calcite crystal according to the properties of that state and is detected, much like the result heads or tails in our coin-tossing example is predetermined by the existence of both sides of the coin at the start. Under such circumstances, the collapse of the wave function would simply represent a sharpening of our knowledge of the state of the photon. Prior to measurement, the photon is in either the $|v\rangle$ or $|h\rangle$ state, and the measurement merely tells us which.

Or does the superposition really reflect the fact that the linear polarization state of the photon is completely *undetermined* prior to measurement? In this case, the collapse of the wave function represents more than just a change in our state of knowledge of the system. In fact, this way of thinking requires a fundamental revision of our conception of the process of measurement compared with classical mechanics. For example, I assume the length of my desk to be a predetermined quantity. Although I accept that I have no precise knowledge of this quantity until I measure it, I do not assume that the very act of looking at my desk to locate its edges in space changes its length from an undetermined to a determined quantity. In classical physics, to have no knowledge of a physical quantity does not imply that it is not determined before a measurement is made.

---

[4] Despite the fact that, in what follows, I will tend to refer to state vectors rather than wave functions, I will always refers to this process as the 'collapse of the wave function'.

[5] Von Neumann, John (1955). *Mathematical foundations of quantum mechanics*. Princeton University Press, Princeton, NJ, p. 357.

The American physicist John Wheeler developed an entertaining baseball analogy that, although developed to make a slightly different point, serves a purpose here. Wheeler painted a mental picture of three baseball umpires discussing aspects of their craft. In judging the plays, the first umpire declares that: 'I calls 'em like I see 'em.' The second declares: 'I calls 'em the way they *are*.' The third, who has clearly taken an introductory course on quantum theory, declares: 'They ain't *nothin'* till I calls 'em.'[6]

We will see in Part III that Niels Bohr and his colleagues in Copenhagen favoured the view that quantum measurement involves an inescapable indeterminism that is an inherent feature of the interaction between a quantum system and a measuring device. Although von Neumann was not part of the Copenhagen school, and rejected the need to distinguish between a quantum world of measured systems and a classical world of measuring devices, it is clear that he too believed that measurement involves an essential discontinuity. Albert Einstein and his colleagues stood firm for a completely deterministic approach. The subsequent arguments between Bohr and Einstein led to the development of a series of important experimental test cases, which we will review in Part IV.

State preparation

The process of quantum measurement is intimately linked with the process of quantum state preparation. In the discussion above, we have assumed that it is possible to obtain a quantum system composed of a collection of identically prepared states and then perform measurements on this system. For example, we can ascribe a 50 per cent probability of detecting left-circularly polarized photons which have been passed through a vertically orientated polarization filter, or which emerge from the vertical channel of a calcite crystal. For this statement to make any sense, it presupposes that we have been able to find a means of generating a collection of identically prepared left-circularly polarized photons in the first place.

State preparation must be distinguished from measurement itself. Measurement is defined as an operation performed on a quantum system which probes the state immediately *before* the measurement and yields a 'result' which can be interpreted from some physically registered event or series of events (such as the click of a Geiger counter, firing of a photomultiplier tube, or, most simply, the precipitation of silver atoms in a photographic emulsion). The measurement process changes the state of the system (and often destroys the system altogether). State preparation, however, is an operation performed on a system consisting of many particles, which forces it into an ensemble of identical states. The states referred to are those of the system *after* the operation has taken place. In doing this, much the same process of collapsing the wave function takes place, and not all particles in the original system necessarily survive. For example, to prepare an ensemble of vertically polarized photons, we could pass a series of randomly polarized photons from a conventional light source through a vertically oriented polarizing filter. Many photons will be absorbed by the filter and therefore will not contribute to the ensemble. We can safely assume that those photons that pass through the filter will be vertically polarized.

---

[6] Bernstein, Jeremy (1991). *Quantum profiles*. Princeton University Press, Princeton, NJ, p. 96.

But how do we know that such particles are all in the 'same' state? We could obviously detect the photons passed through the vertical polarizer, but this only tells us that a certain number of photons was passed—it does not really tell us that they were necessarily all vertically polarized. We do know that they will certainly all pass through a second vertical polarizer, but, in truth, we have only our operational *experience* with the manipulation of the physical properties of quantum systems and the manifestation of this experience in the quantum formalism. In general, we know that systems exposed to certain preparation operations will behave in predictable (but probabilistic) ways when subjected to certain measurement operations. The formalism requires an initial state vector and information regarding the measurement eigenstates. The projection amplitudes for the initial state vector and the various measurement eigenstates then allow us to calculate the probabilities for the various measurement outcomes. Much like the situation with phase factors described above and in Appendix 12, we adopt a convention within the formalism that allows us to prepare quantum states and perform measurements on them. Provided we stick to the 'rules', we can expect that the formalism will provide us with accurate predictions.

Entangled states

In our discussion of the Pauli principle in Chapter 4, we considered quantum systems comprising of two indistinguishable particles and drew some general conclusions from the symmetry properties of the corresponding two-particle state vectors. From the context of that discussion, it would be reasonable to suppose that these considerations apply in situations where the two particles are somehow always 'involved' with each other as, for example, two electrons sharing the same atomic orbital with their spins paired.

But the considerations are, in fact, much more general than this. We can create two-particle quantum states through the mutual interaction or simultaneous creation of two particles in some appropriate quantum process. Provided they have interacted in some way at some time in their history, the two particles *must* be described in terms of a composite state vector until such time that one or both undergo further interactions (such as a measurement).

Consider two particles that, as a result of some kind of interaction or creation process, can each take up two possible quantum states. To keep things simple, we will assume that the particles are photons, and the two possible available states are those of vertical and horizontal polarization, $|v\rangle$ and $|h\rangle$. We impose a restriction that the two photons must always emerge from the interaction in opposite states of linear polarization, that is, one vertical and one horizontal (but not both vertical or both horizontal). We label the photons 1 and 2, and recall from our earlier discussion that we have no practical way of knowing which photon is in which state, even though the photons themselves may have other, very different properties (such as different energies and hence frequencies or wavelengths). The composite two-particle state vector therefore has to be written as a linear combination of the product states $|v\rangle_1|h\rangle_2$ and $|h\rangle_1|v\rangle_2$. We would need to take a symmetric combination of these states because photons are bosons.[7]

---

[7] In fact, this composite state vector is the *tensor product* of the individual single-particle state vectors and refers to a composite Hilbert space which is the tensor product of the Hilbert spaces of the single-particle states. See Isham, Chris J. (1995). *Lectures on quantum theory.* Imperial College Press, London, p. 168.

As a result of the interaction and the creation of a two-particle state, the photons have lost their independence. They are said to be *entangled*, and the striking properties of such states led Einstein and his colleagues in 1935 to launch one of the most deeply disturbing challenges to the interpretation of quantum theory and its implications for physical reality. We examine this challenge in detail in Part III.

For our purposes here, it is sufficient to record the consequences of making measurements on such entangled states. Suppose we set up polarizing filters and detectors designed to discover the state of polarization of each photon. This implies a composite apparatus consisting of a measuring device for photon 1, which we account for using the measurement operator $\hat{M}_1$, and a second device for photon 2 which we account for using the operator $\hat{M}_2$. There are four possible outcomes for the combined measurements, which we can express as four possible measurement eigenstates. These correspond to the situation where both photons are detected to be vertically polarized (which we denote $|\psi_{++}\rangle$, where + represents vertical polarization), photon 1 is detected to be vertically polarized, and photon 2 is detected to be horizontally polarized ($|\psi_{+-}\rangle$, where − represents horizontal polarization), photon 1 is horizontally polarized, and photon 2 is vertically polarized ($|\psi_{-+}\rangle$), and both photons are horizontally polarized ($|\psi_{--}\rangle$). In our example, we have already placed restrictions on these possibilities that will be reflected in the mathematical structure of the two-particle state vector and which prevent the possibilities $|\psi_{++}\rangle$ and $|\psi_{--}\rangle$ from being realized in the measurements. The contribution from these measurement eigenstates is therefore zero, and we find that the composite two-particle state vector can be rewritten in terms of the remaining measurement eigenstates as a linear combination of $|\psi_{+-}\rangle$ and $|\psi_{-+}\rangle$. As before, we have no way of knowing which eigenstate will be realized in the next measurement; we only know that there is a 50:50 chance of *either* $|\psi_{+-}\rangle$ *or* $|\psi_{-+}\rangle$.

We now have a situation exactly analogous to that considered in our discussion of the collapse of the wave function, above. When we make a measurement on either of the particles, the composite state vector is 'collapsed' to one of the two possibilities, $|\psi_{+-}\rangle$ or $|\psi_{-+}\rangle$. Thus, if photon 1 is found to be in a state of vertical polarization, this means that the composite state vector has collapsed to the measurement eigenstate $|\psi_{+-}\rangle$, and we know that photon 2 *must* be in a state of horizontal polarization (whether, in fact, we measure it or not). Similarly, if photon 1 is found to be in a state of horizontal polarization, we know that the composite state vector has collapsed to the eigenstate $|\psi_{-+}\rangle$ and that photon 2 must therefore be in a state of vertical polarization.

This whole process seems on the surface to be relatively innocuous. After all, we set up the problem in such a way as to allow only the possibilities vertical–horizontal and horizontal–vertical, and we should not then be surprised to find this reflected in the possible measurement outcomes. But here is the rub. Suppose I now allow the photons to travel undisturbed in different directions, so that they become separated by some enormous distance. By definition, if they are undisturbed, they are still entangled—still collectively described by the two-particle state vector, which can be written in terms of the basis vectors $|v\rangle$ and $|h\rangle$ or the measurement eigenstates $|\psi_{+-}\rangle$ or $|\psi_{-+}\rangle$. I know that if I make a measurement on photon 1, there is a 50:50 chance that the composite state vector will be instantaneously 'collapsed' at that moment to realise a $|v\rangle$ state. This means that the state for photon 2 must similarly reduce to a $|h\rangle$ state, *no matter how far apart the photons are at the moment the measurement is made on photon 1*. This will be discussed in much greater detail later.

Which way did it go?

Before we leave this chapter, it is worth taking a quick look at some of the curious observations that can be made with photons, observations that we must somehow attempt to interpret in terms of quantum theory. These will serve to whet the appetite for the fun which is to follow in the remaining chapters.

Thomas Young explained his observation of double-slit interference using a wave theory of light. A light beam of moderately high intensity incident on two closely spaced, narrow apertures produces an interference pattern consisting of bright and dark fringes. Now imagine that we reduce the light intensity of the source so that only one photon passes through the double-slit apparatus at a time, to impinge on some photographic film. Such experiments can, and have, been performed in the laboratory. After a significant number of photons have passed through, we find that the interference pattern is clearly visible (the equivalent experiment with electrons was described in Chapter 2; see Fig. 2.1, p. 8).

If we assume that an individual photon must pass through one—and only one—slit, we should be able to repeat the experiment using a detector to discover which one. However, when such an experiment is done, we find that the interference pattern is replaced with a completely different pattern corresponding to the diffraction of light through the remaining open slit. The act of removing the detector and unblocking the second slit restores the interference pattern. We conclude that if a photon does pass through one slit, it must be somehow affected by the second, even though it cannot 'know' in advance that the second slit is open. Feynman rationalized this by saying that the photon somehow takes all paths from its source to its destination on the screen, passing through both slits on the way.

We have seen that a calcite crystal can be used to decompose left circularly polarized light into vertical and horizontal components. If we take an identical crystal, and orient it in the opposite sense, we can use it to recombine the vertical and horizontal components and reconstitute the left circularly polarized light (see Fig. 5.6). That such a reconstitution can be achieved has been proved in careful laboratory experiments.

Now suppose that an individual left circularly polarized photon passes through the first crystal and emerges from the vertical channel. The photon enters the vertical channel of the second crystal. At first glance, there seems to be no way of obtaining a left-circularly polarized photon out of this, and yet this is exactly what is obtained as the light intensity passing through the arrangement is reduced to very low levels. A detector can be used to check that the photon passes through one—and only one—channel of the first crystal. The photon therefore appears to be 'aware' of the existence of the open horizontal channel, and is affected by it. Close the horizontal channel by inserting a stop between the crystals, and the possibility of producing a left circularly polarized photon is lost: a vertically polarized photon emerges.

These two examples illustrate that, if we are right in our assumptions about the behaviour of individual photons, the further assumption that they pass through an apparatus as *localized* particles is wrong. The non-locality implied by the particle's dual wave–particle nature means that all paths are followed and gives rise to effects that seem to contradict common sense. According to the orthodox interpretation of quantum theory, the state vector of a quantum particle is non-local: it 'senses' the entire measuring apparatus and can be affected by open slits or channels in polarization analyzers in ways that a localized particle cannot. This is why the nature of the measuring apparatus is believed to be so important. The act of measurement itself—which begins with the blackening of

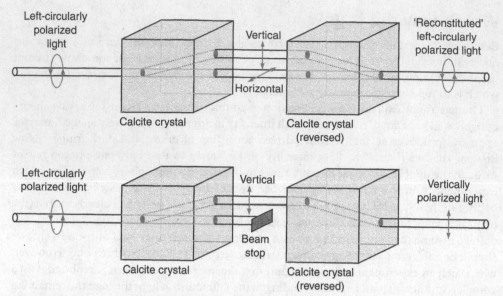

Fig. 5.6 Two calcite crystals placed 'back to back' produce some curious results.

photographic emulsion or the production of an electric current in a photomultiplier—'concentrates' the state vector into a small region of space and hence 'localizes' the quantum particle.

The bomb factory

The importance of non-locality and its role in the measurement process is further highlighted by the following example, based on a 1993 proposal by physicists Avshalom Elitzur and Lev Vaidman.

Suppose you are in charge of quality control in a bomb factory. This factory does not make just any bomb, however. The bombs you make are fitted with a special kind of photosensitive cell and can be detonated by detection of a single photon. Unfortunately, the manufacturing process is poorly refined, and you are aware that in any batch of bombs, there will be some duds. Instead of absorbing incident photons, and triggering an explosion, the photosensitive cells on the dud bombs simply reflect back the photons and the bomb does not go off. Your challenge is to devise a test to identify the dud bombs without triggering too many good ones. At first sight, this seems impossible, as it appears that the only way to tell if the bomb is a dud is to shine light on it and see if it explodes. The good news is that you have taken many advanced courses in quantum theory, and you are very familiar with the phenomenon of quantum non-locality and interference.

The apparatus you devise is shown schematically in Fig. 5.7 and is based on actual experiments devised and performed in 1995 by Paul Kwiat, Harald Weinfurter, Thomas Herzog, Anton Zeilinger at the University of Innsbruck in Austria, and Mark Kasevich at Stanford University in Calfornia. An incident horizontally polarized test photon is passed through an optical switch. This switch can be made transparent for a short time to allow a single photon into or out of the apparatus but otherwise acts as a mirror. The photon then

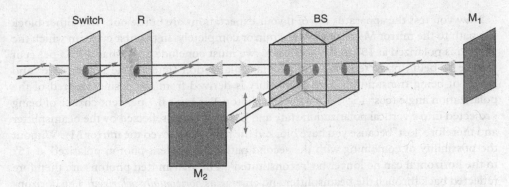

Fig. 5.7 Bomb testing apparatus. A horizontally polarized test photon enters through the optical switch, after which the switch is 'closed' and acts like a mirror, trapping the photon in the apparatus. The polarization of the photon is rotated slightly by the rotator, R, and then passes through a polarizing beamsplitter, BS. Horizontally and vertically polarized components of the photon are reflected back into the beamsplitter by mirrors M_1 and M_2, reconstituting the photon so that it passes back through the rotator and is reflected from the switch. The photon is then rotated further, and the cycle is repeated. After a fixed number of cycles, the switch is opened, and a vertically polarized photon emerges. Blocking the path to M_2 or removing it completely prevents the photon from being reconstituted, and after the same number of cycles, a horizontally polarized photon will emerge from the apparatus.

passes through a polarization rotator, R. This device simply rotates the plane of polarization of the incident photon through an angle, φ, set by the experimenter (see the axis convention given in Fig. 5.2). For the sake of simplicity, we will assume that the rotator only acts on photons travelling one way through the device (from left to right in Fig. 5.7). Photons travelling from right to left pass through with their polarization unchanged. For reasons that will become apparent, you set the rotator to turn the polarization through a small angle, say 15°. After passing through the rotator, the photon then enters an optical device called a polarizing beamsplitter, BS. Horizontally polarized photons are transmitted by the beamsplitter, and vertically polarized photons are reflected. The photon may therefore emerge from the two channels of the beamsplitter—corresponding to transmission or reflection. Two mirrors, M_1 and M_2, placed at equal distances from the beamsplitter, send the photon back into the beamsplitter and 'reconstitute' the incident photon polarized at 15° to the horizontal. (This is exactly analogous to the reconstitution of left-circularly polarized photons from vertically and horizontally polarized photons passed back through a calcite crystal, as shown in Fig. 5.6.)

Now the photon travels backwards through the rotator (which does not affect the polarization) where it encounters the optical switch once more. The switch is now 'closed', meaning that it acts like a mirror. The photon is reflected back into the apparatus, where the plane of polarization is now turned a further 15° (to 30°) by the rotator, and the whole cycle occurs once more. You set the optical switch to open after a time corresponding to six cycles of the photon through the apparatus. After six cycles, the plane of polarization of the test photon has been rotated a total of six times 15°, or 90°. The photon therefore emerges as vertically polarized and can be detected as such. So far, so good.

Now you test the apparatus to see if your expectations are borne out. You either block the path to the mirror M_2 or remove the mirror completely. In the first cycle, in which the photon is polarized at 15° to the horizontal, we must conclude that it has a 93.3 per cent chance of being projected into a horizontal polarization state by the beamsplitter and hence of being transmitted. This probability is derived from the cosine-squared of the polarization angle ($\cos^2 15° = 0.933$). The photon also has a 6.7 per cent chance of being projected into a vertical polarization state and hence of being reflected by the beamsplitter and therefore 'lost' because you have blocked the path or removed the mirror M_2. Without the possibility of combining with the second path, the incident photon polarized at 15° to the horizontal can no longer be 'reconstituted'. The transmitted photons are therefore reflected back through the beamsplitter and emerge as *horizontally polarized*. The horizontally polarized photon passes backwards through the rotator and is reflected by the switch to repeat the cycle. But it is immediately apparent that, as a result of having blocked the path or removed M_2, the photon will *never* be rotated by more than 15°.[8] In fact, after six cycles, the photon has a nearly two-thirds probability (0.933 raised to the sixth power) of emerging horizontally polarized and a little over one-third probability (0.067 raised to the sixth power) of not emerging at all.

You have one further refinement to make before using this apparatus to test the bombs. If you now set the polarization rotator to turn the plane of polarization by no more than 0.01°, and set the optical switch to open after 9000 cycles, in the case where you remove M_2 you can increase the probability of detecting a horizontally polarized photon to 0.9997, reducing the probability of 'losing' the photon to 0.0003. You are now ready. You replace M_2 with a bomb, sealed inside a bomb-proof chamber fitted with a small window through which a single photon can enter and leave. If the bomb is good, this means that any photon incident on the photosensitive cell will be absorbed and therefore cannot be reflected back into the beamsplitter. If the bomb is a dud, the photosensitive cell acts like a mirror and therefore can reflect the photon back into the beamsplitter. The possibility of reconstituting the photon in the beamsplitter therefore depends on whether the bomb is good or not. You open the optical switch and let in a single horizontally polarized photon. After a time corresponding to 9000 cycles, you open the switch and determine the polarization state of the photon that emerges. If it is vertically polarized, you know the bomb is a dud. If it is horizontally polarized, you know the bomb is good. There is a small, 0.03 per cent chance that a good bomb will be exploded by your test apparatus, but you could reduce this possibility even further by reducing the polarization rotation angle and increasing the number of cycles.

It should by now have become clear that we can set up this apparatus in such a way as to reduce the possibility of 'sampling' the reflection path from the beamsplitter to very low levels indeed. And yet the result of the experiment is entirely dependent on this possibility, no matter how small it is. The end result is a near interaction-free measurement: we can tell whether a bomb is good or not by the fact that it does or does not create the *possibility* of a photon path. Even though no photon has interacted directly with the photosensitive

---

[8] This is related to the so-called 'quantum Zeno effect', an inhibition of transitions to some final quantum state as a result of frequent projections of the initial or intermediate states, first elaborated by Baidyanath Misra and George Sudarshan in 1977. In effect, this means that frequent measurements on a quantum system effectively suppress its evolution to some final state—a quantum equivalent of the old adage that 'the watched pot never boils'.

cell of the bomb, we can discover—with certainty—that either the bomb is good or it is a dud. This is *seeing without looking*.

No description of the quantum formalism is possible without reaching for elements of an interpretation. Although I have tried to take care to preface many of the conclusions drawn in this and the previous chapter with the phase 'the orthodox interpretation', the fact is that this interpretation is the one taught to the majority of undergraduate chemists and physicists. It has therefore been important to go through it, if only so that readers can recognize it for what it is. If we are prepared to accept this interpretation, there are a number of logical consequences. Many scientists find these consequences so unacceptable that they reject the theory as somehow incomplete. This was Einstein's view. We will now examine these consequences in detail.

Part III

Meaning

6

The schism

The strange behaviour of photons described in the last chapter immediately raises all sorts of difficult questions about the meaning of quantum theory. We might be encouraged to look for this meaning by going back to the theory's mathematical formalism, perhaps by trying to find out how we might better interpret some of its elements. However, it is a central argument of this book that, no matter where we look, we are always led back to philosophy. At first sight, this might seem to be an odd thing to claim. After all, modern textbooks on quantum physics and chemistry rarely (if ever) discuss philosophy. We accept that the behaviour of photons is strange, but surely it is something that we can at least study experimentally—do we need philosophy in these circumstances? The answer is most certainly yes. Quantum theory directly challenges our understanding of the nature of matter and radiation, and the process of measurement at their most elementary levels, and we cannot go forward unless we adopt some kind of interpretation, some way of trying to make sense of it all. As we will see, this interpretation has to be based on some philosophical position.

I will argue this point by first showing that the orthodox interpretation of quantum theory developed by Niels Bohr and his colleagues in Copenhagen (the one taught by design or default to most modern undergraduate scientists) is derived from a particular philosophical outlook known as *positivism*. It is 'anti-realist' in terms of its approach to unobservable entities and theory construction and use and in terms of its approach to the interpretation of key theoretical concepts. Furthermore, the standard mathematical formalism of quantum theory described in Part II itself reflects this anti-realism, both in the way it is founded (in terms of its postulates) and in the way it is meant to be applied, particularly where measurement is concerned. This must be contrasted with the realist position, which sees attempts at constructing scientific theories and their concepts as evolving an understanding of the world based on a description of real entities capable of real effects and which approximate ever more closely to the 'truth' of some underlying, independent reality.

There are many shades of grey between the extremes of black and white: in this case between naive realism on the one hand and unashamed positivism on the other. It will be very helpful therefore to look briefly at the arguments that shaped the philosophy of science in the twentieth century and which so strongly informed the debate over the interpretation of quantum theory in the 1920s and 1930s. In particular, understanding the nature of the divide between the realist and anti-realist positions will provide a framework for discussion of the great debate between Bohr and Einstein in the next chapter.

Do not be misled into thinking that arguments about philosophy are ultimately futile or irrelevant to important matters of concern to the experimental scientist. That this is not so will be amply demonstrated in Part IV.

The scientific method

When asked to provide a rationale for the ways in which science makes progress, most scientists (and many non-scientists) would cite the 'scientific method'. They would explain that scientists begin by gathering a large body of empirical data from which certain generalizations may be made. These generalizations may be elevated to the status of scientific 'laws'. The generalizations or the laws form a framework that we might refer to as a scientific theory, from which it is possible to deduce predictions of previously unobserved phenomena or the results of new experiments that either verify or falsify the theory. The generalizations or the laws are either thrown out or modified, new predictions are made, and the cycle of hypothesizing and testing continues.

Credit for the articulation of this scientific method is usually given to Francis Bacon who, under James I in the seventeenth century, became successively Attorney-General, Lord Keeper of the Great Seal, Lord Chancellor, Baron Verulam, and Viscount St. Albans. At the height of his influence, he was prosecuted by his bitter rival, Edward Coke (whom he had replaced as Attorney General), for taking bribes in his capacity as Lord Chancellor. Bacon did not deny the charges, but he did rather disingenuously argue that the bribes had not influenced any of his decisions. He was fined and imprisoned, but such was his relationship with the king that his fine was soon returned, and he was released from prison after a few days. He devoted the rest of his life to science and philosophy, his interest in scientific experimentation ultimately proving fatal—he died from a chill caught whilst trying to preserve a chicken by stuffing it with snow.

Much of the content of Bacon's scientific method can actually be found in Aristotle's theory of scientific procedure, although Bacon was aggressively critical of an Aristotelian method which tended to leap to 'immutable truths' from a handful of very limited observations. Nevertheless, Bacon was a highly successful propagandist and considerably raised the profile of organized scientific research as a means of securing new benefits for mankind and reclaiming dominion over nature, often in a highly colourful language that some modern commentators have identified as early mythologizing of the function, purpose, and ultimate goals of science. He argued for a cooperative approach to research which led, some years after his death, to the implementation of many of his projects by the Royal Society. It is perhaps in this sense that Bacon's influence on the ultimate development of modern science is felt most strongly.

Irrespective of Bacon's role, the generation of knowledge and understanding through the process of inductive inference remained a central, unquestioned tenet of the scientific method for several hundred years. This was to change in the early years of the nineteenth century with the rise of a more empiricist outlook and attempts to establish a formally logical foundation for mathematics and philosophy. The problem, first highlighted by the great Scottish philosopher David Hume and elaborated in the Canons of John Stuart Mill, was simply stated. If the laws of science were to be built through inductive inference substantiated or suitably modified as a result of experiment or observation, those laws *could never be certain*.

A relevant example favoured by philosophers of knowledge concerns ravens. After a long series of observations made on ravens all around the world, we might generalize our data into a simple 'law of black ravens'. The law states simply that all ravens are black. This would lead us to predict that the next raven observed (and any number of ravens that might be observed in the future) should conform to this law and therefore be black in colour. But we would have to admit that no matter how many observations we had made, there would always remain the possibility of observing a raven of a different colour, contradicting the law and necessitating its abandonment or revision. The probability of finding a non-black raven might be considered vanishingly small, but it could never be regarded to be zero, and so we could never be certain that the law of black ravens would hold for all future observations. This is a conclusion reinforced by the experiences of Europeans—who might have formulated a similar 'law of white swans'—when they first observed *black* swans (*Cygnus atratus*) around the lakes and rivers of Australia and New Zealand.

The problem of induction

In his book *The problems of philosophy*, the British philosopher Bertrand Russell put it this way[1]:

We are all convinced the sun will rise tomorrow. Why? Is this belief a mere blind outcome of past experience, or can it be justified as a reasonable belief?... If we are challenged as to why we believe that it will continue to rise as heretofore, we may appeal to the laws of motion: the earth, we shall say, is a freely rotating body, and such bodies do not cease to rotate unless something interferes from outside, and there is nothing outside to interfere with the earth between now and tomorrow... But the real question is: Do *any* number of cases of a law being fulfilled in the past afford evidence that it will be fulfilled in the future? If not, it becomes plain that we have no ground whatever for expecting the sun to rise tomorrow... It is to be observed that all such expectations are only *probable*; thus we have not to seek for a proof that they *must* be fulfilled, but only for some reason in favour of the view that they are *likely* to be fulfilled.

Russell generalized the problem to one of belief in what he called (paraphrasing Hume) the 'uniformity of nature', the belief that laws derived from observations of the natural world will continue to apply in the future with no exceptions. His attempted resolution of the problem was to elevate induction to the status of a *principle*.

The problem of induction (and its resolution or otherwise) represented a point of departure for a number of different philosophical approaches. One of these, still popular today, accepts a subjective element in the process of scientific reasoning and acknowledges that we can never achieve certainty, only an improved probability that our theories are right that represents a precise mathematical statement that the theory approaches evermore

---

[1] Russell, Bertrand (1912). *The problems of philosophy*. Oxford University Press, Oxford, pp. 33–34.

closely to some kind of final truth. The *Bayesian* approach was developed by a group of Cambridge philosophers (which included John Maynard Keynes) in recognition of the fact that science appears to deal implicitly with probabilities which can reflect the 'degree of belief' scientists should hold for a particular theory in the light of results of experimental or observational tests. The group drew on a structure first developed by an eighteenth-century clergyman called Thomas Bayes which is based on the principles of probability calculus. In this approach, the scientist assigns a prior probability to a theory, a numerical measure of his belief in its basic truth. This is then tested by experiment or observation, modifying the prior probability to yield a posterior probability. The test provides evidence in favour of the theory if the posterior probability is greater than the prior probability, evidence against if the posterior probability is lower. The posterior probability then becomes the new prior probability in a future test.

What is contentious in this approach is the origin of the early prior probabilities, which, by definition, must be derived from the scientist's subjective assessment of the theory's worth. This assessment is likely to be coloured by the scientist's own value set, with perceptions based on properties such as beauty, simplicity, or empirical adequacy. If it is further accepted that most practising scientists never actually attempt to quantify their degree of belief in a particular theory in the form of a calculated probability, there is much in a form of qualified Bayesianism that would appear to mirror actual scientific practice.

Others, such as Karl Popper, argued that the problem of induction could not be solved and rejected induction altogether as a valid basis on which to build a scientific method. Karl Popper was one of the last century's most influential philosophers of science. Born in Vienna in 1902, he discussed the interpretation of quantum theory directly with its founders—Einstein, Bohr, Schrödinger, Heisenberg, Pauli, *et al.*—and continued the debate into his nineties, adding to a prolific output of writings on the subjects of quantum theory, the philosophy of science, and the evolution of knowledge. This output began in 1934 with the publication in Vienna of his now famous book *Logik der forschung*, first published in English in 1959 as *The logic of scientific discovery*. Popper believed that as there is no rational, objective process by which theories can be inductively inferred (and therefore no objective way of establishing a prior probability), science should proceed by generating hypotheses (referred to by Popper as conjectures) by whatever means seems reasonable and then seek logically to refute them through experimental or observational tests. This is an extension of the 'hypothetico-deductive' approach—the natural alternative to the inductivist view—in which problems are solved through a constant process of creating hypotheses, deducing their consequences, and testing to see which of these consequences are indeed borne out. Popper elevated 'falsifiability' to the status of a principle and made it the cornerstone of his philosophy of science, using it both as a criterion for defining what constitutes a scientific theory and as a scientific method. He argued that we make real progress in science *only when a theory is falsified*.

Popper's account of scientific discovery continues to find many adherents amongst the community of practising scientists, although it has been dismissed by a large cross-section of the community of modern philosophers of science. There are many reasons for this rejection, but the principal reasons concern the basic motivation of scientists for carrying out experiments or observations and their actual behaviour on encountering potentially falsifying data. We will examine these arguments later in this chapter.

A further approach to the problem of induction was to dismiss it as a non-problem and shift attention to more practical issues. This was the view of the philosopher A. J. Ayer, who wrote[2]:

Thus it appears that there is no possible way of solving the problem of induction, as it is ordinarily conceived. And this means that it is a fictitious problem, since all genuine problems are at least theoretically capable of being solved: and the credit of natural science is not impaired by the fact that some philosophers continue to be puzzled by it. Actually, we shall see that the only test to which a form of scientific procedure which satisfies the necessary condition of self-consistency is subject, *is the test of its success in practice*. We are entitled to have faith in our procedure just so long as it does the work which it is designed to do ...

Ayer proposed that it does not matter so much *how* we arrive at our ideas for scientific theories. Induction is as good as, if not better than, many alternatives, and we can continue to trust it for as long as it continues to provide a useful guide to theory-making. Much more important is whether or not the theory's predictions are borne out by experiment or observation.

Ayer wrote these words in his seminal book *Language, truth and logic*, first published in January 1936 when he was just 25. He had not long returned from Vienna where he had become a convert to a radical new wave of philosophical thought that was sweeping through continental Europe. The book established him as the movement's British spokesman. The movement's outlook was marked by a striking rejection of any untestable, unverifiable, and hence purportedly metaphysical statements as non-scientific. It was this outlook that was beginning to dominate philosophical thought at about the time quantum theory was being developed and just as its interpretation was being shaped.

Logical positivism and the rejection of metaphysics

The empiricist tradition can be traced back to Hume, through the French philosopher Auguste Comte and the physicist Ernst Mach. Mach was professor of physics at the Universities of Prague and Vienna from 1867 to 1901. Drawing on and extending the empiricist approach, he argued that scientific activity involves the study of facts about nature revealed to us through our sensory perceptions (perhaps aided by some instrument) and the attempt to understand their interrelationship through observation and experiment. According to Mach, this attempt should be made in the most economical way.

Mach rejected as non-scientific any statements made about the world that are not *empirically verifiable*. What do we mean here by the word empirical? The dictionary definition identifies empirical as purely experimental (i.e. without reference to theory) so that statements which are not experimentally (or observationally) verifiable are rejected as non-scientific. However, this definition does not seem to tell the whole story. Science is certainly not about the mindless collecting of empirical facts about nature; it is about interrelating those facts and making predictions on the basis of some kind of theory. According to Mach, the key question concerns not the way in which the theories themselves are arrived at but rather *how* the theories and the concepts they use should be interpreted and understood.

[2] Ayer, A. J. (1936). *Language, truth and logic*. Victor Gollancz (republished by Penguin Books, London, p. 35). Emphasis added.

Let us take a specific example. The philosophers of ancient Greece developed a cosmological model that placed the earth at the centre of the universe. An essential element in that model was the ideal of the perfect circle, and the motions of the stars around the earth appear to conform to this ideal. However, as seen from the earth, the motions of the planets are far from circular. In about AD 150, the philosopher Ptolemy attempted to explain the observed motions of the planets around the earth by constructing an elaborate theory based on epicycles—combinations of circles in which the ideal was at least preserved. To a certain extent, he succeeded, but as observations became more accurate, he found that he had to add more and more epicycles. Now, Ptolemy's statements about the motions of the planets are empirically verifiable: if we use the theory in the prescribed manner, we would expect to be able to compare them with observations and so verify that they describe the motions of the planets (albeit with limited accuracy).

In fact, we can easily imagine that we could develop a modern refinement of the Ptolemaic system, with a very large number of epicycles, and that with a little computer power, we should be able to make some fairly accurate predictions about the motions of the planets. Does this mean that we should regard the epicycles to be 'real' in the sense that they represent real elements of the dynamics of planetary motion? Perhaps our immediate reaction is to say 'Of course not!'. But why not? Ptolemy's difficulties were created by his assumption that the sun and planets orbit the earth, whereas a much more *economical* theory places the sun at the centre of the solar system, as suggested by Copernicus. But does this system necessarily represent reality any better than Ptolemy's?

Mach's point was that there is no purpose to be served by seeking to describe a reality beyond our immediate senses. Instead, our judgement should be guided by the criteria of verifiability (does the theory agree with experimental observations?) and simplicity (is it the simplest theory that will agree with the experimental observations?). Thus, if both the Ptolemaic and Copernican systems can be developed to the point at which they make identical predictions for the motions of the planets, our choice should be based solely on their relative conceptual or mathematical simplicity. In this case, the Copernican system wins out because it is the simplest.

In constructing a physical theory, we should therefore seek the most economical way of organizing facts and making connections between them. We should not attach a deeper significance to the concepts used in a theory (such as epicycles) or the entities they describe if these are not in themselves observable and hence subject to empirical verification. According to Mach, only those elements that we can perceive actually exist, and there is no point in searching for a physical reality that we cannot perceive: we can know only what we experience. Mach's criterion of what constituted a verifiable statement was particularly stringent. It led him to reject the concepts of absolute space and absolute time, and to side with Ludwig Boltzmann's opponents in rejecting the reality of atoms and molecules.

Speculations that are intrinsically not verifiable, that involve some kind of appeal to the emotions or to faith, are not scientific. However, these speculations, which belong to a branch of philosophy called metaphysics (literally, 'beyond physics'), are not rejected outright. They are recognized as a legitimate part of the process of developing an attitude towards life, but they are perceived to have no place in science. This emphasis on verifiability and an unmerciful attitude to the elimination of metaphysics from science is a philosophical position generally known as positivism, a name first coined by Comte.

Mach placed particular emphasis on the correct use of language, calling it 'the most wonderful economy of communication'. His views were enormously influential in the development of a new school of philosophical thought that emerged in Vienna in the early 1920s. Centred around Moritz Schlick, professor of philosophy at the University of Vienna, Rudolf Carnap, Otto Neurath, and others, the 'Vienna Circle' extended the positivist outlook through the use of formal logical analysis. They drew their inspiration from a wide variety of sources, particularly the work of the physicists Mach, Boltzmann, and Einstein. Philosophically, their particular brand of positivism was foreshadowed in the work of David Hume and Auguste Comte, and they were greatly influenced by the analytical approaches of their contemporaries Bertrand Russell in Cambridge and Ludwig Wittgenstein (a former student of Russell's) in Vienna.

The Vienna Circle began with the contention that the only true knowledge is scientific knowledge and that, in order to be meaningful, a scientific statement has to be both formally logical and verifiable. The foundations of their philosophy, which is sometimes known as logical positivism, was logical analysis, the criterion of verifiability and a strict demarcation between what are accepted to be scientific statements and metaphysical statements.

Scientists might think it rather obvious that science has to be logical. But the rigorous application of formal logic actually leads to an exhaustive analysis of the use of language and the meaning of words. This is necessary to rule out tautological, self-contradictory or meaningless statements. At times, logical positivism appears more like philology than philosophy. Of course, we would never accept mathematical statements that use undefined terms or are self-contradictory: why should we expect less from language?

Most importantly, the use of logical analysis leads to the elimination of all metaphysical statements as meaningless. With one stroke, the logical positivists eliminated from philosophy centuries of 'pseudostatements' about mind, being, reality, and God, reassigning them to the arts alongside poetry and music. The views of the Vienna Circle came to dominate the philosophy of science in the middle of the last century.

The implications for the development of physical theories are clear. Theories are merely instruments[3] for making connections between observations or the results of experiments in the most economical way possible. If they describe the behaviour of entities that we cannot directly perceive, then the entities themselves are merely convenient theoretical devices and are no more real than Ptolemy's epicycles. Statements that the theory makes about the physical world are logical statements connecting the concepts of the theory but they are *not* statements about some underlying independent reality as such a reality is metaphysical and therefore meaningless. This does not necessarily mean that there is no such thing as reality, but it does mean that we have to temper our expectation. Theories describe elements of an *empirical reality*—the reality manifested as effects that we can directly perceive and hence verify—but do not expect to be able to go beyond this empirical level. To do so is to engage in meaningless speculation.

The Copenhagen interpretation

We saw in Chapter 2 that Schrödinger and Heisenberg adopted very different positions with regard to the interpretation of quantum theory. Schrödinger was a realist: he believed

---

[3] Such an outlook is also sometimes called *instrumentalism*.

that his wave mechanics provided part of a description of an underlying independent reality. Heisenberg took a fairly uncompromising positivist stance, insisting that his matrix mechanics served its purpose as nothing more than an algorithm through which the results of experimental observations could be correlated and new predictions made. When Schrödinger demonstrated that the two approaches are mathematically equivalent, physicists were presented with a clear choice. This was more than just a choice between two equivalent mathematical formalisms: it was a choice between different philosophies.

Schrödinger's wave mechanics was the more popular, because of its instinctive (a positivist would say emotional or metaphysical) appeal. It held the promise that its further development might reveal a little more of that underlying independent reality, perhaps one of wave fields and their superpositions. In October 1926, Bohr invited Schrödinger to join with him and Heisenberg in Copenhagen to debate the issues, but Schrödinger remained stubbornly unconvinced by their arguments. Their failure to persuade Schrödinger made Bohr and Heisenberg more determined than ever to find a radical new interpretation of quantum theory.

However, Bohr and Heisenberg themselves had different, deeply held views. As we have seen, they argued bitterly over the interpretation of the theory. Bohr believed that some form of wave interpretation was essential and rejected an approach based on point-particles. But Heisenberg identified waves with his rival, Schrödinger. In February 1927, Bohr departed to Norway for a skiing holiday, leaving Heisenberg in Copenhagen to marshal his thoughts and write his now famous paper on the uncertainty principle; a paper which he believed would completely demolish Schrödinger's wave field idea. When Bohr returned, he launched into Heisenberg's finished paper, treating it much like he would treat a first draft of one of his own papers. Heisenberg was dismayed: he wanted to publish his paper as quickly as possible to gain the upper hand in the debate with Schrödinger. Eventually, Wolfgang Pauli stepped in to referee the ensuing argument between Bohr and Heisenberg on the interpretation of the uncertainty principle, and was probably himself the architect of the dualistic wave–particle approach. A consensus was reached, which was not so much a 'united front' on the question of interpretation but more of an amalgamation and compromise of the very different views of the group.

These three physicists (see photograph) developed what became known as the *Copenhagen interpretation of quantum theory*. Its foundations are the uncertainty principle, wave–particle duality, Born's probabilistic interpretation of the wave function, and the identification of eigenvalues as the measured values of observables. It is the interpretation that pervades the very core of the mathematical formalism developed subsequently by John von Neumann (although the Copenhagen view differed from von Neumann's regarding some of the subtler aspects of quantum measurement and von Neumann was never part of the Copenhagen school). It is an interpretation that is so well entrenched in physics that many students are surprised to discover that there are alternatives. We can now admit that the references to quantum theory's orthodox interpretation, made many times in Part II, are actually references to the Copenhagen interpretation.

The Copenhagen interpretation requires that we consider very carefully the methods by which we acquire knowledge of the physical world. It shifts the focus of scientific activity from the objects of our studies to the relationships between those objects and the instruments we use to reveal their behaviour: the instrument takes centre stage, alongside the object, and the distinction between them is considerably blurred.

Between them Bohr, Heisenberg, and Pauli formulated an interpretation of the new quantum theory founded on the dual wave–particle properties of quantum entities. This came to be known as the Copenhagen interpretation. This photograph is from the Niels Bohr archive, courtesy AIP Emilio Segrè Visual Archives.

According to this interpretation, it is not meaningful to regard a quantum particle as having *any* intrinsic properties independent of some measuring instrument. Although we may speak of electron spin, velocity, orbital angular momentum, and so on, these are properties we have assigned to an electron for convenience—each property becomes 'real' only when the electron interacts with an instrument specifically designed to reveal that property. Agreement between theory and experiment allows us to do no more than interpret these concepts as elements of an empirical reality. These concepts help us to correlate and describe our observations, but they have no meaning beyond their use as a means of connecting the object of our study with the instrument we use to study it.

Thus, when we make a statement such as 'This photon has vertical polarization', we should also make reference to (or at least be aware of) the experimental arrangement by which we have come by that knowledge. We might modify our statement thus: 'This photon was generated in such-and-such a way and was transmitted through a polarizing filter with its axis of maximum transmission oriented vertically with respect to some laboratory reference frame. Its passage through the filter was confirmed by the generation of a blackened spot on a piece of photographic film. This photon therefore combined with the instrument to reveal properties we associate with vertical polarization.' Note the emphasis on the past: in making the measurement, the state of the photon was certainly changed irreversibly.

Bohr insisted that we can say nothing at all about a quantum particle without making very clear reference to the nature of the instrument which we use to make measurements on it. If our instrument is a double-slit apparatus, and we study the passage of a photon through it, we know that we can understand the physics of the photon–instrument interaction using the wave concept as expressed in the photon's wave function or state vector. If our instrument is a photomultiplier or a piece of photographic film, we know that the photon–instrument interaction can be understood in terms of a particle picture. We can design instruments to demonstrate a quantum particle's wave-like properties *or* its particle-like properties, but we cannot demonstrate both simultaneously. According to the Copenhagen interpretation, this is not because we lack the ingenuity to conceive of such an instrument, but because such an instrument is inconceivable.

As scientists, we perhaps find it difficult to resist the temptation to conjure up a mental picture of an individual photon existing in some kind of predetermined state independently of our measurements. But according to the Copenhagen interpretation, such a mental picture would be at best unhelpful and at worst positively misleading.

Complementarity

Bohr summarized his views in a lecture delivered to a meeting of physicists on 16 September 1927 at Lake Como in Italy. It was during this lecture that he introduced his concept of *complementarity*. This concept went through many refinements and restatements but now tends to be presented in terms of wave–particle duality. Bohr argued that although the wave picture and the particle picture are mutually exclusive, they are not contradictory, but complementary. For Bohr, complementarity lay at the heart of the strange nature of the quantum world. The uncertainty principle becomes merely a mathematical statement expressing the limits imposed on our ability to make measurements based on complementary concepts of classical physics. Specifying the position and momentum of a particle with absolute precision might seem to be possible in principle but becomes problematic as soon as we consider the quantum entity's wave-like characteristics. The mathematical formalism of quantum theory then becomes an attempt to repackage complementary wave and particle descriptions in a single, all-encompassing theory. This does not imply that the theory is wrong or somehow incomplete. On the contrary, it is the best we can do and goes as far as we can go.

Because of the emphasis placed on the importance of the observer or observing instrument, many physicists and philosophers have accused the Copenhagen interpretation of depending on the choices made by the subject, that is, they accuse the interpretation of being *subjective*. Clearly, the subject (the observer) appears to exercise remarkable powers over reality, with the freedom to choose what kind of reality is to be exposed in a measurement. In the language of quantum measurement described in Chapter 5, a simple reorientation of a polarizing filter changes instantaneously the measurement eigenstates of a quantum system, thereby changing the nature of the reality that can be examined. The whole process of expanding the state vector in terms of the measurement eigenstates then becomes a subjective process—what is written down depends upon the subject's personal preferences, not on the independent, objectively real properties of the object under study.

This charge of subjectivism is somewhat unfair. In many of his most oft-quoted statements, Bohr insisted that he was searching for objectivity. But his was the weaker objectivity that we can associate with positivism rather than the strong objectivity of the realist. The state vector of the Copenhagen interpretation might not reflect an objectively real behaviour, but the results of individual measurements are objectively real, and the information communicated by one physicist to another about the methods used to analyse the state vector in terms of the measuring instrument (using the language of classical physics) means that the experiment can be repeated, the experiences shared, and the interpretation understood.

According to postulate 4, the time-dependent Schrödinger equation predicts that the state vector evolves in a way that is quite deterministic. Once the initial conditions have been established, the future behaviour of the state vector is predictable through the quantum-mechanical laws of motion. However, this determinism does not apply to our classical conception of space–time. It is not a determinism in any 'normal', classical sense of the word. In order to deal in a practical way with the state vector, it has to be projected into a form that we can recognize within the reference frame provided by classical physics—we must make a measurement on it. But the act of measurement destroys the continuity provided by the time-dependent Schrödinger equation. Heisenberg again[4]:

Our actual situation in research work in atomic physics is usually this: we wish to understand a certain phenomenon, we wish to recognise how this phenomenon follows from the general laws of nature. Therefore, that part of matter or radiation which takes part in the phenomenon is the natural 'object' in the theoretical treatment and should be separated in this respect from the tools used to study the phenomenon. This again emphasises a subjective element in the description of atomic events, since the measuring device has been constructed by the observer, and we have to remember that what we observe is not nature in itself but nature exposed to our method of questioning.

This quotation nicely captures the anti-realist flavour of the Copenhagen interpretation.

Although they were not particularly concerned to devote much time to the elaboration of their philosophy, the members of the Copenhagen school of physicists were nevertheless aware that their interpretation created considerable problems for the understanding of what constitutes knowledge and the methods of its acquisition. Heisenberg in particular made it his business to raise awareness of these issues in his many public addresses on the subject. Eventually, the philosophers began to take note, and, in correspondence spanning the years 1930–1932, Moritz Schlick of the Vienna school of logical positivists sought advice and direction from Heisenberg on quantum theory's implications for causality and the philosophy of knowledge.[5] Clearly, the positivists of the Vienna Circle found much resonance with what the Copenhagen physicists were saying.

We should recognize what we are dealing with here. The Copenhagen interpretation essentially states that in quantum theory, we have reached the limit of what we can know. To try to go beyond this limit is pointless (how can we ever hope to know something that is unknowable?). The argument is that any attempt to introduce a new concept to describe an underlying independent reality inevitably involves a reworking of familiar classical concepts and a descent into metaphysics. We always return to the idealized concepts that summarize the fullest extent of our knowledge—waves and particles.

[4] Heisenberg, Werner (1989). *Physics and philosophy*. Penguin, London, pp. 45–46.

[5] Cassidy, David C. (1992). *Uncertainty: the life and science of Werner Heisenberg*. W. H. Freeman, New York, pp. 255–256.

It is interesting to note that in some branches of physics, scientists have long since given up making such attempts. The families of quarks described in Chapter 3 are now generally accepted as elements of an empirical reality—they contribute to reproducible effects that can be observed in large-scale experiments without themselves ever being revealed (indeed, by definition they are 'confined' inside larger particles). The names given to the different types of quarks are intentionally abstract: they are intended only to provide an economical means of communicating their properties and conceptual status in a somewhat abstract theory. This is Bohr's philosophy writ large.

There is no quantum world

Bohr had developed his own distinctive philosophy even before he became a physicist. Interestingly, Bohr's emphasis was also on the use of language, and he is quoted as saying[6]:

Our task is to communicate experience and ideas to others. We must strive continually to extend the scope of our description, but in such a way that our messages do not thereby lose their objective or unambiguous character.

This sentiment was translated through to Bohr's scientific papers, the drafting of which would involve seemingly endless searching for just the right words or phrases that would communicate exactly what Bohr meant to say.

However, this emphasis on language went far beyond wordplay. It transcended forms of written and verbal communication and included the sum of human experience. Bohr argued that we, as experimental scientists, design, perform, interpret, and communicate the results of our experiments using the concepts of *classical* physics. We understand how large-scale laboratory instruments work only in terms of classical physics. The effect of an event occurring at the level of an individual quantum particle must be somehow amplified, or otherwise turned into some kind of macroscopic signal (such as a deflection of a pointer on a voltage scale) so that we can perceive and measure it. Our perceptions function at the level of classical physics, and the only concepts with which we are entirely familiar, and for which we have a highly developed language, are classical concepts.

In his book *Physics and philosophy*, published in 1962, Heisenberg wrote that the Copenhagen interpretation of quantum theory actually rests on a paradox. This is the paradox of describing quantum phenomena in terms of idealized classical concepts. We only know of waves and particles—these are the concepts we have inherited from the experiences registered in our daily lives and from a long tradition of classical physics. This interpretation requires that we accept that we can never 'know' quantum concepts: they are simply beyond human experience and are therefore metaphysical. A quantum entity is neither a wave nor a particle. Instead, we substitute the appropriate classical concept—wave or particle—as and when necessary.

---

[6] Petersen, Aage, in French, A. P. and Kennedy, P. J. (eds.) (1985). *Niels Bohr: a centenary volume*. Harvard University Press, Cambridge, MA, p. 301.

Compare the statement credited to Bohr[7]:

There is no quantum world. There is only an abstract quantum physical description. It is wrong to think that the task of physics is to find out how nature is. Physics concerns what we can say about nature.

with the following comment on logical positivism by A. J. Ayer[8]

The originality of the logical positivists lay in their making the impossibility of metaphysics depend not upon the nature of what could be known but upon the nature of what could be said.

There is little doubting the positivism of the young Heisenberg, but was Bohr himself a positivist? Here, we must admit shades of grey. A careful analysis of Bohr's philosophical influences and his writings on the Copenhagen interpretation and complementarity suggests that, philosophically, he was closer to the tradition known as *pragmatism* than to positivism.[9] Pragmatism, founded by the American philosopher Charles Sanders Pierce, traces its lineage to Hegel and has many of the characteristics of positivism in that they both roundly reject metaphysics. There are differences, however. The positivist doctrine is one of 'seeing is believing', and so what we can know is therefore limited by what we can observe. The pragmatist doctrine admits a more practical (or, indeed, pragmatic) approach to the reality of entities—such as electrons—whose properties and behaviour are described by theories and which produce secondary observable effects but which themselves cannot be seen. According to the pragmatist, what we can know is limited not by what we can see but by what we can *do*. It seems logical that the father of modern atomic theory would want to accept the reality of atoms. But Bohr placed limits on what a theory of the internal structure of the atom could say. He argued that we live in a classical world, and our experiments are classical experiments. Go beyond these concepts, and you cross the threshold between what you can know and what you cannot.

Positivist or pragmatist, the most important feature of Bohr's philosophy is that it was principally anti-realist. It denied that quantum theory has anything meaningful to say about an underlying physical reality that exists independently of our measuring devices. It denied the possibility that further development of the theory could take us closer to some as yet unrevealed truth.

The aim and structure of physical theory

The grip that the logical positivists held on the philosophy of science was strong and lasted 30 years, but it was also doomed. One by one, the key pillars of logical positivist doctrine crumbled, and the outlook, with its trenchant rejection of metaphysics, was largely discredited.

The building blocks of logical positivism were the criterion of verifiability, the use of this criterion to provide a strict demarcation between science and metaphysics and an analytical structure based on a formal theory of logic. Popper had attacked the principle

---

[7] Petersen, Aage, In French, A. P. and Kennedy, P. J. (eds.) (1985). *Niels Bohr: a centenary volume*. Harvard University Press, Cambridge, MA, p. 305.

[8] Ayer, A. J. (ed.) (1959). *Logical positivism*. The Library of Philosophical Movements. The Free Press of Glencoe, New York, p. 11.

[9] Murdoch, Dugald (1987). *Niels Bohr's philosophy of physics*. Cambridge University Press, Cambridge, pp. 231–232.

of induction and the criterion of verifiability, arguing that theories can *never* be completely empirically verified, but they can be falsified. There are, in fact, problems with both criteria that have to do with the nature of theory adaptation in the light of potentially falsifying data. This is best illustrated by reference to an example from the history of planetary astronomy.

The planet Uranus was discovered by William Herschel in 1781. Its influence on planetary dynamics was subsequently incorporated into the calculations as an additional contribution to the series of inter-planet gravitational perturbations to the principal source of gravitational force in the solar system—the sun. When this was done, the predicted orbit of Uranus was *not* in agreement with the observed orbit. What happened? Was this example of disagreement between theory and observation taken to falsify the basis of the calculations, and hence the entire structure of Newtonian mechanics on which the calculations were based? No. If Newton's laws of motion are taken as the foundation of the theory of classical mechanics, it must be admitted that these laws properly apply only in idealized, model situations, and so *they are never tested directly by observation or experiment*.[10]

This seems an outrageous claim. But if we think about how the laws are actually applied to practical situations, such as the calculation of planetary orbits, we are forced to admit that no such application is possible without a whole series of so-called *auxiliary assumptions* or hypotheses. And, when faced with potentially falsifying data, the tendency of most scientists is *not* to throw out an entire theoretical structure (especially one that has stood the test of time for several hundred years), but instead to tinker with the auxiliary assumptions. In a series of articles published in 1904–1905 with the general title *The aim and structure of physical theory*, the French scientist, philosopher and historian Pierre Duhem argued[11]:

In sum, the physicist can never subject an isolated hypothesis to experimental test, but only a whole group of hypotheses; when the experiment is in disagreement with his predictions, what he learns is that at least one of the hypotheses constituting this group is unacceptable and ought to be modified; but the experiment does not designate which one should be changed.

This is indeed what happened in the case of the predicted orbit of Uranus. The auxiliary assumption that was challenged was the (unstated) one that the solar system consists of just seven planets. John Adams and Urbain Leverrier independently proposed that this assumption be abandoned in favour of the introduction of an as yet unobserved eighth planet beyond Uranus that was perturbing its orbit. In 1846, the German astronomer Johann Galle discovered the new planet, subsequently called Neptune, less than one degree from its predicted position.[12]

Note that this does not necessarily mean that a theory can never be falsified by observation or experiment, but it does mean that falsifiability is not a robust criterion for a scientific method. This much is revealed in an exactly analogous example which had dramatically different consequences. Emboldened by his success, in 1859, Leverrier challenged the same auxiliary assumption in attempting to solve the problem of the anomaly

---

[10] See, for example, Cartwright, Nancy (1983). *How the laws of physics lie*. Oxford University Press, Oxford, especially pp. 54–73.

[11] Duhem, P. (1954). *The aim and structure of physical theory*. English translation by Wiener, Philip P. of the second French edition of 1914, Princeton University Press, Princeton, NJ, p. 187.

[12] Irregularities in the orbit of Neptune were, in turn, taken as evidence for the existence of a ninth planet by Percival Lowell in 1905. Although Lowell's predictions were inaccurate, a ninth planet—Pluto—was discovered by Clyde Tombaugh in 1930.

in the perihelion of Mercury, by proposing another as yet unobserved planet—which he called Vulcan—between the sun and Mercury itself. No such planet could be found. As Duhem states, in the face of disagreement, at least one of the hypotheses constituting the group must be modified, but the experiment does not state which. In fact, the hypotheses that form the central core of Newtonian mechanics proved to be untenable in this case: Einstein's general theory of relativity (of which Newton's gravity is a limiting case) correctly predicts the advance in the perihelion of Mercury.

Duhem's argument, suitably extended by Willard Quine (though perhaps not extended quite as far as Quine originally proposed)[13] is known to philosophers of science as the *Duhem–Quine thesis*.

Although Popper used his criterion of falsifiability to distinguish science from metaphysics, this was not done for the purposes of rejecting metaphysics *per se*. In seeking to discredit the positivist doctrine, Popper argued that even metaphysical statements can be verified in principle, and, he claimed, not only can unverifiable metaphysical concepts or statements be meaningful but also they support the progressive development of science. In fact, a very powerful argument against the possibility of ever making observations devoid of theoretical structures or concepts, and hence of ever truly separating science from metaphysics, predates the Vienna Circle's particular brand of positivism by about 20 years. In *The aim and structure of physical theory*, Duhem denied that it is ever possible to make an observation that is not itself laden with theoretical concepts. He went on to argue that without a theoretical interpretation, it is, in fact, impossible to use experimental apparatus of any kind[14]:

Go into this laboratory; draw near this table crowded with so much apparatus: an electric battery, copper wire wrapped in silk, vessels filled with mercury, coils, a small iron bar carrying a mirror. An observer plunges the metallic stem of a rod, mounted with rubber, into small holes; the iron oscillates and, by means of the mirror tied to it, sends a beam of light to a celluloid ruler, and the observer follows the movement of the light beam on it. There, no doubt, you have an experiment; by means of the vibration of this spot of light, this physicist minutely observes the oscillations of the piece of iron. Ask him now what he is doing. Is he going to answer: "I am studying the oscillations of the piece of iron carrying this mirror"? No, he will tell you that he is measuring the electrical resistance of a coil. If you are astonished, and ask him what meaning these words have, and what relation they have to the phenomena he has perceived and which you at the same time perceived, he will reply that your question would require some very long explanations, and he will recommend that you take a course in electricity.

Duhem's is a particularly powerful argument. As scientists gradually come to terms with theoretical concepts and unobservable entities, these things begin to enter into the scientists' common language. So common do some of these things become that scientists quickly lose sight of their status and cease questioning their own degree of belief in them. They become an intrinsic part of the scientist's world. This does not necessarily make them any more real,[15] but it does make virtually impossible the task of separating

---

[13] Gillies, Donald (1993). *Philosophy of science in the twentieth century*. Blackwell, Oxford, pp. 112–116.

[14] Duhem, P. (1954). *The aim and structure of physical theory*. English translation by Wiener, Philip P. of the second French edition of 1914, Princeton University Press, Princeton, NJ, p. 145.

[15] Duhem was something of an anti-realist in that he held firmly to the need to separate physics from metaphysics and sided with Mach on the non-existence of atoms. However, he denied that he was a positivist (see Louis De Broglie's foreword to the 1954 Princeton University Press edition of *The aim and structure of physical theory*, especially pp. ix–x).

scientific statements from statements containing precisely these metaphysical concepts. I say virtually impossible, because it is obviously not totally impossible to do this: it is just that no practising scientist would ever use the language that resulted from such an exercise.

The remaining foundation of the Vienna Circle's brand of positivism was the use of the techniques of symbolic logic for clarifying philosophical issues. Bertrand Russell had earlier embarked on an ambitious project to reduce all mathematics to logic. He had drawn on the theories of logic developed by Giuseppe Peano and Gottlob Frege, and his attempt culminated in the publication of his *Principia mathematica* with Alfred North Whitehead between 1910 and 1913. Unfortunately, Russell's project was subsequently shown in 1931 by the Czech mathematician Kurt Gödel to be by definition unachievable. Put simply, Gödel showed that any formalized logical system based on a consistent set of axioms will allow the possibility of constructing propositions that are formally *undecidable* (i.e. they cannot be proved by reference to the axioms). Such undecidable propositions can, however, be proved 'informally' through so-called meta-mathematical arguments *outside* the axiomatic structure. It is not possible to escape the problem by expanding the number of axioms, as Gödel's arguments can be repeated for the larger system. This is Gödel's famous 'incompleteness theorem', which demonstrates the rather humbling fact that it is impossible to establish the internal logical consistency of a large class of deductive systems, including many important systems of mathematics (such as elementary arithmetic). Most philosophers abandoned logicism following the publication of Gödel's proof. Interestingly, Gödel was numbered among the members of the Vienna Circle.

By the mid-1930s, the Vienna Circle was in any case on the point of breaking up. Many in the Circle were Jewish and politically left-leaning, making them targets first under the rule of the Austrian Fascist Party, which seized power in 1934, and then subsequently under Nazism following Hitler's invasion of Austria in 1938. Schlick was shot dead in 1936 by a Nazi student, whose thesis (on ethics, of all things) Schlick had failed.

The failure of the logical positivist program led, perhaps inevitably, to a somewhat more ameliorative attitude to metaphysics, leading Ayer to observe: 'The metaphysician is treated no longer as a criminal but as a patient...'.[16] But the positivist program is far from dead. It is very much alive in the hands of Bas van Fraassen, a leading proponent of *constructive empiricism*. This outlook retains the negative attitude towards metaphysics and denies that scientific theories develop towards a literally true representation of some independent reality. Instead, van Fraassen claims that science aims to give us theories which are empirically adequate, and acceptance of a theory requires us to believe *only* that it is empirically adequate.

Gone from the program is the verifiability criterion and the obsession with logic and the analysis of language. We should take theories literally, we should take as true what theories say about things we can observe, but we do not need (and should not need) to believe that unobservable entities are 'real' and that the theories themselves are true.

Social constructivism and incommensurability

The realist believes that scientific theories take us ever closer to 'the truth'. The trouble is that as soon as we attempt to define and understand the nature of truth, truth itself

---

[16] Ayer, A. J. (ed.) (1959). *Logical positivism*. The Library of Philosophical Movements. The Free Press of Glencoe, New York, p. 8.

becomes as slippery as reality. The starting point for most discussions about truth is the work of Polish philosopher Alfred Tarski. At considerable risk of over-simplification, Tarski's approach amounts to what we might refer to as a *correspondence* theory of truth. Basically, it says that a statement is true if, and only if, it corresponds to the facts. The statement 'snow is white', is therefore true if, and only if, snow is, in fact, white, and we have a common understanding of the meaning of the words 'snow', 'is', and 'white'. This seems like common sense, if somewhat trivial, but begs some questions about what should be regarded as 'facts' in the first place.

We could be a little less ambitious and argue that theories are approximations to the truth, that is, they have an element of 'truth-likeness', or what Popper referred to as *verisimilitude*. As science progresses and our theories become more refined, we could therefore expect that they would tend to *converge* on 'the truth'. However, whether correspondence or convergence, there is nothing in these theories of truth that helps us differentiate between the realist and the anti-realist positions. Both lay claim to interpretations of a physical world that is true, but their standards of acceptance of what constitute relevant facts are very different.

There is yet another view. What if the theoretical concepts, descriptions, interpretations of the facts, and, indeed, what is accepted as true are all *conventions* constructed by the community of people engaged in science in any particular generation? This position, which generally goes by the name *social constructivism*, was famously advocated by Thomas Kuhn in his book *The structure of scientific revolutions*. Kuhn differentiated between two types of scientific activity. 'Normal' science is routine puzzle-solving within the framework of concepts, descriptions, meanings, and interpretations accepted (mostly without question) by the community as 'the truth'. This framework, which Kuhn called a *paradigm*,[17] represents the scientist's 'world view': it shapes their value system, supports their value judgements, and so determines what is regarded as good and bad science. 'Revolutionary' science occurs as a result of an accumulation of data that throw increasing doubt on the veracity of the existing paradigm, leading ultimately to a crisis in science. The paradigm is changed not because it is falsified by the evidence, or indeed because a replacement paradigm is verified, but because a sufficient number within the community become persuaded that the new paradigm is better than the old. This process is no less than a socio-cultural revolution and, as such, is often characterized by bitter argument as the community breaks into rival factions. Indeed, Kuhn saw parallels with the process of political revolution.

Social constructivism is profoundly anti-realist. In fact, Kuhn failed to find any convincing arguments that paradigm-shifts promote any real progressive evolution of theories towards some final truth, even a collective truth forged from convention. For example, he saw progression in the theories of Aristotle through Newton to Einstein, but only as instruments for puzzle-solving in normal science. In terms of what the theories actually

---

[17] Kuhn is famous also for his introduction of this term, which is used in *The structure of scientific revolutions* in at least two different senses. One is sociological—the paradigm represents the entire approach to puzzle-solving and the belief and value systems of the community of scientists. The other is exemplary—the paradigm is *the* observation, experiment, or model that defines the 'rules' and shapes the scientists' approach to puzzle-solving within the boundaries of normal science. See Kuhn, Thomas S. (1970). *The structure of scientific revolutions* (2nd edn.). University of Chicago Press, Chicago, p. 135.

say, he found Einstein's general theory of relativity closer to Aristotle than either of them is to Newton.

Rather, the change in paradigm reflects for Kuhn not so much a progressive development but a change in language and description, usually with a carry-over of many of the terms and concepts of the old paradigm to the new, but equally usually with a fundamental change in the *meaning* of those terms and concepts. In the midst of crisis, the meanings have become so divergent that rival factions struggle to communicate with each other effectively. They are using the same words, but the words now mean different things to different scientists. The paradigms have become *incommensurable*—they no longer have any 'common measures' by which they can be unambiguously compared, a concept applied to the philosophy of science by Kuhn and Paul Feyerabend in the early 1960s.

Social constructivism is a knight's move from relativism, the view that scientific theories are relative to the society or culture from which they emerge. Taken to its extreme, relativism challenges our preconceptions of reason, rationalism, and Western cultural ideals of progress.

The rational character of reality

Faced with the onslaught from positivism (or its modern embodiment, constructive empiricism), pragmatism, social constructivism, and numerous other anti-realist stances that unfortunately will have to go undiscussed in this book, is *any* rational defence of realism possible?

The philosopher Ian Hacking had not spent much time pondering the merits or otherwise of realism until a friend described to him the details of some experiments he was involved in to try to observe the fractional electric charges characteristic of quarks. The experiment was a modern equivalent of J. A. Millikan's fabled oil-drop experiment, with oil drops replaced by balls of niobium cooled to below their superconducting transition temperature. Hacking asked how they changed the charge on the niobium ball, and described his friend's reply[18]:

"Well at that stage", said my friend, "we spray it with positrons to increase the charge or with electrons to decrease the charge." From that day forth I've been a scientific realist. *So far as I'm concerned, if you can spray them then they are real.*

Mach's views on space and time greatly influenced the young Einstein, whose admiration for Mach's work on mechanics never diminished. Indeed, the young Einstein's approach to his science and particularly the special theory of relativity was markedly positivist in nature. But by the time he was formulating the general theory of relativity, he had grown wary and more circumspect, and his philosophical outlook shifted much more towards a realist position. Of course, the general theory was to resolve one the greatest conundrums of Newton's mechanics. The action at a distance implied in Newton's theory of gravitation that had caused Newton much personal discomfort was replaced in Einstein's by a balancing of forces of attraction and motion acted out in a four-dimensional space–time curved by the presence of massive bodies. Gravitation does not reach out and exert forces on distant masses, as if by magic. Instead, the masses themselves distort the space–time

---

[18] Hacking, Ian (1983). *Representing and intervening*. Cambridge University Press, Cambridge, p. 22. The emphasis is Hacking's.

around them, and the motions of distant masses are affected by this distortion in a way that can be explained perfectly well by classical mechanics.

But it is not enough simply to claim that Einstein was a realist: here again, we must deal with shades of grey. The discoverer of special relativity was hardly a naive realist—he did not believe that constant refinement of our theoretical models of the physical world took us any closer to some absolute truth and conceded that our notions of physical reality could never be final. But he did hold fast to some key concepts of the realist tradition, including a belief that the purpose of science is to refine our description of a reality that is conceptualized as independent of ourselves and our measuring devices. Einstein was convinced that a proper, realist theory should provide a formal description of the behaviour of real entities interacting through entirely causal (not probabilistic) laws in a comprehensible space–time representation. Of these three criteria, observer-independence and causality were fundamentally important to Einstein; the space–time representation was perhaps secondary.[19]

Einstein was a realist, but he inherited particularly from Mach a strongly empiricist core to his realist philosophy which led him to emphasize the importance of verification and economy. As to the justification for his realism, he never attempted to rationalize his motives with scientific arguments, preferring instead to acknowledge that no scientific rationale could be given[20]:

I have no better expression than the term "religious" for this trust in the rational character of reality and in its being accessible, to some extent, to human reason. Where this feeling is absent, science degenerates into senseless empiricism.

Why, then, did Einstein choose realism if the only justification that can be found for it is an appeal to faith? Einstein's answer was simple. The existence of an observer-independent reality founded on causal laws had been a largely unstated belief of scientists for hundreds of years and had remained unquestioned until the advent of quantum theory. It had been the unspoken drive behind all of the most significant discoveries in science. And, Einstein believed, unlike 'sterile empiricism', this passion to comprehend reality was much more likely to drive fruitful theory development in the future.

Much of the debate about the realism of entities and the realism of theories concentrates on our interpretation of theoretical descriptions. But, as Hacking points out, science has two aims: theory *and* experiment. Theories represent, says Hacking, and experiments intervene[21]:

I suspect there can be no final argument for or against realism at the level of representation. When we turn from representation to intervention, to spraying niobium balls with positrons, anti-realism has less of a grip ... The final arbitrator in philosophy is not how we think but what we do.

---

[19] Fine, Arthur (1986). *The shaky game: Einstein, realism and the quantum theory* (2nd edn.). University of Chicago Press, Chicago, p.103.

[20] Solovine, M. (ed.) (1956). *Albert Einstein: lettres à Maurice Solovine.* Gauthier-Villars, Paris, quoted in Fine, Arthur (1986). *The shaky game: Einstein, realism and the quantum theory* (2nd edn.). University of Chicago Press, Chicago, p.110.

[21] Hacking, Ian (1983). *Representing and intervening.* Cambridge University Press, Cambridge, p. 31.

The schism: realism *versus* anti-realism

I do not think it is weak-minded to recognize in actual scientific practice elements of *all* the different philosophical positions considered in this chapter. Scientists deal with observations or the results of experiments which reveal facts about nature. They will employ whatever appropriate methods are available to obtain generalizations from bodies of empirical data, provided the methods are reliable and will stand up to the scrutiny of their scientific peers. They will not be overly troubled by the fact that some philosophers are puzzled by the problem of induction. Their generalizations are then tested through new observations or new experiments. The best (quite often the simplest and therefore most economical) theory is the one which accounts for all the known facts and can be used to make predictions, the accuracy of which can then be readily tested. Extreme metaphysical speculations (about the existence of God, for example) do not usually play any part in the scientists' routine activity, although most scientists, when prompted, will certainly have a developed and highly distinctive metaphysical outlook that makes them complete as human beings.

During their formal education, most scientists learn to use the *methods* extolled as virtuous by the positivists, in order to maintain a sceptical approach and avoid woolly thinking. Young scientists are instructed by their teachers as to what qualifies as science, what 'doing science' means and how it should ideally be conducted. They learn to adopt a pragmatic attitude to science in which philosophy—and particularly metaphysics—appears to play no part. This attitude is the source of most of the debates about the value content of science, and particularly the concern that science is governed by its own rules and not subject to the value system of the society in which it is embedded.

However, whilst complete belief might be suspended pending a consensus on the evidence, for many scientists, the stuff of their theories—atoms, electrons, photons, and so on—are quite 'real'. Those that stop to think about it assume these entities to have an existence independent of the instruments used to produce the effects their theories are supposed to explain. Many do not stop to think about it. It would, perhaps, be very difficult for high-energy physicists to justify the financial investments needed to build the Large Hadron Collider if they were not convinced of the reality of the objects on which they wished to make measurements. This entity realism carries over into scientific theorizing, since, as we learn from Duhem, it is virtually impossible to separate observation and experiment from theoretical statements concerning observable and unobservable entities *and their properties*. However, this entity realism is not all or nothing. Scientists learn to deal with ambiguity very early in their careers and are more than willing to live with an entire spectrum of degrees of belief, reflecting the state of scientific knowledge of their time. Are unicorns real? No. Is the chair on which you are sitting real? Yes. Are electrons real? Yes. Are quarks real? Maybe.[22] Are superstrings real? Even more maybe. Furthermore, it is a simple fact that most practising scientists seek to uncover an independent physical reality lying beneath the phenomena they observe: to explain *why* the world is the way it is, which goes beyond merely registering the fact that

[22] When he first proposed the existence of quarks, Murray Gell-Mann sought to avoid any philosophical wrangling over their status by referring to them as 'mathematical', reasoning that it was going to be hard to call them 'real' if they were permanently confined. Numerous commentators overlooked Gell-Mann's careful definitions and claimed that he did not really believe that quarks existed. He did. See Gell-Mann, Murray (1994). *The quark and the jaguar.* Little Brown, London, p. 182.

instrument A will give effect B under conditions C. If you doubt this, read their scientific publications.

There is no doubting the socio-cultural context in which science is conducted, debates are held, and consensus is reached. Yes, it is true that observation and experiment themselves are human practices and, as such, are not infallible generators of necessarily unambiguous empirical facts or confirmations.[23] Yes, it is true that scientific laws or core theoretical concepts are only tested together with relevant auxiliary assumptions and, when something does not work out, scientists will draw on their own judgement and experience when choosing which assumptions to abandon. Yes, it is also true that, once a community achieves agreement that a new theory (or paradigm) is superior to the old, history tends to be rewritten to show science (falsely) as an entirely logical, linear progression.

But the real situation in science is what makes science so successful and powerful. It is a mistake to think that new theories (or new paradigms) necessarily replace the old. Theories are indeed discarded when it becomes clear that they no longer provide adequate descriptions of aspects of reality. So, we no longer teach phlogiston theory to the new generation of chemists. However, theories are *not* discarded when only their domain of applicability is shown to be limited. We will happily teach Newtonian mechanics to the new generation of physicists, whilst pointing out its limitations and restricted range of application. Physicists will happily *use* Newtonian mechanics to solve problems that are recognized to be within its range of application. Outside this range, they will reach for special or general relativity or quantum mechanics, as appropriate. These theories may well be incommensurable, creating additional heat during a crisis, but the barriers are not insurmountable: when the crisis is over, scientists learn quickly how to translate from one theoretical language to another.

Does Newtonian mechanics provide a completely truthful description of reality? No, it does not. It does, however, appear to provide a description with a high level of truth-likeness when applied in certain circumstances (e.g. in calculations of the gravitational dynamics of simple systems such as the earth and the moon). Does the general theory of relativity provide a completely truthful description of reality? No, it does not. It provides a description with a higher level of truth-likeness than Newtonian mechanics when applied in certain circumstances (e.g. in calculations of the perihelion of Mercury). Real progress is made in science because it has the ability to *assimilate new and better descriptions of the world into the body of accepted scientific knowledge*. Against a background of scientific values and practices passed, with modification, from generation to generation, science *grows*. Will we ever arrive at a final theory whose statements encapsulate absolute truths? I very much doubt it and sincerely hope not.

The anti-realist questions the realist's use of truth concepts and the usefulness of searching too hard for an independent reality. However, there is a distinction to be made. Realist scientists might be convinced that there is an independent reality 'out there' which is probed through observation and experiment. An anti-realist might accept that there are

---

[23] See, for example, Broad, William and Wade, Nicholas. (1982). *Betrayers of the truth*. Oxford University Press, Oxford and Collins, Harry and Pinch, Trevor. (1993). *The Golem*. Cambridge University Press, Cambridge.

elements of an empirical reality which are probed in this way but points out that the realist view involves a logical contradiction, since we obviously have no way of observing an observer-independent reality, and hence we can never hope to verify that such a reality exists. We have no means of acquiring knowledge of the physical world except through observation and experiment, and so the face of the reality we expose is, of necessity, a face that is dependent on the observer for its appearance: we see only the reality exposed to our method of questioning. The anti-realist argues that, since we cannot verify the existence of an observer-independent reality, such a reality is metaphysical and therefore quite without meaning. The logical contradiction implied in the realist's view is sidestepped only by an appeal to the emotions or to faith.

A scientist might typically adopt the *methods* of the anti-realist but the *outlook* of the realist. If this position seems a little confusing and ill thought out, it is perhaps because scientists rarely spend time working out where they stand on these philosophical issues. Indeed, a pragmatic scientist might have little time for what seems like a kind of philosophical nit-picking. However, it is very difficult to avoid these issues in quantum theory. A quantum system exhibits properties we associate with waves and particles. Its behaviour appears to be determined by the kind of instrument we use to probe its properties, and, what is more, the mathematical formalism of quantum theory is set up in such a way to make this instrument dependence explicit. One kind of instrument will tell us that a quantum entity is a wave. Another kind will tell us that it is a particle. All we can know is the *empirical* reality—here, the quantum entity behaves like a wave, here it behaves like a particle. Is it therefore meaningful to speculate about what the quantum entity *really is?*

Should we accept that, in quantum theory, we have reached the end of the road? Should we deny that there is a way forward through just the kind of metaphysical speculation that can introduce concepts which begin life as abstract mathematical constructions (such as atoms and quarks) into elements of reality? Can we afford not to push science along apparently 'meaningless' paths? What if, despite appearances, we have not reached the limit after all? What if there *is* something more to discover about reality if only we have the wit to ask the right questions? Whatever our personal thoughts on this matter, we should admit that it goes against the grain of human nature not to *try*.

Quantum theory is the most important fundamental theory of science that is truly anti-realist both in its mathematical formalism and in its orthodox interpretation. In this regard, it represents a marked break with the past. Discussions of its interpretation have consequently dominated the philosophical and, increasingly, the scientific literature ever since. Commenting on these, Karl Popper noted[24]:

One remarkable aspect of these discussions was the development of a split in physics. Something emerged which may be fairly described as a quantum orthodoxy: a kind of party, or school, or group, led by Niels Bohr, with the very active support of Heisenberg and Pauli; less active sympathizers were Max Born and P. Jordan and perhaps even Dirac. In other words, all the greatest names in atomic theory belonged to it, except two great men who strongly and consistently dissented: Albert Einstein and Erwin Schrödinger.

The Copenhagen school of physicists was convinced that its interpretation of quantum theory was the only sensible interpretation. Other physicists disagreed, however. As we

---

[24] Popper, Karl R. (1982). *Quantum theory and the schism in physics*. Unwin Hyman, London, pp. 99–100.

have seen, Schrödinger refused to bow to pressure from Bohr and Heisenberg to reconsider his position. Einstein was never comfortable with quantum theory's implications for causality. These two were only the most eminent and directly involved of the physicists who were unhappy with what the Copenhagen school was saying. Einstein in particular confronted Bohr head on in a now famous debate on the meaning of quantum theory.

7

A bolt from the blue

The confrontation between Einstein and Bohr over the interpretation of quantum theory has been painted in the past as a direct conflict between realism and positivism. However, more recent attempts to uncover the philosophical positions of both Einstein and Bohr have revealed that these were not so black and white. Einstein's realism was conditioned by the strong empiricist influence in his own early experiences as a theorist and the growing wariness that led him eventually to reject Machian positivism. His was a realism born of a sense of necessity: a belief that a realist stance is the only one that leads to productive theorizing about the physical world. Bohr's philosophy went through some quite marked changes, principally in response to Einstein's challenges. The father of modern atomic theory could hardly subscribe to an uncompromising positivist position that rejected the reality of atoms. But the reality of the inner workings of those atoms was, Bohr contended, beyond the reach of a knowledge founded on complementary classical concepts of waves and particles. As Einstein's challenges became deeper and more subtle, so Bohr's position shifted to that of a distinctive anti-realism. Add to this a rhetoric in Bohr's speeches and published writings that insisted on the logical inevitability of the Copenhagen interpretation, that insisted on its status as the only feasible interpretation, and that countenanced no other alternative, and you have dogma in the making. Heisenberg called it 'Kopenhagener Geist der Quantentheorie', the Copenhagen spirit of quantum theory which, he declared in 1930, 'has directed the entire development of modern atomic physics'.[1] Einstein frequently referred to the Copenhagen doctrine as 'Talmudic'.

The disagreement between Einstein and Bohr had a profound effect on the older generation of physicists that, over the course of 20 years of the most exciting developments in the history of their science, had developed a particularly deep fondness and respect for its two most distinguished members. Paul Ehrenfest was close to both protagonists and reported intimate details of their discussions. His son, Paul Jr. (called Pavlik by his

---

[1] See the preface to Heisenberg, Werner (1930). *The physical principles of the quantum theory*. University of Chicago Press, Chicago (republished in 1949 by Dover, New York).

parents) took many of the photographs that have since come to symbolize their intellectual struggle. Concerned by the widening gulf between them, Ehrenfest felt he had to make a choice between the approaches to interpretation offered by Einstein and Bohr, and sometime around mid-1927, he explained, in tears, to Samuel Goudsmit that he had to agree with Bohr.[2]

Although invited, Einstein did not attend the meeting of physicists at Lake Como in September 1927 at which Bohr first presented his ideas about complementarity. He was nevertheless very active in the debate in private correspondence with a number of leading physicists. At the same time, he was developing his own ideas about the interpretation of quantum theory, based on the statistical properties of large collections of particles. From the very beginning, Einstein believed that quantum theory lacked some key ingredients and that, in a very significant sense, it was 'incomplete'. He compared it with the theory of light before the advent of light quanta. Quantum theory, he believed, was perhaps a 'correct theory of statistical laws', but it provided 'an inadequate conception of individual elementary processes.'[3]

Finally, on 24 October 1927, Einstein, Bohr, and many other leading physicists assembled in Brussels for the fifth Solvay Conference, on 'Electrons and Photons'.

The fifth Solvay Conference

The fifth Solvay Conference was opened by Hendrik Lorentz. There followed presentations from W. L. Bragg and Arthur Compton. Louis de Broglie then presented his 'theory of the double solution' (which we cover in detail in a later chapter). Born and Heisenberg described matrix mechanics, the uncertainty relations, and the probabilistic interpretation, and argued provocatively that quantum theory in its then current form was 'complete', with no further modification of its basic physical and mathematical hypotheses required. Schrödinger delivered a paper on wave mechanics. Lorentz then opened the meeting for general discussion, expressing his personal reservations that the probabilistic interpretation was being advanced as a basic postulate of quantum theory (an 'input') rather than as a consequence (an 'output') of physical constraints or theoretical considerations, before calling on Bohr to address the meeting. Bohr described his concept of complementarity, much as he had first presented it at Como, but directing his statements to Einstein who was hearing this argument for the first time. Einstein did not respond immediately.

The discussion on those comments he eventually did make spilled over into the dining room of the hotel in which all the conference participants were staying and which became the scene of one of the most important scientific debates ever witnessed, as Einstein directly challenged Bohr over the meaning of quantum theory. At stake was the interpretation of quantum theory and its implications for the way we attempt to understand the

---

[2] Pais, Abraham (1982). *Subtle is the Lord: the science and life of Albert Einstein*. Oxford University Press, Oxford, p. 443. Ehrenfest was a particularly tragic figure. He suffered from bouts of depression all his life, and, as it became ever more difficult for him to follow developments in his beloved physics with any understanding, his thoughts turned to suicide. On 25 September 1933, in the waiting room of a mental institute in Amsterdam where his youngest son Vassily was undergoing treatment for Downs syndrome, he shot and killed his son before turning the gun on himself. Einstein's memorial for Ehrenfest is amongst his most moving writings.

[3] Einstein, Albert, letter to Sommerfeld, Arnold. 9 November 1927. Quoted in Fine, Arthur (1986). *The shaky game: Einstein, realism and the quantum theory* (2nd edn.). University of Chicago Press, Chicago, p. 29.

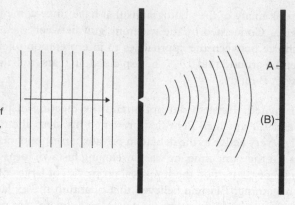

Fig. 7.1 Simple electron or photon diffraction experiment cited by Einstein in his debate with Bohr. Reprinted by permission of Open Court Publishing Company, a division of Carus Publishing Company, Peru, IL, from (1949). The Library of Living Philosophers, Vol. VII, *Albert Einstein: philosopher–scientist*, edited by Paul Arthur Schilpp, p. 212.

physical world. The outcome of the debate would determine the directions of the future development of quantum physics.

This debate has been described in great detail by Bohr himself in a book published in 1949 in celebration of Einstein's 70th birthday. Einstein began by expressing his general reservations about quantum theory by reference to an experiment involving the diffraction of a beam of electrons or photons through a narrow aperture, as shown in Fig. 7.1. The diffraction pattern appears on a second screen and is recorded (using photographic film, say). According to the Copenhagen interpretation, quantum theory is a complete theory of individual processes, the behaviour of each individual quantum particle is described by an appropriate wave function or state vector, and it is the properties of the state vector that give rise to the diffraction pattern. However, at the moment the state vector impinges on the second screen, it 'collapses' instantaneously, producing a localized spot on the screen which indicates 'a particle struck here'.

Einstein objected to this way of looking at the process. Suppose, he said, that the particle is observed to arrive at position A on the second screen (see Fig. 7.1). In making this observation, we learn not only that the particle arrived at A, but also that it definitely did not arrive at position B. What is more, we learn of the particle's non-arrival at B instantaneously with the observation of its arrival at A. Before observation, the probability of finding the particle is, supposedly, 'smeared out' over the whole screen. Einstein believed that the collapse of the wave function implies a peculiar 'action at a distance'. The particle, which is somehow distributed over a large region of space, becomes localized instantaneously, the act of measurement appearing to change the physical state of the system far from the point where the measurement is actually recorded. Einstein felt that this kind of action at a distance violated the postulates of special relativity. 'If one works only with Schrödinger waves,' commented Einstein, 'the interpretation...of [the modulus-square of the wave function], I think, contradicts the postulate of relativity.'[4]

There is an alternative description, however. What if the state vector represents a probability amplitude not for a single quantum particle but for a large collection of identically prepared particles (an ensemble) which is described in terms of a single state vector? According to this view, each individual particle passes through the aperture along a

[4] Quotation from Jammer, Max (1974). *The philosophy of quantum mechanics.* Wiley-Interscience, New York, p. 116.

defined, localized path, to arrive at the second screen. There are many such paths possible, and the diffraction pattern thus reflects the statistical distribution of large numbers of particles, each following different but defined paths. This distribution is related to the modulus-square of the wave function or state vector, which expresses the probability density of one (of many) particles rather than a probability density for each individual particle.

We should note that we cannot choose between these possibilities by observing what happens to an individual quantum particle. Both descriptions say that one particle passes through the aperture to arrive at one specific location on the second screen. In the first description, the point of arrival is determined at the moment the particle interacts with the detector, with a probability given by the modulus-square of the state vector. In the second description, the point of arrival is determined by the actual path that the particle follows, which is in turn obtained from a statistical probability given by the modulus-square of the state vector. In both cases, we see the diffraction pattern only when we have detected a large number of particles.

Is quantum mechanics consistent?

Einstein chose to attack the Copenhagen interpretation of quantum theory by attempting to show that, as a result of its incompleteness, one of the theory's principal foundations—the physical meaning of the uncertainty relations—is inconsistent. The debate took the form of a series of puzzles, developed by Einstein as hypothetical experiments. These 'thought' experiments were not intended to be taken too literally as practical experiments that could be carried out in the laboratory. It was enough for Einstein that the experiments could be conceived and carried out in principle.

Einstein asked the assembled audience what might happen if a quantum particle passed through an apparatus such as that shown in Fig. 7.1 under conditions where the transfer of momentum between the particle and the first screen is carefully controlled and observed. A particle hitting the screen as it passes through the aperture would be deflected, its path beyond being determined by the conservation of momentum. Now imagine that we insert another screen—one with two slits—between the first screen and the detector (Fig. 7.2). If we control the transfer of momentum between the particle and the first screen, we should be able to discover towards which slit in the second screen the particle is deflected. If the particle is ultimately detected, we can deduce from our measurements that it passed through one or other of the two slits, and we have thus determined the particle's approximate trajectory through the apparatus. We can now leave the apparatus to detect a large number of particles—one after the other—from which we expect to see a double-slit interference pattern. Thus, Einstein concluded, we can demonstrate the particle-like (defined trajectory) and wave-like (interference) properties of quantum particles simultaneously, in contradiction to Bohr's notion of complementarity, thus proving that the Copenhagen interpretation is internally inconsistent.

Bohr's reaction was to take the thought experiment a stage further. He sketched out in a pseudo-realistic style the kind of apparatus that would be needed to make the measurements to which Einstein referred. His purpose was not to try to imagine how the experiments could be done in practice but primarily to focus on what he saw to be flaws in Einstein's arguments.

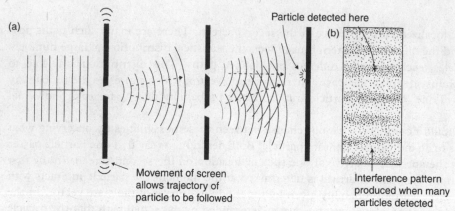

(a)

Particle detected here

(b)

Movement of screen
allows trajectory of
particle to be followed

Interference pattern
produced when many
particles detected

Fig. 7.2 (a) Controlling and observing the momentum transferred
between a quantum particle and the first screen allows the
approximate trajectory of the particle to be traced through a
double-slit apparatus. (b) After many particles have passed through the
apparatus, the double-slit interference pattern should be visible.
Reprinted by permission of Open Court Publishing Company, a division
of Carus Publishing Company, Peru, IL, from (1949). The Library of
Living Philosophers, Vol. VII, *Albert Einstein: philosopher–scientist*,
edited by Paul Arthur Schilpp, p. 216.

Fig. 7.3 Hypothetical instrument
designed by Bohr to demonstrate
how the measurement of the
momentum transfer to the first
screen might be made. Reprinted
by permission of Open Court
Publishing Company, a division of
Carus Publishing Company, Peru,
IL, from (1949). The Library of
Living Philosophers, Vol. VII,
*Albert Einstein: philosopher–
scientist*, edited by Paul Arthur
Schilpp, p. 220.

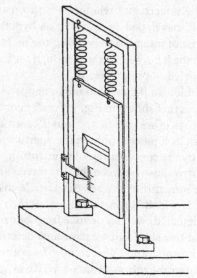

Controlling and observing the transfer of momentum from the quantum particle to
the first screen requires that the screen be capable of movement in the vertical plane.
Observing the recoil of the screen in one direction or the other as the particle passed
through the aperture would then allow the experimenter to draw conclusions about the
direction in which the particle had been deflected. Bohr envisaged a screen suspended by
two weak springs, as shown in Fig. 7.3. A pointer and scale inscribed on the screen allow
the measurement of the amount of movement of the screen, and hence the momentum

imparted to it by the particle. The fact that Bohr had in mind a macroscopic apparatus presents no problem, provided we assume that the apparatus is sufficiently sensitive to allow observation of individual quantum events. This sensitivity is important, as we will see.

Bohr had to demonstrate the consistency of the uncertainty principle, and hence of his concept of complementarity, when applied to the analysis of this kind of thought experiment. He argued that controlling the transfer of momentum to the screen in the way Einstein suggested *must* imply a concomitant uncertainty in the screen's position in accordance with the uncertainty principle. If we measure the screen's momentum in the vertical plane with a certain precision, there must result an uncertainty in the position of the screen such that the product of the uncertainties is greater than or equal to $h/4\pi$.

Bohr was able to show that the resulting uncertainty in the position of the aperture in the first screen destroys the *phase coherence* of the waves as they spread out beyond the double slits in the second screen, and the interference pattern is 'washed out'. Controlling the transfer of momentum from the particle to the first screen allows us to follow the trajectory of the particle through the apparatus, but prevents us from observing interference effects, in accordance with the complementary nature of the wave and particle descriptions.

Bohr's argument rests on the assumption that controlling and measuring the momentum transferred to the first screen sufficiently precisely to determine the particle's future direction automatically leads to an uncertainty in the screen's position. Why should this be? Bohr's answer was that, in order to read the scale inscribed on the first screen sufficiently accurately, it has to be illuminated. This illumination involves the scattering of photons from the screen and hence an uncontrollable transfer of momentum, preventing the momentum transfer from the quantum particle to be measured precisely. We can measure the latter with precision only if we reduce the illumination completely, but then we cannot determine the position of the pointer against the scale. Bohr concluded[5]:

we are presented with a choice of *either* tracing the path of a particle *or* observing interference effects, which allows us to escape from the paradoxical necessity of concluding that the behaviour of an electron or a photon should depend on the presence of a slit in the [second screen] through which it could be proved not to pass. We have here to do with a typical example of how the complementary phenomena appear under mutually exclusive experimental arrangements and are just faced with the impossibility, in the analysis of quantum effects, of drawing any sharp separation between an independent behaviour of atomic objects and their interaction with the measuring instruments which serve to define the conditions under which the phenomena occur.

We should note here that Bohr's counterargument rests on exactly the same concept of 'disturbance' that he had criticized earlier in Heisenberg's original explanation of the origin of uncertainty. Admittedly, the rationale for Bohr's criticism had centred on Heisenberg's struggle with the basic theory of the resolving power of his hypothetical gamma-ray microscope, but he had nevertheless strongly rejected the need for Heisenberg to invoke the Compton effect as the source of a disturbance so large in relation to the properties being measured that uncertainty in position and momentum inevitably resulted. Yet here he is arguing that 'any reading of the scale, in whatever way performed, will involve an

[5] Bohr, N. in Schilpp, P. A. (ed.) (1949). *Albert Einstein: philosopher–scientist*. The Library of Living Philosophers, Open Court, La Salle, IL, p. 217. The italics are Bohr's.

uncontrollable change in the momentum of the [second screen]', leading to a 'reciprocal relationship between our knowledge of the position of the slit and the accuracy of the momentum control'. Although Bohr's counterargument won the day in the eyes of the majority of physicists assembled in Brussels, the basis of it was eventually to prove untenable, as we shall see.

Einstein did not give up. He produced further thought experiments that we do not have room to consider fully here. He could not shake his deeply felt misgivings about the Copenhagen interpretation and forced Bohr to defend it. The fifth Solvay Conference ended with Bohr having successfully argued for the logical consistency of the Copenhagen interpretation, but he had failed to convince Einstein that it was the only interpretation.

The photon box experiment

As described in Chapter 3, the years 1927–1930 witnessed further dramatic breakthroughs in the development and refinement of quantum theory. Dirac had successfully accommodated the requirements of special relativity and, as by-products of his linearization approach, had produced a theoretical description of electron spin and predicted the existence of anti-matter (though it was not clear by 1930 just what was described by the other two solutions of Dirac's relativistic quantum theory of the electron).

The debate on interpretation recommenced at the sixth Solvay Conference, which was held in Brussels during 20–25 October 1930. Although the conference was devoted to the physics of magnetism, there was keen interest in the discussions on the interpretation of quantum theory that took place between the conference's formal proceedings. Einstein described his latest and most ingenious thought experiment, a further development of one that he had originally used in discussions at the fifth Solvay Conference. This is the 'photon box' experiment.

Suppose, said Einstein, that we build an apparatus consisting of a box which contains a clock mechanism connected to a shutter. The shutter closes a small hole in the box. We fill the box with photons and weigh it. At a predetermined and precisely known time, the clock mechanism triggers the opening of the shutter for a very short time interval, and a single photon escapes from the box. The shutter closes. We reweigh the box, and, from the mass difference and special relativity ($E = mc^2$), we determine the precise energy of the photon that escaped. By this means, we have measured precisely the energy and time of passage of a photon through a small hole, in contradiction to the energy–time uncertainty relation.

Bohr's immediate reaction has been described by Léon Rosenfeld[6]:

During the whole evening he was extremely unhappy, going from one to the other and trying to persuade them that it couldn't be true, that it would be the end of physics if Einstein were right; but he couldn't produce any refutation.

Bohr experienced a sleepless night, searching for the flaw in Einstein's argument that he was convinced must exist. By breakfast the following morning, he had his answer.

Again, Bohr produced a sketch of the apparatus that would be required to make the measurements in the way Einstein had described them, and this is shown in Fig. 7.4.

---

[6] Rosenfeld, L. (1968) *Proceedings of the fourteenth Solvay conference*. Interscience, New York.

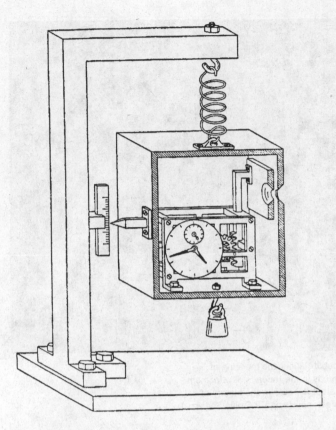

Fig. 7.4 Photon box experiment. Hypothetical instrument designed by Bohr to show how the measurements suggested by Einstein might be carried out. Reprinted by permission of Open Court Publishing Company, a division of Carus Publishing Company, Peru, IL, from (1949). The Library of Living Philosophers, Vol. VII, *Albert Einstein: philosopher–scientist*, edited by Paul Arthur Schilpp, p. 227.

The whole box is suspended by a spring and fitted with a pointer so that its position can be read on a scale affixed to the support. A small weight is added to bring the pointer to the zero on the scale. The clock mechanism is shown inside the box, connected to the shutter. After the release of one photon, the small weight is replaced by another, heavier weight so that the pointer is returned to the zero of the scale. The weight required to do this can be determined independently with arbitrary precision. The difference in the two weights required to balance the box gives the mass lost through the emission of one photon, and hence the energy of the photon.

Let us focus on the first weighing, before the photon escapes. Obviously, we will have set the clock mechanism to trigger the shutter at some predetermined time, and the box will be sealed. The actual reading of the clock face is, of course, not possible, since this would involve an exchange of photons—and hence energy—between the box and the outside world. To weigh the box, we must select a weight that just sets the pointer to the zero of the scale. However, to make a precise position measurement, the pointer and scale will again need to be illuminated, and, following Bohr's earlier arguments, this implies an uncertainty in the momentum of the box. How does this affect the weighing? The uncontrollable transfer of momentum to the box causes it to jump about unpredictably. Although we can fix the box's instantaneous position against the scale, the sizeable interaction during the act of measurement means that the box will not stay in that position. Bohr argued that we can increase the precision of measurement of the *average* position by allowing ourselves a long time interval in which to perform the whole balancing procedure.

Bohr and Einstein continued their debate about the meaning of
quantum theory through critical periods in the theory's development.
This was to become one of the greatest debates in the history of
science. Photograph by Paul Ehrenfest, courtesy AIP Emilio Segrè
Visual Archives.

This will give us the necessary precision in the weight of the box. Since we can anticipate the need for this, we can set the clock mechanism so that it opens the shutter after the balancing procedure has been completed.

Now comes Bohr's *coup de grâce*. According to Einstein's general theory of relativity, a clock moving in a gravitational field is subject to time-dilation effects: the very act of weighing a clock effectively changes the way it keeps time. This phenomenon is responsible for the red shift in the frequency of radiation emitted from the sun and stars. Because the box is jumping about unpredictably in a gravitational field (owing to the act of measuring the position of the pointer), the rate of the clock is changed in a similarly unpredictable manner. This introduces an uncertainty in the exact timing of the opening of the shutter which depends on the length of time needed to weigh the box. The longer we make the balancing procedure (the greater the ultimate precision in the measurement of the energy of the photon), the greater the uncertainty in its exact moment of release. Bohr was able to show that the relationship between the uncertainties of energy and time is in accord with the uncertainty principle. This response was hailed as a triumph for Bohr and for the Copenhagen interpretation of quantum theory. Einstein's own general theory of relativity had been used against him.

However, Einstein remained stubbornly unconvinced, although he did change the nature of his attacks on the theory. Instead of arguing that the theory is inconsistent, he began to develop arguments that he believed demonstrated its incompleteness in a much more fundamental way. When discussing the photon box experiment, Einstein conceded that it now appeared to be 'free of contradictions', but in his view, it still contained 'a certain unreasonableness'.

We should not leave the photon box experiment without noting that many physicists, including Bohr, have since examined it again in considerable detail, and it continues to generate comment in the current scientific literature. Some have rejected Bohr's response completely, denying that the uncertainty principle can be 'saved' in the way Bohr maintained. Others have rejected Bohr's response but have given alternative reasons why the uncertainty principle is not invalidated. Despite these counterproposals, the prevailing view in the physics community at the time appears to be that Bohr won this particular round in his debate with Einstein. However, Bohr appears to have been quite unprepared for Einstein's next move.

Is quantum mechanics complete?

Bohr had successfully defended the Copenhagen interpretation by arguing that the inevitable and sizeable disturbance of the observed system in any physical measurement process effectively precludes knowledge with a precision greater than that allowed by the uncertainty principle. According to Bohr, this disturbance places a fundamental limit on our ability to acquire knowledge in a quantum world, forcing us to be satisfied with statistical predictions derived from our use of complementary classical concepts. At first sight, there seems to be no argument against this position. A measurement of any kind will always involve physical interaction on the same scale as the system being measured, and it is a pointless descent into naivety to talk about the properties of quantum systems *only* in the absence of measurement. Einstein somehow had to find a way around this.

There were still some clues in the photon box experiment. What if, Einstein reasoned in 1931, we use the experiment not to confront the consistency of the uncertainty relations (which is what he had done at the Solvay Conference) but to use it instead to derive a logical paradox from what he saw to be the theory's lack of completeness? As before, we set the clock to trigger the release of one photon, but this time, we synchronize it with a second, external clock. We fill the box with photons. Einstein now accepted that his own general theory of relativity precluded precise knowledge of the moment of release, but perhaps this was not really the point after all. Let the released photon travel to a fixed mirror placed a long distance from the box (say, half a light year away). Whilst the photon is on its round-trip journey of one year, we can now *choose* what measurement we wish to make. We could open the box and compare the two clocks. They will no longer be synchronous because of the effect of weighing the box on the rate of the clock inside, but we can now correct for this by reference to the external clock, and so we can draw a retrodictive conclusion as to the exact timing of release of the photon. We know how long it will take on its journey, and so we can calculate its exact time of arrival back in the laboratory. Alternatively, we could choose to keep the box sealed and reweigh it. We could take as long as we needed over this second weighing, and, as before, this would tell us the exact energy of the released photon.

We make the further assumption that a photon half a light year distant cannot be affected by the decision we make in the laboratory to measure *either* the time of release *or* the energy of the photon. In other words, the photon is not in any way *disturbed* by measurements we make over three thousand billion miles away. We are left to conclude that the photon must possess simultaneously exact values of both energy and time. As there is nothing in quantum theory that tells us about the simultaneous exact values of these complementary observables, we further conclude that the theory provides an incomplete description of individual quantum systems.

Adolf Hitler came to power in Germany on 30 January 1933, and Einstein took up a position at the Institute for Advanced Study in Princeton and permanent residence the following October. Looking around for bright young mathematicians with whom to work, Einstein was drawn to a Russian, Boris Podolsky, and an American, Nathan Rosen. Einstein had already worked with Podolsky and had published a paper jointly with him and Richard Tolman in 1931, arguing that the uncertainty relations implied an uncertainty in past events as well as in future events. Rosen had recently completed a Ph.D. with John C. Slater at the Massachusetts Institute of Technology.

In the meantime, the photon box experiment had evolved further, and in May 1935, Einstein, Podolsky, and Rosen (EPR) published a paper in the journal *Physical Review*, entitled 'Can quantum-mechanical description of physical reality be considered complete?'. The abstract reads as follows[7]:

In a complete theory there is an element corresponding to each element of reality. A sufficient condition for the reality of a physical quantity is the possibility of predicting it with certainty, without disturbing the system. In quantum mechanics in the case of two physical quantities described by non-commuting operators, the knowledge of one precludes the knowledge of the other. Then either (1) the description of reality given by the wave function in quantum mechanics is not complete or (2) these two quantities cannot have simultaneous reality. Consideration of the problem of making predictions concerning a system on the basis of measurements made on another system that had previously interacted with it leads to the result that if (1) is false then (2) is also false. One is thus led to conclude that the description of reality as given by a wave function is not complete.

It must be stated at the outset that the EPR paper appears to have been written largely by Podolsky, and there is much in the language and nature of the argumentation employed that Einstein appears later to have regretted. In particular, the criterion of physical reality as described in the paper exposed EPR to a weakness that was to be exploited by Bohr in his counterargument, as we will see below. All the more disappointing, perhaps, as the main challenge presented by EPR does not require this (or any) criterion of physical reality, though it does rest on the assumption that however reality is defined, it is defined to be *local*, as we will see.

Einstein had needed to find a physical situation in which it was possible in principle to acquire knowledge of the state of a quantum particle without disturbing it in any way, thereby denying Bohr the escape route that he had used to such great effect in Einstein's earlier challenges. The modified photon box experiment, in which the experimenter is allowed to choose which of two complementary observables to measure exactly, was a first step towards the solution. The EPR thought experiment took the approach a stage further. It involves measurements made on one of two quantum particles that have interacted at some time in their history and moved apart. We will denote these as particle A and particle B. The position and momentum of particle A are complementary observables, and we cannot measure one without introducing an uncertainty in the other in accordance

---

[7] Einstein, A., Podolsky, B., and Rosen, N. (1935). *Physical Review*, **47**, 777.

with Heisenberg's uncertainty principle. Similar arguments can be made for the position and momentum of particle B.

However, if we now consider the *difference* between the positions of particles A and B and the *sum* of their momenta, it is relatively straightforward to show that these are quantities whose operators commute (see Appendix 14). There is therefore no restriction in principle on the precision with which we can measure the difference between the positions of the particles and the sum of their momenta.

A reasonable definition of reality

EPR allowed themselves what seems at first sight to be a fairly reasonable definition of physical reality[8]:

If, without in any way disturbing a system, we can predict with certainty (i.e. with a probability equal to unity) the value of a physical quantity, then there exists an element of physical reality corresponding to this physical quantity.

The purpose of this statement is to make clear that for each particle considered individually, the measurement of one physical quantity (the position of B, say) with certainty implies an infinite uncertainty in its momentum. Therefore, according to EPR's definition of reality, under these circumstances the position of particle B is an element of physical reality, but the momentum is not. Obviously, by choosing to perform a different measurement, we can (in the language of EPR) establish the reality of the momentum of particle B but not its position. The Copenhagen interpretation of quantum theory insists that we can establish the reality of one or the other of two complementary observable physical quantities but not both simultaneously.

But we can show that the difference in the positions of particles A and B and the sum of their momenta are quantities whose operators commute. The Copenhagen interpretation says that we can therefore establish the physical reality of these quantities simultaneously. It is enough for the EPR argument that these quantities are simultaneously real in principle, although their actual determination might require a physical measurement.

Now suppose we allow the two particles to interact and move a long distance apart. We perform an experiment on particle A to measure its position with certainty. We know that the difference in position must be a physically real quantity, and so in principle we can deduce the position of particle B also with certainty. We therefore conclude that the position of B must be an element of physical reality according to the EPR definition. However, suppose instead that we choose to measure the momentum of particle A with certainty. We know that the sum of the momenta must be physically real, and so in principle we can deduce the momentum of particle B with certainty. We conclude that it too must be an element of physical reality. Thus, although we have not performed any measurements on particle B following its separation from A, we can in principle establish the reality of *either* its position *or* its momentum from measurements we *choose* to perform on A which, *by definition, do not disturb B*.

The Copenhagen interpretation denies that we can do this. We are forced to accept that if this interpretation of quantum theory is correct, the physical reality of either the

[8] Einstein, A., Podolsky, B., and Rosen, N. (1935) *Physical Review*, 47, 777.

position or momentum of particle B is determined by the nature of the measurement we choose to make on a completely different particle an arbitrarily long distance away. EPR argued that 'No reasonable definition of reality could be expected to permit this.'

As presented above, the EPR argument is based on a hypothetical experiment and is concerned with matters of principle. At the time the argument was developed, it was unimportant that the proposed experiment is difficult, if not impossible, to perform. However, we will see in Part IV that the experimental study of the behaviour of quantum particles that have interacted and moved apart is made much more practicable if their spin properties are probed rather than their positions and momenta.

Spooky action at a distance

The EPR thought experiment strikes right at the heart of the Copenhagen interpretation. If the uncertainty principle applies to an individual quantum particle, then it appears that we must invoke some kind of action at a distance if the reality of the position or momentum of particle B is to be determined by measurements we choose to perform on A.

Whether it involves a change in the physical state of the system or merely some kind of communication, the fact that this action at a distance must be exerted instantaneously on a particle an arbitrarily long distance away from our measuring device suggests that it violates the postulates of special relativity, which restricts any signal to be communicated no faster than the speed of light. EPR did not believe that such action at a distance is necessary: the position and momentum of particle B are defined all along, and, as there is nothing in the wave function or state vector which tells us how these quantities are defined, quantum theory is incomplete. EPR concluded[9]:

While we have thus shown that the wave function does not provide a complete description of physical reality, we left open the question of whether or not such a description exists. We believe, however, that such a theory is possible.

Einstein attacks quantum theory

The EPR argument actually found its way into the popular press even before it appeared in *Physical Review*. The 4 May 1935 edition of *The New York Times* carried an article entitled 'Einstein attacks quantum theory', which provided a non-technical summary of the main arguments, with extensive quotations from Podolsky. Einstein deplored the article and the publicity surrounding it. The newspaper article was followed in the same edition by a report by physicist Edward Condon, who noted that the arguments raised a 'point of doubt' but who also identified the reality criterion as the argument's most significant weakness.

Bohr first heard of the EPR argument from Léon Rosenfeld, who was at that time working with Bohr in Copenhagen. Rosenfeld later reported that[10]:

... this onslaught came down upon us like a bolt from the blue. Its effect on Bohr was remarkable ... as soon as Bohr had heard my report of Einstein's argument, everything else was abandoned: we have

[9] Einstein, A., Podolsky, B., and Rosen, N. (1935). *Physical Review*, 47, 777.

[10] Rosenfeld, L. in Rozenthal, S. (1967). *Niels Bohr; his life and work as seen by his friends and colleagues.* North-Holland, Amsterdam.

to clear up such a misunderstanding at once. We should reply by taking up the same example and showing the right way to speak about it. In great excitement, Bohr immediately started dictating to me the outline of such a reply. Very soon, however, he became hesitant. 'No, this won't do, we must try all over again . . . we must make it quite clear.' So it went on for a while, with growing wonder at the unexpected subtlety of the argument.

Others were devastated by EPR. Pauli was furious. Dirac believed that they would have to start all over again because Einstein had proved that it did not work.[11]

Bohr's reply to the EPR argument was published in *Physical Review* in October 1935. He chose to use the same title that EPR had used in May and the abstract reads as follows[12]:

It is shown that a certain 'criterion of physical reality' formulated in a recent article with the above title by A. Einstein, B. Podolsky and N. Rosen contains an essential ambiguity when it is applied to quantum phenomena. In this connection a viewpoint termed 'complementarity' is explained from which quantum-mechanical description of physical phenomena would seem to fulfill, within its scope, all rational demands of completeness.

Bohr's paper is essentially a summary of complementarity and its application to quantum theory. He rejects the argument that the EPR thought experiment creates serious difficulties for the Copenhagen interpretation and stresses once again the importance of taking into account the necessary interactions between the objects of study and the measuring devices. He wrote:

From our point of view we now see that the wording of the above-mentioned criterion of physical reality proposed by Einstein, Podolsky and Rosen contains an ambiguity as regards the meaning of the expression 'without in any way disturbing a system' . . . there is essentially the question of an influence on the very conditions which define the possible types of predictions regarding the future behaviour of the system. Since these conditions constitute an inherent element of the description of any phenomenon to which the term 'physical reality' can be properly attached, we see that the argumentation of the mentioned authors does not justify their conclusion that quantum-mechanical description is essentially incomplete.

Many in the physics community seemed to accept that Bohr's paper put the record straight on the EPR experiment. I find Bohr's wording really rather vague and unconvincing. Bohr inevitably targeted the reality criterion as the Achilles heel in the EPR argument, emphasizing once again the important role of the measuring instrument in defining the elements of reality that we can observe. Thus, setting up an apparatus to measure the position of particle A with certainty, from which we can infer the position of particle B, excludes the possibility of measuring the momentum of A and hence inferring the momentum of B. If there is no mechanical disturbance of particle B (as EPR assume), its elements of physical reality must be defined by the nature of the measuring device we have selected for use with particle A. This is classic anti-realism.

The EPR argument pushed Bohr from his previous, rather ambiguous philosophy to a fixed anti-realist position. It was at this point that Bohr dropped the use of 'disturbance' as a counter-argument and focused instead on the nature of the experimental arrangement itself precluding the type of reality that could be exposed. Mechanical disturbance could no longer serve Bohr's purpose. Rather, this was a matter of the essential inseparability

[11] From a quotation in Beller, Mara. (1999). *Quantum dialogue*. University of Chicago Press, Chicago, p. 145.
[12] Bohr, N. (1935). *Physical Review*, **48**, 696.

of observed system and measuring device, with the way the measuring device is set up effectively 'defining' the (statistical) measurement outcomes. In the language of quantum measurement in Chapter 5, we would say that the possible measurement eigenstates are defined by the way we set up the apparatus. Measuring the position of A requires that we set up an apparatus whose measurement eigenstates are the possible positions of A and hence (by inference) the possible positions of B. This same apparatus cannot be used to measure the momentum of A, and so cannot be used to infer the momentum of B. In the context of EPR's definition, this combination of apparatus and observed system has measurement eigenstates that make it meaningless to ascribe a reality to the momentum of B.

However, Bohr failed to respond to the real challenge posed by EPR. What EPR had created in their thought experiment was a two-particle state that allows for correlations to be established between quantum particles over potentially vast distances. Making a measurement, involving a notional collapse of the wave function, implies a 'spooky' action at a distance that appears to violate the basic postulates of special relativity.

Does such a measurement necessarily imply an action at a distance? Certainly, if we could somehow delay our choice of measuring instrument (position versus momentum) until almost the last moment, then in principle the information available to us about a particle some considerable distance away changes instantaneously. We are left to wonder how particle B is supposed to 'know' what physical property—position or momentum—it is supposed to reveal as a result of measurements made on A. An action at a distance will be required if the measurement performed on A changes the physical state of B or results in some kind of communication to B of particle A's changed circumstances.

Now if the two-particle wave function or state vector reflects only our state of knowledge of the quantum system, then its collapse would not necessarily seem to affect the system's physical properties. However, the problem remains that the collapse of the wave function requires that those physical properties become manifest in the quantum system where before they were not defined. The physical properties of particle B suddenly become 'real' (i.e. measurable), where before they were not.

There is no mechanism for this in the Copenhagen interpretation of quantum theory.

Einstein separability

In June 1935, Schrödinger wrote to congratulate Einstein on the EPR paper. He wrote[13]:

I was very happy that in the paper just published in [Physical Review] you have evidently caught dogmatic [quantum mechanics] by the coat-tails... My interpretation is that we do not have a [quantum mechanics] that is consistent with relativity theory, i.e., with a finite transmission speed of all influences. We have only the analogy of the old absolute mechanics... The separation process is not at all encompassed by the orthodox scheme.

Schrödinger's reference to the 'separation process' highlights the essential difficulty that the EPR argument creates for the Copenhagen interpretation. According to this interpretation, the wave function or state vector for the two-particle quantum state does not separate as the particles themselves separate in space–time. Instead of dissolving into two

[13] Schrödinger, Erwin, letter to Einstein, Albert, 7 June 1935, quoted in Moore, Walter (1989). *Schrödinger: life and thought*. Cambridge University Press, Cambridge, p. 304.

completely separate state vectors, one associated with each particle, the two-particle state vector is 'stretched' out and, when a measurement is made, collapses instantaneously despite the fact that it may be spread out over a considerably large distance. In his 1935 article in *Die Naturwissenschaften*, Schrödinger identified the two particles to have become *entangled* as a result of their interaction[14]:

If two separated bodies, each by itself known maximally, enter a situation in which they influence each other, and separate again, then there occurs regularly that which I have just called *entanglement* of our knowledge of the two bodies.

EPR's definition of physical reality requires that the two particles are considered to be isolated from each other, that is, they are no longer described by a single two-particle state vector at the moment a measurement is made. The reality thus referred to is sometimes called 'local reality', and the ability of the particles to separate into two locally real independent physical entities is sometimes referred to as 'Einstein separability'. Under the circumstances of the EPR thought experiment, the Copenhagen interpretation denies that the two particles are Einstein separable and therefore denies that they can be considered to be locally real (at least, before a measurement is made on one or other of the particles, at which point they both become localized).

Entangled states and Schrödinger's cat

Einstein responded immediately to Schrödinger's letter of 7 June 1935[15]:

All physics is a description of reality; but this description can be 'complete' or 'incomplete'. To begin with, the sense of this expression is even a problem itself. I will explain with the following analogy:
  In front of me stand two boxes, with lids that can be opened, and into which I can look when they are open. This looking is called 'making an observation'. In addition there is a ball, which can be found in one or the other of the two boxes when an observation is made.
  Now I describe the situation thus: the probability that the ball is in the first box is $\frac{1}{2}$. Is this a complete description?

(1) NO. A complete description is: the ball is in the first box (or is not). This is the way to express the characterization of the state by a complete description.

(2) YES. Before I open the box the ball is not in *one* of the two boxes. This existence in a definite box first occurs when I open one of the boxes. In this way arises the statistical character of the world of experience or its empirical system of laws [*Gesetzlichkeit*]. The state *before* the box is opened is completely described by the number $\frac{1}{2}$.

The Talmudic philosopher doesn't give a straw for 'reality', a bogy of naiveté, and explains both statements as only different ways of expression.
  I bring in the *separation principle*. The second box is independent of anything that happens to the first box. If one holds fast to the separation principle, only the Born description is possible, but now it is incomplete.

---

[14] Schrödinger, E. (1935). *Naturwissenschaften*, **23**, 807–812, 823–828, 844–849.
[15] Einstein, Albert, letter to Schrödinger, Erwin, 19 June 1935, quoted in Moore, Walter (1989). *Schrödinger: life and thought*. Cambridge University Press, p. 304. For a slightly different translation (and the original German text), see Fine, Arthur (1986). *The shaky game: Einstein, realism and the quantum theory* (2nd edn.). University of Chicago Press, Chicago, p. 71.

Their correspondence continued through the summer of 1935. Schrödinger was still pursuing the idea that the quantum wave function—and its statistical interpretation—reflected an underlying physical reality of waves, wave packets, and their superpositions. Einstein insisted that the wave function was inadequate as a complete description of reality and reflected only the statistical probabilities of ensembles of systems. In seeking to persuade Schrödinger to this point of view, Einstein developed yet another thought experiment, one that was ultimately to lead Schrödinger to present one of the most famous paradoxes of quantum theory. This thought experiment consisted of a charge of gunpowder that could spontaneously combust[16]:

In the beginning the ψ-function characterizes a reasonably well-defined macroscopic state. But, according to your equation, after the course of a year this is no longer the case at all. Rather, the ψ-function then describes a sort of blend of not-yet and of already-exploded systems. Through no art of interpretation can this ψ-function be turned into an adequate description of a real state of affairs; [for] in reality there is just no intermediary between exploded and not-exploded.

Schrödinger eventually came to be persuaded to Einstein's point of view, and summarized his own position in a series of three articles published in *Die Naturwissenschaften*, entitled 'The present situation in quantum mechanics', from which I have already quoted. However, despite the broad scope of subject matter covered in these articles, they will forever be known by a single paragraph that appears in a section headed 'Are the variables really blurred?'. This paragraph introduces the famous paradox of Schrödinger's cat.

In our discussion of quantum measurement in Chapter 5, the notion of the collapse of the wave function was presented without reference to the point in the measurement process at which the collapse is meant to occur. It might be assumed that the collapse occurs at the moment the microscopic quantum system interacts with the macroscopic measuring device. But is this assumption justified? After all, a macroscopic measuring device is composed of microscopic entities—molecules, atoms, protons, neutrons, and electrons. We could argue that the interaction takes place on a microscopic level and should, therefore, be treated using quantum mechanics, as John von Neumann had assumed when developing his original quantum theory of measurement. The problem is that the collapse is itself not contained in *any* of the mathematical apparatus of quantum theory: we simply cannot obtain a collapse of the state vector using Schrödinger's continuous, deterministic equation of motion from which the time evolution of the interaction between a quantum system and a measuring device is deduced. It is impossible to separate the combined description of quantum system plus measuring device into its constituent parts except by invoking an indeterministic collapse, or von Neumann's 'projection postulate'.

Schrödinger drew on the ball-in-the-box and the exploding-gunpowder examples that he had discussed extensively in his correspondence with Einstein to show up the apparent absurdity of this situation by shifting the focus from the microscopic world of subatomic particles to the macroscopic world of cats and human observers. The essential ingredients are shown in Fig. 7.5. A cat is placed inside a steel chamber together with a Geiger tube containing a small amount of radioactive substance, a hammer mounted on a pivot and a phial of prussic acid. The chamber is closed. From the amount of radioactive substance used and its known half-life, we expect that within 1 h, there is a probability of $\frac{1}{2}$ that

[16] Einstein, Albert, letter to Schrödinger, Erwin, 8 August 1935, quoted in Fine, Arthur (1986). *The shaky game: Einstein, realism and the quantum theory* (2nd edn.). University of Chicago Press, Chicago, p. 78.

Fig. 7.5 Schrödinger's cat.

one atom has disintegrated. If an atom does indeed disintegrate, the Geiger counter is triggered, releasing the hammer, which smashes the phial. The prussic acid is released, killing the cat.

Prior to actually measuring the disintegration, the state vector of the atom of radioactive substance must be expressed as a linear superposition of the measurement eigenstates, corresponding to the physical states of the intact atom and the disintegrated atom. However, as we can see from Appendix 15, treating the measuring instrument as a quantum object and using the equations of quantum mechanics leads us to an infinite regress. We create a superposition of the two possible outcomes of the measurement.

But what about the cat? These arguments would seem to suggest that we should express the state vector of the system-plus-cat as a linear superposition of the products of the state vectors describing a disintegrated atom and a dead cat and of the state vectors describing an intact atom and a live cat. In fact, the state vector of the dead cat is in turn a shorthand for the state corresponding to the triggered Geiger counter, released hammer, smashed phial, released prussic acid, and dead cat. Prior to measurement, the physical state of the cat is therefore 'blurred'—it is neither alive nor dead but some peculiar combination of both states. We can perform a measurement on the cat by opening the chamber and ascertaining its physical state. Do we suppose that, at that point, the state vector of the system-plus-cat collapses, and we record the observation that the cat is alive or dead as appropriate?

Although obviously intended to be somewhat tongue in cheek, Schrödinger's para-dox nevertheless brings our attention to an important difficulty that must be confronted. The Copenhagen interpretation says that elements of an empirical reality are defined by the nature of the experimental apparatus we construct to perform measurements on a quantum system. It insists that we resist the temptation to ask what physical state a par-ticle (or a cat) was actually in prior to measurement, as such a question is quite without meaning. As far as Copenhagen is concerned, Schrödinger's cat is indeed blurred: it is meaningless to speculate on whether it is really alive or dead until the box is opened.

However, this anti-realist interpretation sits uncomfortably with some scientists, par-ticularly those with a special fondness for cats. Einstein saw the paradox as yet further evidence for the basic incompleteness of quantum theory[17]:

your cat shows that we are in complete agreement concerning our assessment of the character of the current theory. A ψ-function that contains the living as well as the dead cat just cannot be taken as a description of a real state of affairs. To the contrary, this example shows exactly that it is reasonable to let the ψ-function correspond to a statistical ensemble that contains both systems with live cats and those with dead cats.

Summary

The sustained attack by Einstein and Schrödinger on the 'Kopenhagener Geist', fast becoming an entrenched, anti-realist interpretation of quantum theory, was intended primarily to highlight the theory's considerable conceptual and philosophical problems. These problems were (and still are today) derived from the unique role of measurement in quantum theory and relate to the following.

The interpretation of quantum probability. Modern quantum theory remains today an incon-gruous mix of Copenhagen (which insisted on the completeness of quantum theory when applied to individual systems) combined with an ensemble interpretation. Unlike classical probabilities, quantum probabilities do *not* reflect our ignorance of the intricate details of some underlying physical reality. They are rather an expression of the likelihood that interaction between the quantum system and the measuring device will 'create' or 'realize' specific outcomes. These outcomes are not generally identified with individual systems but rather apply to the distribution of results of repeated measurements on a collection of identically prepared systems. However, unlike Einstein's ensembles, these systems cannot be considered to consist of individual entities that themselves possess individually realized states until they are exposed to a measuring device.

The collapse of the wave function. The transition from potentialities to measurement actu-alities is simply not described at all in the theory's mathematical framework. It is rather postulated as a way of breaking the infinite regress implied when the measuring device itself is considered to be composed of quantum particles and therefore subject to quantum theory's basic laws, and was introduced by John von Neumann as a 'projection postulate'.

---

[17] Einstein, Albert, letter to Schrödinger, Erwin, 4 September 1935, quoted in Fine, Arthur (1986). *The shaky game: Einstein, realism and the quantum theory* (2nd edn.). University of Chicago Press, Chicago, p. 84.

The Copenhagen interpretation is completely silent on the question of where in the process this collapse is supposed to take place, leaving us to ponder the fate of cats suspended in states of quantum superposition.

Entangled states, locality, and action at a distance. The essence of the EPR argument was a demonstration of 'spooky' action at a distance implied by the entanglement of two quantum particles that have interacted and moved apart. The Copenhagen interpretation insists that the theory is complete and that, therefore, the two particles continue to be described by a single two-particle state vector, no matter how far apart they travel. The particles have lost their individuality and their locality in space–time. When we make a measurement on one particle, the state vector collapses instantaneously, forcing the second particle to realize a specific state, though it may be half way across the universe. EPR insisted that 'no reasonable definition of reality could be expected to permit this'.

Some have accepted the EPR argument that quantum theory is incomplete. They have set about searching for an alternative interpretation or an alternative theory, one that allows us to attach physical significance to the properties of particles without the need to specify the nature of the measuring instrument, one that allows us to define an independent reality and that reintroduces strict causality. Even though searching for such a theory might be engaging in meaningless metaphysical speculation, they believe that it is a search that has to be undertaken. Each of these alternatives attempts to resolve one or more of the problems outlined above, with varying degrees of success. We will review them in detail in Part V.

Before undertaking this review, it is important first to examine the simplest of the potential solutions of quantum theory's great conundrums. If we take Einstein at face value, we could argue that quantum probabilities are in principle no different from the probabilities of classical physics. Just as Boltzmann's statistical mechanics provides us with a means of identifying relations between physical quantities without having to deal with the enormously complex, but nonetheless real, 'hidden' motions of atoms and molecules, so perhaps quantum theory provides us with statistical relations between similarly local hidden variables associated with real individual, localized quantum particles moving in a real space–time frame. By invoking hidden variables, we might be able to eliminate the collapse of the wave function, entanglement and spooky action at a distance. Hidden variable theories are compelling not least because of their conceptual simplicity, and we examine them in detail in the next chapter.

8

Bell's theorem and local reality

If we reject the 'spooky' action at a distance that seems to be required in the Copenhagen interpretation of quantum theory, and which is highlighted by the EPR thought experiment, we must accept the EPR argument that the theory is somehow incomplete. In essence, this involves the rejection of the first postulate of quantum theory: the state of a quantum mechanical system is *not* completely described by the wave function or state vector.

Those physicists who, in the 1930s, were uncomfortable with the Copenhagen interpretation were faced with two options. Either they could scrap quantum theory completely and start all over again, or they could try to extend the theory somehow to reintroduce strict causality, local reality, or both. There was a general recognition that quantum theory was too good to be consigned to history's waste bin of scientific ideas. The theory seemed to do a good job of rationalizing the available experimental information on the physics of the microscopic world of quantum entities, and its predictions had been shown to be consistently correct. What was needed, therefore, was some means of adapting the theory to bring back those aspects of classical physics that it appeared to lack.

Einstein had hinted at a statistical interpretation. In his opinion, the squares of the wave functions or state vectors of quantum theory represented statistical probabilities obtained by averaging over a large number of real particles. The Copenhagen interpretation of the EPR experiment insists that the reality of the physical states that can be measured is defined by the nature of the interaction between two quantum particles and the nature of the experimental arrangement. A completely deterministic, locally real version of quantum theory demands that the physical states of the particles be established at the moment of their interaction and that the particles separate as individually real entities in those physical states. The physical states of the particles are prescribed by the physics of their interaction and independent of how we choose to set up the measuring instrument, and so no reference to the nature of the latter is necessary, except to define how the independently real particles interact with it. The instrument thus probes an observer-independent reality.

Quantum theory in the form taught to undergraduate students of chemistry and physics tells us nothing about such physical states. This is either because they have no basis in reality (the anti-realist position as embodied in the Copenhagen interpretation) or because the theory is incomplete (EPR argument). One way in which quantum theory can be made 'complete' in this sense is to introduce a new set of variables. These variables determine which physical states will be prescribed as a result of a quantum process (such as an emission of a photon or a collision between two quantum entities). As these variables are not revealed in laboratory experiments, they are necessarily 'hidden'.

Of course, hidden-variable theories of one form or another are not without precedent in the history of science. Any theory which rationalizes the behaviour of a system in terms of parameters that are for some reason inaccessible to experiment is a hidden-variable theory. These variables have often later become 'unhidden' through the application of new experimental technologies. The obvious example is again Boltzmann's use of the 'hidden' motions of real atoms and molecules to construct a statistical theory of mechanics. Mach's opposition to Boltzmann's ideas was based on the view that such hidden variables are entirely metaphysical, and introducing them unnecessarily complicates a theory and takes science forward no further.

Einstein on hidden variables

We should note that, although the introduction of hidden variables in quantum theory appears to be consistent with Einstein's belief that quantum theory is somehow incomplete, it is quite clear that he himself came to reject such an approach. Understanding his reasons requires another short diversion into the history of quantum physics.

Einstein took just such an approach himself in May 1927, when he read a paper to the Prussian Academy of Sciences in Berlin entitled 'Does Schrödinger's wave mechanics determine the motion of a system completely or only in the sense of statistics?' In this paper, he reconfigured conventional quantum theory (in the form of the time-independent Schrödinger equation), to associate specific particle directions with individual kinetic energy and velocity terms in the equation. Schrödinger's wave mechanics deals only with position and momentum, so these principal 'directions' take the form of hidden variables in Einstein's modification. By using the wave function to determine a unique velocity at each point in a multi-dimensional configuration space, Einstein could ensure that the specification of an initial state serves to specify a completely deterministic trajectory for the system, with surfaces of constant phase of the wave function representing 'wavefronts' in configuration space.

Einstein's modification was essentially a synthesis of classical wave and particle descriptions, with the wave function of Schrödinger's quantum mechanics taking the role of a 'guiding field' (Führungsfeld), guiding physically real point particles. In this scheme, the wave function is responsible for all the wave-like effects, such as diffraction and interference, but the particles maintain their integrity as local, physically real entities. There are obvious parallels here with what became known as de Broglie's 'double solution' and subsequent 'pilot wave' variations of quantum theory, which we will examine later in a little more detail.

Whilst Einstein was excited by his result at the beginning of May 1927, his enthusiasm for it had evaporated by the end of that month. He began to have doubts about attaching physical significance to a wave function defined in multidimensional configuration space. Perhaps most importantly, he noted that his hidden-variables scheme still allowed the peculiar entanglement of individual systems when described in terms of a composite wave function. In other words, although the particles might remain localized, they are influenced by non-local correlations in the guiding wave function, and consequently they cannot be regarded to be Einstein-separable. This was to become the basis of the EPR argument some 8 years later. Einstein realized that his scheme had not solved the problem of 'spooky' action at a distance. Einstein withdrew his paper before it could be published. It survives in the Einstein Archives as a handwritten manuscript.[1]

This experience probably led Einstein to conclude that his initial belief—that quantum theory could be 'completed' through a more direct fusion of classical wave and particle concepts—was misguided. He later asserted that such attempts were 'too cheap' and expressed the opinion that a complete theory could emerge only from a much more radical revision of the entire theoretical structure. This is the likely reason that he was very cool towards subsequent attempts at devising both local and non-local hidden variables alternatives to standard quantum theory. He appears, rather, to have been convinced that solutions to the conceptual problems of quantum theory would be found in an elusive grand unified field theory, the search for which took up most of his intellectual energy in the last decades of his life.

A simple example

Einstein's objections notwithstanding, local hidden variables theories do in principle appear to offer a very direct route to the resolution of many of the conceptual and philosophical problems of quantum theory and, for this reason, cannot be ignored. It will serve our purpose here to illustrate the kinds of properties required of such a theory by considering a very simple, if somewhat contrived, example.[2]

A photon is in a state of left-circular polarization described by the state vector $|L\rangle$. We know that its interaction with a linear polarization analyser, and its subsequent detection, will reveal the photon to be in a state of vertical or horizontal polarization. Suppose, then, that the photon is completely described by $|L\rangle$ *supplemented* by some hidden variable λ, which prescribes which state of linear polarization will be observed experimentally. By definition, λ itself is inaccessible to us through experiment, but its value somehow controls the way in which the photon interacts with the analyser.

We could imagine that λ has all the properties we would normally associate with a linear polarization vector. We presume that λ (or its projection) can take up any angle in the plane perpendicular to the direction of propagation, as shown in Fig. 8.1(a). In a large ensemble of N left-circularly polarized photons, there would be a distribution of λ values over the N photons spanning the full 360° range. Thus, photon 1 has a λ value which we characterize in terms of the angle φ_1 it makes with the vertical axis, photon 2 has a λ value characterized by φ_2 and so on until we reach photon N, which has a λ value characterized by φ_N. These angles are randomly distributed in the range 0–360°.

We now need to suppose further that these λ values control the way in which the photons interact with the polarization analyser. A simple mechanism is as follows. If the angle that λ makes with the vertical axis lies within ±45° of that axis, then the photon passes through the vertical channel of the analyser and is detected as a vertically polarized photon, as

[1] Belousek, Darrin W. (1996). *Studies in the History and Philosophy of Modern Physics*, **27**, 437–461.
[2] This example is adapted from Rae, Alastair (1986) *Quantum physics: illusion or reality?* Cambridge University Press, Cambridge, pp. 34–36.

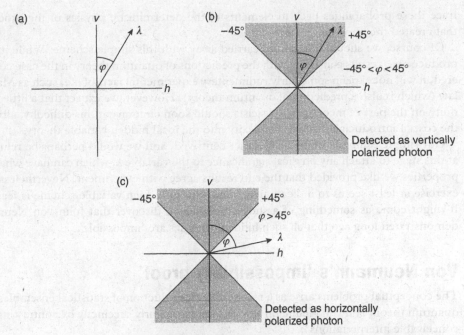

Fig. 8.1 Simple example of a local hidden variable. (a) The hidden vector λ determines the behaviour of a circularly polarized photon when it interacts with a linear polarization analyser. (b) If λ lies within $\pm 45°$ of the vertical axis of the analyser, it passes through the vertical channel. (c) If it lies within $\pm 45°$ of the horizontal axis, it passes through the horizontal channel.

illustrated in Fig. 8.1(b). If, however, λ makes an angle with the vertical axis which lies outside this range, then the photon passes through the horizontal channel of the analyser and is detected as a horizontally polarized photon, Fig. 8.1(c). We would need to suggest that the photon retains some 'memory' of its original circular polarization if we are to avoid the kinds of problems described in Chapter 5, and which arise when two calcite crystals are placed 'back to back'. In fact, why not suppose that, when the linear polarization properties of the photon become revealed, its circular polarization properties become hidden, controlled by another hidden variable.

In this scheme, the probability of detecting a photon in a state of vertical polarization becomes equal to the probability that the photon has a λ value within $\pm 45°$ of the vertical axis. If there is a uniform probability that the λ value lies in the range 0–360°, then the probability that it lies within $\pm 45°$ of the vertical axis is clearly $\frac{1}{2}$. Similarly, the probability of detecting a photon in a state of horizontal polarization is also $\frac{1}{2}$. Thus, this simple hidden variable theory predicts results consistent with those of quantum theory for the case of a beam of left-circularly polarized photons interacting with a linear polarization analyser.

Note that while we are still referring here to probabilities, unlike those of quantum theory, these are now classical statistical probabilities, averaged over a large number of photons which individually possess clearly defined and predetermined properties. If the hidden variable approach were proved to be correct, we would presumably be able to trace these probabilities back to elements of the deterministic physics of the processes that created the photons.

Of course, we should not get too carried away with this simple scheme. While it does produce *some* results consistent with the predictions of quantum theory in the case considered, it will not explain some fairly rudimentary experimental facts of life, such as Malus's law (which is also a prediction of quantum theory). However, we expect that a little ingenuity on the part of theoretical physicists should soon circumvent this difficulty, albeit at the cost of introducing further complexity into the local hidden variable theory.

Our simple example is obviously rather contrived, and we would perhaps be reluctant at this stage to attach any physical significance to the variable λ, which can have whatever properties we like provided that the end results agree with experiment. Nevertheless, this exercise at least seems to indicate that some kind of hidden variable scheme is feasible. It might come as something of a shock therefore to discover that John von Neumann demonstrated long ago that all such hidden variables are 'impossible'.

Von Neumann's 'impossibility proof'

The conceptual problems arise as a result of the introduction of statistical ensembles into quantum theory and von Neumann sets this up particularly succinctly by contrasting two conceivable interpretations[3]:

I. The individual systems . . . of our ensemble can be in different states, so that the ensemble . . . is defined by their relative frequencies. The fact that we do not obtain sharp values for the physical quantities in this case is caused by our lack of information: we do not know in which state we are measuring, and therefore cannot predict the results.

II. All individual systems . . . are in the same state, but the laws of nature are not causal. Then the cause of the dispersions is not our lack of information, but is nature itself, which has disregarded the 'principle of sufficient cause'.

A hidden variable extension of quantum theory is specifically designed to meet the needs of Case I, above. An ensemble of N quantum particles, all described by some state vector $|\psi\rangle$, contains particles with some distribution of λ values. For an individual particle, the value of λ prescribes its behaviour during the measurement process. Suppose that the result of a measurement (with measurement operator $\hat{M}$) is one of two possibilities, R_+ and R_-. The ensemble N can then be divided into two sub-ensembles, which we denote N_+ and N_-. The sub-ensemble N_+ consists of those particles with λ values that prescribe the result R_+ for each particle. The sub-ensemble N_- similarly contains only those particles predisposed to give the result R_-. Referring to our simple example given above, N_+ would contain all those photons with λ values characteristic of vertical polarization, and N_- would contain those photons with λ values characteristic of horizontal polarization.

When we perform measurements on particles in the sub-ensemble N_+, we know that we should always obtain the result R_+. For this sub-ensemble, R_+ is an *eigenvalue* of the measurement operator, and this result is obtained with *unit probability*, as demanded by the deterministic physics supposedly operating through the hidden variables. The sub-ensemble therefore has the property that the expectation value for the result of operating

[3] Von Neumann, John (1955). *Mathematical foundations of quantum mechanics*. Princeton University Press, Princeton, NJ, p. 302.

on the state vector with the square of the measurement operator, $\langle \hat{M}^2 \rangle$, is equal to the square of the expectation value for that same result, $\langle \hat{M} \rangle^2$. (In this case, both $\langle \hat{M}^2 \rangle$ and $\langle \hat{M} \rangle^2$ are equal to R_+^2—see Appendix 16.) The difference between these two quantities is the *variance* (the square of the standard deviation, a measure of the predicted spread of results around the expected value), and an ensemble with zero variance is said to be *dispersion-free*. Hidden variable theories demand that such dispersion-free ensembles exist, since they represent the situation where each particle in the ensemble is expected to yield a measurement outcome with unit probability in an entirely deterministic way. The purpose of hidden variable theories is to ensure exactly this. Von Neumann's proof rests on the demonstration that such dispersion-free ensembles are, in fact, impossible, and hence no hidden variable theory can reproduce the results of quantum theory.

Von Neumann's proof is included in his book *Mathematical foundations of quantum mechanics*. The proof is quite complicated, and we will deal with it here only in a superficial manner. Interested readers are advised to consult Appendix 16 and the more advanced texts given in the Bibliography. The proof is based on a number of postulates, one of which merits our attention. Imagine that we apply a second measurement, with operator $\hat{L}$ and outcomes S_+ and S_-, simultaneously with the first measurement with operator $\hat{M}$. We can use the same line of argument as above to identify within the sub-ensemble N_+ a sub-sub-ensemble (which we denote N_{++}), which consists of those particles with λ values or combinations of values that prescribe the results R_+ *and* S_+ for each particle. Von Neumann postulated that the expectation value for the combined measurement $\hat{M}$ plus $\hat{L}$ can be obtained as a linear combination of the expectation values of the operators $\hat{M}$ and $\hat{L}$ applied separately, whether or not these operators commute. This is known as von Neumann's 'additivity' postulate. It is relatively straightforward to show that this postulate is indeed valid for operators in quantum theory.

But now, according to von Neumann, we have a problem. If we apply the combined measurement $\hat{M}$ plus $\hat{L}$ on the particles that form the sub-sub-ensemble N_{++}, then the additivity postulate leads us to an expectation value $\langle \hat{M} + \hat{L} \rangle$ given as the sum $R_+ + S_+$. By definition, these results must each be obtained with unit probability. However, although the expectation values of non-commuting quantum mechanical operators are additive, their eigenvalues clearly are not. If they were, then an appropriate choice of measurement operators would allow us simultaneously to measure the position and momentum of a quantum particle with arbitrary precision, or mutually exclusive electron spin orientations, or simultaneous linear and circular polarization states of photons. This clearly conflicts with experiment. Von Neumann therefore concluded that dispersion-free ensembles (and hence hidden variables) are impossible. He went on to state[4]:

It should be noted that we need not go any further into the mechanism of the 'hidden parameters', since we now know that the established results of quantum mechanics can never be re-derived with their help...It is therefore not, as is often assumed, a question of re-interpretation of quantum mechanics—the present system of quantum mechanics would have to be objectively false, in order that another description of the elementary processes than the statistical one be possible.

Von Neumann's proof certainly discouraged the physics community from taking the idea of hidden variables seriously, although a few (notably Schrödinger and de Broglie)

[4] Von Neumann, John (1955). *Mathematical foundations of quantum mechanics*. Princeton University Press, Princeton, NJ, p. 324.

were not put off by it. Others began to look closely at the proof and became suspicious. A few questioned the proofs' correctness. The physicist Grete Hermann suggested that von Neumann's proof is circular—that it presupposes what it is trying to prove in its premises. She argued that the additivity postulate, whilst certainly true for quantum states in ordinary quantum theory, cannot be automatically assumed to hold for states described in terms of hidden variables. Since von Neumann's proof rests on the general non-additivity of eigenvalues, it collapses without the additivity postulate.

In his book *The philosophy of quantum mechanics*, published in 1974, Max Jammer examined Hermann's arguments and concluded that the charge of circularity is not justified. He noted that the additivity postulate was intended to apply to all operators, not just non-commuting operators (which would give rise to non-additive eigenvalues). However, for commuting operators, the case against dispersion-free ensembles is not proven by von Neumann's arguments. Jammer wrote: 'What should have been criticized, instead, is the fact that the proof severely restricts the class of conceivable ensembles by admitting only those for which [the additivity postulate] is valid.'[5]

It is also worth noting an objection raised by the physicist John S. Bell (who we will meet again later in this chapter). Bell argued that von Neumann's proof applies to the simultaneous measurement of two complementary physical quantities. But such measurements require completely incompatible measuring devices that cannot therefore be applied simultaneously, so nobody should be surprised that the corresponding eigenvalues are not additive.

It gradually began to dawn on the physics community that hidden variables were not impossible after all. But about 20 years passed between the publication of von Neumann's proof and the resurgence of interest in hidden variable theories. By that time, the anti-realist Copenhagen interpretation was well entrenched as the *only* interpretation of quantum physics, and those arguing against it were in a minority.

Bohm's version of the EPR experiment

Work on hidden variable solutions to the conceptual problems of quantum theory did not exactly stop after the publication of von Neumann's proof, but then it hardly represented an expanding field of scientific activity. Two decades passed before David Bohm, a young American physicist, began to take more than a passing interest in the subject.

Bohm was born in Pennsylvania and completed his doctorate under Robert Oppenheimer at the Berkeley Radiation Laboratory in California. With Oppenheimer's help, he moved to Princeton in 1947 and attended the Shelter Island conference on the problems of quantum theory with Oppenheimer that same year. Two years later, he became embroiled in the Communist witch-hunt that was subsequently to become associated with the name of Senator McCarthy. Bohm, a Communist himself, was brought before the House Un-American Activities Committee and interrogated on his dealings with Steve Nelson, who was believed to have passed atomic secrets to the Soviet Union and who was a close friend of Oppenheimer's wife, Kathryn. When asked if he knew Nelson, Bohm pleaded the Fifth Amendment. Bohm was arrested in December 1950 and charged with contempt of Congress. Under pressure from Princeton University's

---

[5] Jammer, Max (1974). *The philosophy of quantum mechanics.* John Wiley, New York.

conservative benefactors, the University's administration terminated Bohm's contract and Oppenheimer, probably sensing the tide of anti-Communist sentiment that would lead eventually to his own condemnation and withdrawal of his security clearance in 1954, advised Bohm to leave the country. Bohm worked in Brazil and Israel before moving to Bristol University in England in 1957. He eventually took the chair in Theoretical Physics at Birkbeck College in London.[6]

Bohm made his first, all-important contributions to the debate over the interpretation of quantum theory in 1951. In February of that year, he published a book in which he presented a discussion of the EPR thought experiment. At that stage, he appeared to accept Bohr's response to EPR as having settled the matter in favour of the Copenhagen interpretation. He wrote: 'Their [EPR's] criticism has, in fact, been shown to be unjustified, and based on assumptions concerning the nature of matter which implicitly contradict the quantum theory at the outset.'[7] But the subtle nature of the EPR argument, and the apparently natural and common-sense assumptions behind it, encouraged Bohm to analyse the argument in some detail. In this analysis, he made extensive use of a derivative of the EPR thought experiment that ultimately led other physicists to believe that it could be brought down from the lofty heights of pure thought and put into the practical world of the physics laboratory. It is this aspect of Bohm's contribution that we will consider here.

Bohm's work on the EPR argument set him thinking deeply about the problems of the Copenhagen interpretation. He was very soon tinkering with hidden variables, and his first papers on this subject were submitted to the journal *Physical Review* in July 1951, only 4 months after the publication of his book. However, Bohm's hidden variables differ from those we have considered so far (and with which we will stay in this chapter) in that they are non-local. We examine Bohm's non-local hidden variable theory in Chapter 11.

Bohm considered a molecule consisting of two atoms in a quantum state in which the total electron spin angular momentum is zero. A simple example would be a hydrogen molecule with its two electrons spin-paired in the lowest (ground) electronic state. We suppose that we can dissociate this molecule in a process that does not change the total angular momentum to produce two equivalent atomic fragments. The hydrogen molecule is split into two hydrogen atoms. These atoms move apart but, because they are produced by the dissociation of an excited molecule with no net spin and, by definition, the spin does not change, the spin orientations of the electrons in the individual atoms remain opposed.

The spins of the atoms themselves are therefore correlated. Measurement of the spin of one atom (say atom A) in some arbitrary laboratory frame allows us to predict, with certainty, the direction of the spin of atom B in the same frame. Viewed in terms of classical physics or via the perspective of local hidden variables, we would conclude that the spins of the two atoms are determined by the nature of the initial molecular quantum state and the method of dissociation. The atoms move away from each other with their spins fixed in unknown but opposite orientations, and the measurement merely tells us what these orientations are.

In contrast, the two atoms are described in quantum theory by a single wave function or state vector until the moment of measurement. The atoms are entangled. If we choose to measure the component of the spin of atom A along the laboratory z-axis, our observation

[6] Hiley, B. J. (1992). *Professor David Bohm. The Independent*, 30 October, 12. See also Horgan, John (1993). *Last words of a quantum heretic. New Scientist*, 27 February, p. 38.
[7] Bohm, David (1951). *Quantum theory*. Prentice-Hall, Englewood Cliffs, NJ, p. 611.

that the state vector is projected into a state in which atom A has its angular momentum vector aligned in the $+z$ direction (say) means that atom B must have its angular momentum vector aligned in the $-z$ direction. But what if we choose, instead, to measure the x or y components of the spin of atom A? No matter which component is measured, the physics of the dissociation demand that the spins of the atoms must still be correlated, and so the opposite results must always be obtained for atom B. If we accept the definition of physical reality offered by EPR, we must conclude that all components of the spin of atom B are elements of reality, since it appears that we can predict them with certainty without in any way disturbing B.

However, the state vector specifies only one spin component, associated with the magnetic spin quantum number m_s. This is because the operators corresponding to the three components of the spin vector in Cartesian coordinates do not commute (the components are complementary observables). Thus, either the wave function or the state vector is incomplete, or EPR's definition of physical reality is unjustified. The Copenhagen interpretation says that no spin component of atom B 'exists' until a measurement is made on atom A. The result we obtain for B will depend on how we choose to set up our instrument to make measurements on A. This is entirely consistent with EPR's original argument, couched in terms of the complementary position–momentum observables of two correlated quantum particles. However, the measurement of the spin component of an atom (or an electron) is much more practicable than the measurement of the position or momentum of an atom. Some physicists saw that further elaborations of Bohm's version of the EPR experiment could be carried out *in the laboratory*, and not just in the mind.

Correlated photons

Suppose an atom in an electronically excited state emits two photons in rapid succession as it returns to the lowest-energy, 'ground' state. Suppose also that the total electron orbital and spin angular momentum of the atom in the excited state is the same as that in the ground state. Conservation of angular momentum demands that the net angular momentum carried away by the photons is zero.

We know from our discussion on photon spin in Chapter 5 that all photons are bosons, possess a spin quantum number $s = 1$ and can have 'magnetic' spin quantum numbers $m_s = \pm 1$ corresponding to states of left- and right-circular polarization. The net angular momentum of the photon pair can be zero only if the photons are emitted with opposite values of m_s, that is, in opposite states of circular polarization. This scheme is exactly analogous to Bohm's version of the EPR experiment, but we have now replaced the creation of a pair of atoms with opposite electron spin orientations with the creation of a pair of photons with opposite spin orientations (or circular polarizations). We discuss how this can be achieved in practice in the next chapter.

The experimental arrangement drawn in Fig. 8.2 is designed not to measure the circular polarizations of the photons but, instead, measures their vertical and horizontal polarizations. A photon moving to the left (photon A) passes through polarization analyser 1 (denoted PA_1). This analyser is oriented vertically (which we denote as orientation a) with respect to some arbitrary laboratory frame. The derivation of the quantum-theoretical correlations between the photon spins for this particular arrangement is given in Appendix 17. We find that the detection of photon A in a state of vertical polarization means that when B passes through polarization analyser 2 (PA_2, which also has orientation a),

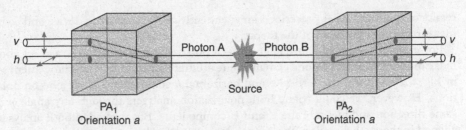

Fig. 8.2 Experimental arrangement to measure the polarization states of pairs of correlated photons.

quantum theory predicts that it *must* be measured also in a state of vertical polarization. This polarization state of B will be 180° out of phase with the corresponding state of A, because the net angular momentum of the pair must be zero, but such phase information is not recovered from the measurements. Similarly, the measurement of A in a state of horizontal polarization implies that B *must* be measured also in a state of horizontal polarization. Therefore, we can predict, with certainty, the vertical versus horizontal polarization state of B from measurements that we make on A.

Another, very succinct, way of expressing the correlation between these measurement outcomes is via the expectation value for the joint measurement, which is derived in Appendix 17 and which, for convenience, we abbreviate as $E(a, a)$ with the symbols in parentheses indicating the orientations of the two analysers. Assigning values of ± 1 to the measurement results depending on whether they are vertical $(+)$ or horizontal $(-)$ allows us to deduce a value of $+1$ for $E(a, a)$ for the arrangement depicted in Fig. 8.2, meaning that the results are perfectly correlated: for every A photon detected in a state of vertical polarization, we expect a B photon to be detected in a state of vertical polarization, and similarly for horizontal polarization states (see Appendix 17).

According to the Copenhagen interpretation, we know only the joint quantum-theoretical probabilities that the photons will be detected in vertical or horizontal polarization states; the polarization directions are not prescribed by any property that the photons possess prior to measurement. In contrast, according to a local hidden variable theory, the behaviour of each photon is governed by a hidden variable, which imposes the correlation and precisely defines the polarization directions (along any axis) that each individual photon possesses at the time the photon is created. Each photon therefore follows a prescribed path through the measuring device.

If the discussion above has seemed reasonable so far, we must acknowledge one important point about it. Although there are some properties of the two-photon state vector that depend only on the nature of the physics of the two-photon emission and the atomic quantum states involved, quantum theory can provide meaningful predictions only when couched in terms of the measurement eigenstates of the apparatus (see Appendix 17). There are no 'intrinsic' states of the quantum system. Even the initial two-photon state vector is only meaningful if we relate it to some kind of experimental arrangement. Of course, quantum theory tells us nothing whatsoever about the 'real' polarization directions of the photons (these are properties that supposedly have no basis in reality). Consequently, the only way of treating the problem is in relation to the measuring device. It is worth emphasizing this point once more: the very formalism of quantum theory is structured in precisely the way demanded by the Copenhagen interpretation, which denies the reality of quantum states prescribed independently of the measuring device and so places measurement at the heart of the theory.

For example, we could have aligned each polarization analyser as described above to make measurements along one of many quite arbitrary directions. The arrangement shown in Fig. 8.2 measures the vertical v and horizontal h components of the photon polarizations. However, we could rotate both polarization analysers through any angle φ in the same direction and measure the v' and h' components. But, provided both analysers are aligned in the same direction, the observed results would be just the same. All polarization components are therefore possible, but only in an incompletely defined sense. To obtain a complete specification, the photons must interact with a device that defines the direction in which the components are to be measured and simultaneously excludes the measurement of all other components. Definiteness in one direction must lead to indefiniteness (more correctly, an 'indefinability') in all other directions.

Bohm closed his discussion of his version of the EPR experiment with the comment[8]:

Thus, we must give up the classical picture of a precisely defined spin variable associated with each atom, and replace it by our quantum concept of a potentiality, the probability of whose development is given by the wave function ... Thus, for a given atom, *no* component of the spin of a given variable exists with a precisely defined value, until interaction with a suitable system, such as a measuring apparatus, has taken place.

Bohm's reference to 'potentialities'—the potential inherent in a quantum system to produce a particular result—suggests that he may already have been thinking about non-local hidden variables, despite his outward adherence to the Copenhagen interpretation. He also noted that the mathematical formalism of quantum theory did not contain elements that provide a one-to-one correspondence with the actual behaviour of quantum particles. 'Instead', he wrote, 'we have come to the point of view that the wave function is an abstraction, providing a mathematical reflection of certain aspects of reality, but not a one-to-one mapping.'

He further concluded that: ' . . . no theory of mechanically determined hidden variables can lead to *all* of the results of the quantum theory.'

Quantum *versus* hidden variable correlations

For the arrangement described in Fig. 8.2, a local hidden variable theory can obviously be configured to predict an expectation value, $E(a, a)$, which is entirely consistent with the quantum theoretical prediction of perfect correlation. So what? If we have no means to test one version of the theory *versus* the other, then arguably the difference between them is one of philosophical preference—realism *versus* anti-realism.

But there are other arrangements where this is not the case. The correlations between the results observed for a beam of photons passing through a polarization analyser is not changed as we rotate the analyser. In principle, the measurement eigenstates corresponding to vertical *versus* horizontal polarization refer only to the direction 'imposed' on the quantum system by the apparatus itself—we need to use the notation v' and h' only when one analyser orientation differs from the other. The correlations do not depend on whether we orient the apparatus along the laboratory z-axis, x-axis, or, indeed, any axis. However, important differences arise when two sets of apparatus are used to make measurements

[8] Bohm, David (1951). *Quantum theory*. Prentice-Hall, Englewood Cliffs, NJ, p. 621.

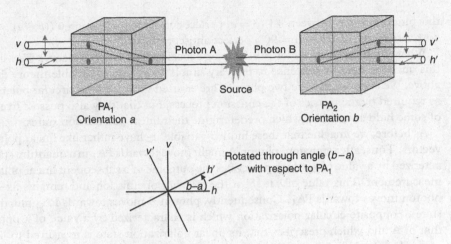

Fig. 8.3 Same arrangement as shown in Fig. 8.2, but with one of the polarization analysers oriented at an angle with respect to the vertical axis of the other.

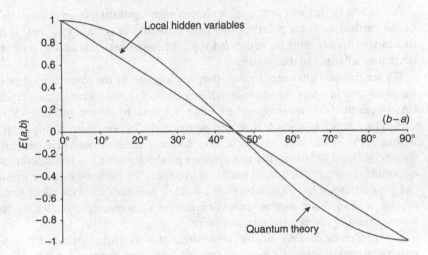

Fig. 8.4 Correlation between the photon polarization states predicted by quantum theory, plotted as a function of the angle between the vertical axes of the analysers. The equivalent prediction derived using a simple hidden variable theory is shown for comparison.

on correlated pairs of quantum particles, since the two sets of measurement eigenstates need not refer to the *same direction*.

If we rotate PA_2 through some angle with respect to PA_1, as shown in Fig. 8.3, then the measurement eigenstates change to reflect the directions of the new vertical v' and horizontal h' axes of PA_2. Denoting the new orientation of PA_2 as b defines the angle between the vertical axes of the analysers as the difference, $(b-a)$. As we might expect, the projection probabilities for the various measurement outcomes now depend on this angle, with the probability of obtaining joint $++$ and $--$ results given by $\frac{1}{2}\cos^2(b-a)$ and the probability for $+-$ or $-+$ results given by $\frac{1}{2}\sin^2(b-a)$ (see Appendix 18). Consequently, the expectation value of the joint measurement $E(a,b)$ is given by the cosine of twice the angle $(b-a)$. The function $\cos 2(b-a)$ is plotted against $(b-a)$ in Fig. 8.4. Note how this function varies between $+1$ ($b = a$, perfect correlation), through 0 (($b - a$) = 45°, no correlation) to -1 (($b - a$) = 90°, perfect anti-correlation).

What are the predictions for $E(a, b)$ using a local hidden variable theory? We will answer this question here by reference to the very simple local hidden variable theory described above. We suppose that the two photons are emitted with opposite circular polarizations, as required by the physics of the emission process, but that they also possess fixed values of some hidden variables which predetermine their linear polarization states.

As before, we imagine that these hidden variables behave rather like linear polarization vectors. Thus, after emission, photon A might move towards PA_1 in a quantum state characterized by a value of λ, which prescribes the outcome of a subsequent linear polarization measurement. This value of λ is set at the moment of emission and remains fixed as the photon moves towards PA_1. Consequently, photon B moves towards PA_2 in a quantum state of opposite circular polarization which is characterized by a value of λ opposite to that of A and which prescribes that its linear polarization state is measured to lie in the same plane as that of A. As with A, the λ value of B is set at the moment of emission and remains fixed as it moves towards PA_2. Its value is not changed on detection of photon A, as demanded by Einstein separability.

According to this simple theory, a photon with λ pointing in any direction within $\pm 45°$ of the vertical axis of a polarizer will pass through the vertical channel. If it lies outside this range, then it must lie within $\pm 45°$ of the horizontal axis and so passes through the horizontal channel of the polarizer.

We set the two analysers so that they are aligned in the same direction ($b = a$). For simplicity, we imagine the situation where photon A passes through the vertical channel of PA_1 ($+$ result). This means that the λ value of A must have been within $\pm 45°$ of the vertical axis. The λ value of photon B, which points in the opposite direction, must therefore lie within $\pm 45°$ of the vertical axis of PA_2 and so passes through the vertical channel, as shown in Fig. 8.5. Hence, the two photons produce a joint $++$ result, consistent with the quantum theory prediction of perfect correlation. In fact, we can see immediately that the properties we have ascribed to the hidden variables will not allow joint $+-$ or $-+$ results, and so this theory is entirely consistent with quantum theory for this particular arrangement of the polarization analysers.

Now, let us rotate PA_2 through some angle ($b - a$) with respect to PA_1. We denote the new polarization axes of PA_2 as v' and h'. Again, we assume for the sake of simplicity that photon A gives a $+$ result. This has the same implications for the hidden variable of photon B as before, that is, λ points in the opposite direction and lies within $\pm 45°$ of the v-axis. However, for photon B to give a $+$ result, λ must lie within $\pm 45°$ of the new v'-axis (see Fig. 8.5). Clearly the probability of obtaining a joint $++$ result with this new arrangement will depend on the probability that λ for photon B lies within $\pm 45°$ of *both* the v- *and* v'-axes (doubly shaded area shown in Fig. 8.5). The hidden variable equivalent of $E(a, b)$ is derived in Appendix 18 and illustrated alongside the quantum-theoretical prediction in Fig. 8.4. Note that when ($b - a$) = 0°, 45°, and 90°, both versions of the theory predict $E(a, b)$ = $+1$, 0, and -1, respectively. The simple local hidden variable theory is therefore consistent with the quantum theory predictions at these three angles. However, the two theories predict very different results at all other angles. The greatest difference between them occurs at an angle of $22\frac{1}{2}°$, where quantum theory predicts that $E(a, b)$ has the value $\cos 45°(= 1/\sqrt{2})$, and the local hidden variable theory predicts $E(a, b) = \frac{1}{2}$.

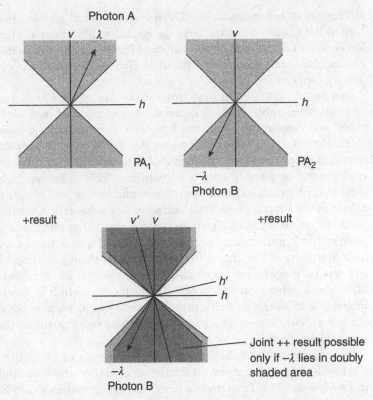

Fig. 8.5 Origin of correlations based on a simple hidden variable theory.

This appears to be merely a confirmation of Bohm's contention, quoted above, that 'no theory of mechanically determined hidden variables can lead to all the results of the quantum theory'. But you might not yet be satisfied that the case is proven. After all, the local hidden variable theory we have described here is a very simple one. Might it not be possible to devise a more complicated version that could reproduce all the results of quantum theory? More complicated local hidden variable theories are indeed possible, but, in fact, *none* can reproduce all the predictions of quantum theory. The truth of this statement is demonstrated in a celebrated theorem devised by John S. Bell.

Bell's theorem

Bohm's early work on the EPR experiment and non-local hidden variables reawakened the interest of a small section of the physics community in these problems. Many dismissed Bohm's work as 'old stuff, dealt with long ago', but for some, his approach served to heighten their own unease about the interpretation of quantum theory, even if they did not necessarily share his conclusions. One physicist who became very suspicious was John S. Bell.

Bell was born in Belfast and worked first as a laboratory assistant in the Physics Department of Queen's University in Belfast before securing financial support to study for degrees in experimental physics and mathematical physics. Following periods at the United Kingdom's Atomic Energy Research Establishment at Harwell and Birmingham University's Department of Mathematical Physics, he moved to the European Centre for Nuclear Research (known as CERN) in Geneva, Switzerland, where he worked in the Theory Division for 30 years.[9]

In a paper submitted to the journal *Reviews of Modern Physics* in 1964 (but not actually published until 1966), Bell examined, and rejected, von Neumann's 'impossibility proof' and similar arguments that had been used to deny the possibility of hidden variables. However, in a subsequent paper, Bell demonstrated that under certain conditions, quantum theory and local hidden variable theories predict different results for the same experiments on pairs of correlated particles. This difference, which is intrinsic to *all* local hidden variable adaptations of quantum theory and is independent of the precise nature of the adaptation, is summarized in Bell's theorem. Questions about local hidden variables immediately changed character. From being rather academic questions about philosophical preferences, they became practical questions of profound importance for quantum theory. The choice between quantum theory and local hidden variable theories was no longer a matter of taste, it was a matter of *correctness*. Given the profound affect these papers were to have on subsequent research in fundamental physics, it is interesting to note that Bell's principal work was in particle physics and the design of particle accelerators—nagging away at the meaning of quantum theory was very much a spare-time activity.

We will derive Bell's theorem through the agency of Dr Bertlmann, a real character used by Bell for a discussion on the nature of reality which was published in the *Journal de Physique* in 1981. I can find no better introduction than to use Bell's own words[10]:

The philosopher in the street, who has not suffered a course in quantum mechanics, is quite unimpressed by Einstein–Podolsky–Rosen correlations. He can point to many examples of similar correlations in everyday life. The case of Bertlmann's socks is often cited. Dr Bertlmann likes to wear two socks of different colours. Which colour he will have on a given foot on a given day is quite unpredictable. But when you see [Fig. 8.6] that the first sock is pink you can be already sure that the second sock will not be pink. Observation of the first, and experience of Bertlmann, gives immediate information about the second. There is no accounting for tastes, but apart from that there is no mystery here. And is not this EPR business just the same?

We can suppose that Dr Bertlmann happens to be a physicist who is very interested in the physical characteristics of his socks. Imagine that he has secured a research contract from a leading sock manufacturer to study how his socks stand up to the rigours of prolonged washing at different temperatures. Bertlmann decides to subject his left socks (socks A) to three different tests:

1. test *a*, washing for 1 h at 0°C;

2. test *b*, washing for 1 h at 22.5°C;

3. test *c*, washing for 1 h at 45°C.

[9] Rubbia, Carlo (1990) *J.S. Bell. The Independent*, 5 October, 17.
[10] Bell, J. S. (1981) *Journal de Physique*, Colloque C2 (suppl. au numero 3), **42**.

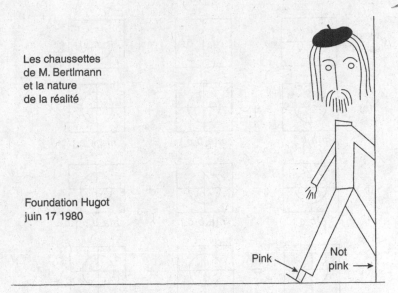

Les chaussettes
de M. Bertlmann
et la nature
de la réalité

Foundation Hugot
juin 17 1980

Pink

Not
pink →

Fig. 8.6 Bertlmann and the nature of reality. Reprinted with permission from Bell, J. S. (1981) *Journal de Physique (Paris), Colloque C2* (suppl. au numero 3), **42**, 41–61.

He is particularly concerned about the numbers of socks A that survive intact (+ result) or are destroyed (− result) by prolonged washing at these different temperatures.[11] Being a theoretical physicist, he knows that he can discover some simple relationships between the numbers of socks passing or failing these tests without actually having to perform the tests using real socks and real washing machines. This makes his study inexpensive and therefore attractive to his research sponsors.

Bertlmann defines the 'space' of possible results in the form of a simple square. Socks that pass test a give a result denoted a_+ and constitute the upper half of the space (see Fig. 8.7). Socks that fail test a give a result denoted a_- and constitute the lower half of the space. Socks that pass or fail test b give results denoted b_+ and b_- and constitute the left and right halves of the space. Socks that pass test c give results denoted c_+ and take up a circular space in the centre of the square: those that fail c are denoted c_- and lie outside the circle. Having set up these definitions, Bertlmann goes on to investigate their relationships. Denoting the number of socks that pass test a and fail test b as $n[a_+b_-]$, he reasons that $n[a_+b_-]$ can be written as the sum of the numbers of socks which belong to two subsets, one in which the individual socks pass test a, fail b, and pass c and one in which the socks pass test a, fail b, and fail c. This is summarized in terms of the spaces occupied by these results in Fig. 8.7(a). This same reasoning can obviously be applied to any combination of sets and subsets. Thus, $n[b_+c_-]$ is equal to the sum of the numbers of socks which pass test a, fail b, and pass c and which pass test a, fail b, and fail c, as illustrated in Fig. 8.7(b). Bertlmann now adds $n[a_+b_-]$ and $n[b_+c_-]$ together, and collects the four subsets implied by the sum into two groups. The first group is constituted by those socks that pass a, pass b, and fail c and those that pass a, fail b, and fail c, as shown

[11] This derivation is based on an example originally used by Bell. See Bell, J. S. (1981). *Journal de Physique (Paris), Colloque C2* (suppl. au numero 3), **42**, 41–61. This article is reproduced in Bell, J. S. (1987) *Speakable and unspeakable in quantum mechanics*. Cambridge University Press, Cambridge, p. 139.

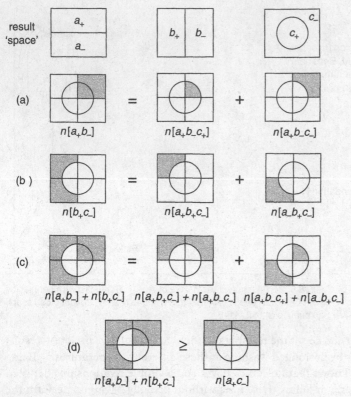

Fig. 8.7 The different possible results of experiments on Bertlmann's socks can be divided into three kinds of result 'spaces'. The space of results in which the sock passes test a is denoted a_+, failure is denoted a_-. Similarly, b_+ and b_- and c_+ and c_- denote pass and fail results for tests b and c. We can illustrate these spaces using a simple square. (a) The number $n[a_+b_-]$ lies in the result space illustrated on the left and can be written as the sum of two subsets, illustrated on the right. (b) Similarly, the number $n[b_+c_-]$ can be written as the sum of two subsets. (c) Adding $n[a_+b_-]$ and $n[b_+c_-]$ together gives results that occupy the space shown on the left. The four subsets implied by the sum can be organized into two groups, as illustrated. The first group is the set $n[a_+c_-]$. (d) It follows, therefore, that the sum of $n[a_+b_-]$ and $n[b_+c_-]$ must be greater than or equal to $n[a_+c_-]$. Based on d'Espagnat, Bernard (1979) *Scientific American*, **241**, 132.

in Fig. 8.7(c). The second group is constituted by those socks that pass a, fail b, and pass c and those that fail a, pass b, and fail c. He then notes that the first group is simply the set $n[a_+c_-]$. It follows from this that the sum of $n[a_+b_-]$ and $n[b_+c_-]$ must therefore be *greater than or at least equal to* $n[a_+c_-]$, as shown in Fig. 8.7(d). This follows simply from the fact that the number in the second group of subsets must be zero or greater (it cannot be negative). This inequality is derived in Appendix 19.

It is at this stage that Bertlmann notices the flaw in his reasoning which readers will, of course, have spotted right at the beginning. Subjecting one of the socks A to test a will necessarily change irreversibly its physical characteristics such that, even if it survives the test, it may not give the result for test b that might be expected of a brand-new sock. And, of course, if the sock fails test b, it will simply not be available for test c. The numbers $n[a_+b_-]$ are theoretically interesting, perhaps, but they have no practical relevance.

But then Bertlmann remembers that his socks always come in *pairs*. He assumes that, apart from differences in colour, the physical characteristics of each sock in a pair are identical. A test performed on the right sock (sock B) can be used to predict what the result of the same test would be if it was performed on the left sock (sock A), even though the test on A is not actually carried out. He must further assume that *whatever test he chooses to perform on B in no way affects the outcome of any other test he might perform on A*, but this seems so obviously valid that he does not give it a second thought.

Bertlmann now devises three different sets of experiments to be carried out on three samples containing the same total number of pairs of his socks. In experiment 1, for each pair, sock A is subjected to test a, and sock B is subjected to test b. If sock B fails test b, this implies that sock A would also have failed test b had it been performed on A. The number of pairs of socks for which A passes test a and B fails test b, $N_{+-}(a, b)$, must be equal to the (theoretical) number of socks A which pass test a and fail test b, that is, $N_{+-}(a, b)$ must be equal to $n[a_+b_-]$.

In experiment 2, for each pair, sock A is subjected to test b, and sock B is subjected to test c. The same kind of reasoning allows Bertlmann to deduce that $N_{+-}(b, c)$ must be equal to $n[b_+c_-]$. Finally, in experiment 3, for each pair, sock A is subjected to test a, and sock B is subjected to test c, for which it follows that $N_{+-}(a, c)$ must be equal to $n[a_+c_-]$. It is clear where this is leading. From the arguments given above, Bertlmann concludes that the sum of $N_{+-}(a, b)$ and $N_{+-}(b, c)$ must be greater than or equal to $N_{+-}(a, c)$.

Bertlmann can now generalize this result for any batch of pairs of socks. By dividing each number by the total number of pairs of socks (which he can presume is the same for each experiment), he arrives at the relative frequencies with which each joint result was obtained. He identifies these relative frequencies with probabilities for obtaining the results for experiments yet to be performed on any batch of pairs of socks that, statistically, have the same properties. He is led to the inescapable conclusion that the sum of the probabilities $P_{+-}(a, b)$ and $P_{+-}(b, c)$ must be greater than or equal to the probability $P_{+-}(a, c)$. This is Bell's inequality.

This result has nothing whatsoever to do with quantum physics, philosophy, realism, anti-realism, or hidden variables. It is a simple result of numerical relationships. But follow the above arguments through once more, replacing socks with photons, pairs of socks with pairs of correlated photons, washing machines with polarization analysers, and temperatures with polarizer orientations, and you will arrive again at Bell's inequality.

Our three tests now refer to polarization analysers set with their vertical axes oriented at $a = 0°$, $b = 22\frac{1}{2}°$, and $c = 45°$ These different arrangements can be summarized, as shown in Table 8.1.

Table 8.1 Different arrangements of polarization analysers

Experiment	Photon A PA$_1$ orientation	Photon B PA$_2$ orientation	Difference in orientations
1	$a(0°)$	$b(22\frac{1}{2}°)$	$(b - a) = 22\frac{1}{2}°$
2	$b(22\frac{1}{2}°)$	$c(45°)$	$(c - b) = 22\frac{1}{2}°$
3	$a(0°)$	$c(45°)$	$(c - a) = 45°$

The quantum theoretical prediction for the probability $P_{+-}(a, b)$ is, as we saw above, given by the expression $\frac{1}{2}\sin^2(b-a)$. Using quantum theory without hidden variables, we would therefore expect to obtain the result $P_{+-}(a, b) = 0.073$ in experiment 1. Similarly, $P_{+-}(b, c) = 0.073$ in experiment 2, and $P_{+-}(a, c) = 0.250$ in experiment 3. We know from Bell's inequality that the sum of $P_{+-}(a, b)$ and $P_{+-}(b, c)$ must be greater than or equal to the probability $P_{+-}(a, c)$, so we conclude that 0.146 is greater than or equal to 0.250.

If you are encountering this result for the first time, it is perhaps advisable to break off here, take some tea (or something stronger!) and read through this section once more from the beginning. *This result is inescapable.* For the particular experimental arrangements of the polarization analysers described here, quantum theory predicts results that violate Bell's inequality.

Perhaps one of the most important assumptions we made in the reasoning which led to this inequality was that of Einstein separability or local reality of the photons. We assumed that it is possible to carry out measurements on photon B without in any way disturbing photon A. It is therefore an inequality that is quite independent of the nature of any local hidden variable theory that we could possibly devise. We conclude therefore that *quantum theory is incompatible with any local hidden variable theory and hence local reality.* You might wish to confirm for yourself that the simple local hidden variable theory described earlier, for which the predicted probabilities are derived in Appendix 18, does indeed conform to Bell's inequality for the same set of polarizer orientations.

Perhaps we should not really be entirely surprised by this result, after all. The predictions of quantum theory are based on the properties of a two-particle state vector which, before collapsing into one of the measurement eigenstates, is supposedly 'delocalized' over the whole experimental arrangement. The two particles are, in effect, always in 'contact' prior to measurement and therefore can exhibit a degree of correlation that is impossible for two Einstein separable particles. However, Bell's inequality provides us with a straightforward test. If experiments like those described here are actually performed, the results will allow us to choose between quantum theory and a whole range of theories based on local hidden variables.

Generalization of Bell's inequality

Before we get too carried away with these inequalities, we should remember what it is we are supposed to be measuring here. We are proposing an experiment in which some source (yet to be specified) emits a pair of photons correlated so that they have no net angular momentum. The photons move apart, and each enters a polarization analyser oriented at some angle to the arbitrary laboratory vertical axis. The photons are detected to emerge from the vertical or horizontal polarization channels of these analysers, and the results of coincident measurements are compared with the predictions of quantum theory and local hidden variable theories.

Unfortunately, nothing in this life is ever easy. Bertlmann's derivation of Bell's inequality is based on an important assumption. Remember that he had supposed that, with the exception of colour, each member of any given pair of his socks possesses identical physical characteristics so that the result of any test performed on sock B would automatically imply the same result for A. This, in turn, implies that if we perform the same test on both socks simultaneously, we expect to observe identical results, or a perfect correlation. In the language of the equivalent experiments with photons, if we orient PA_1 and PA_2 so that their vertical axes are parallel, we expect to obtain perfect correlation, or $E(a, a) = +1$. Alas, in the 'real' world, there are a number of limitations in the experimental technology of polarization measurements that prevent us from observing perfect correlation. And any effect that reduces the physicist's ability to measure these correlations below the maximum permitted by Bell's inequality will render the experiments inconclusive.

Firstly, real polarization analysers are not 'perfect'. They do not transmit all the photons that are incident on them (through one or other of the two channels), and they often 'leak', that is, horizontally polarized photons can occasionally pass through the vertical channel, and vice versa. Worse still, the transmission characteristics of the analysers may depend on their orientation. Secondly, detectors such as photomultipliers are quite inefficient, producing measurable signals for only a small number of the photons actually generated. Finally, the analysers and detectors themselves must be of limited size, and so they cannot 'gather' all of the photons emitted, even if they are emitted in roughly the right directions. Experimental factors such as these limit the numbers of pairs that can be detected successfully and will also lead to some pairs being detected 'incorrectly'; for example, a pair which should have given a ++ result actually being recorded as a +− result. These limitations always serve to reduce the extent of correlation between the photons that can be observed experimentally.

There is a way out of this impasse. It involves a generalization of Bell's inequality to include a fourth experimental arrangement, and was first derived by John F. Clauser, Michael A. Horne, Abner Shimony, and Richard A. Holt in a paper published in *Physical Review Letters* in 1969. A derivation is provided in Appendix 20. Denoting the four different orientations of the polarization analysers as a, b, c, and d, this generalized form of Bell's inequality can be written in terms of a specific combination of the expectation values $E(a, b)$, $E(a, d)$, $E(c, b)$, and $E(c, d)$. When subject to the same assumptions of Einstein separability of the photons, this combination can be shown to be less than or equal to 2 (see Appendix 20).

The advantage of this generalization is that nowhere in its derivation is it necessary to rely on a perfect correlation between the measured results for any combination of polarizer orientations. The inequality applies equally well to non-ideal cases. There is more. The implication of the hidden variable approach we have adopted so far is that the λ values are established and fixed at the moment the photons are emitted, and the outcomes of the measurements therefore prescribed. However, there is nothing in the derivation of the general inequality which says this must be so. The only assumption needed is one of locality—measurements made on photon A do not affect the possible outcomes of any subsequent measurements made on B and *vice versa*. The generalized form of Bell's inequality actually provides a test for all classes of locally realistic theories, not just those theories that happen also to be deterministic. It is no longer essential to suppose that the λ values of photons A and B remain determined as they propagate towards their respective analysers.

The photons must still be correlated (no net angular momentum), but their λ values could vary between emission and detection. All that is required for the inequality to be valid is that there should be no 'communication' between the photons at the moment a measurement is made on one of them. As we can arrange for the analysers to be a long distance apart (or *space-like separated*, to use the physicists' term), this requirement essentially means no communication faster than the speed of light. Recall once again

Table 8.2 Specific polarizer orientations

Experiment	Photon A PA$_1$ orientation	Photon B PA$_2$ orientation	Difference in orientations
1	$a(0°)$	$b(22\frac{1}{2}°)$	$(b - a) = 22\frac{1}{2}°$
2	$a(0°)$	$d(67\frac{1}{2}°)$	$(d - a) = 67\frac{1}{2}°$
3	$c(45°)$	$b(22\frac{1}{2}°)$	$(b - c) = -22\frac{1}{2}°$
4	$c(45°)$	$d(67\frac{1}{2}°)$	$(d - c) = 22\frac{1}{2}°$

that Einstein suspected that the Copenhagen interpretation of quantum theory might necessarily lead to a violation of the postulates of special relativity.

The specific polarizer orientations we will want to consider further are summarized in Table 8.2. For these sets of analyser orientations, the combination of expectation values given in Appendix 20 predicts a value of 2.828, in clear violation of the inequality.

This exercise merely confirms once more that quantum theory is not consistent with local reality. Correlations between the photons can be greater than is possible for two Einstein separable particles, since the reality of their physical properties is supposedly not established until a measurement is made. The two particles are in 'communication' over large distances, since their behaviour is governed by a common two-particle state vector. Quantum theory *demands* a 'spooky action at a distance' that violates special relativity. The question is now this: is it right?

Part IV

Experiment

9
Quantum non-locality

It is probably reasonable to suppose that the derivation, in the late 1960s and early 1970s, of an equation that is demonstrably violated by a quantum theory then over 40 years old should have settled the matter one way or the other, once and for all. Correlated quantum particles are everywhere in physics and chemistry, the simplest and most obvious example being the helium atom, an understanding of the spectroscopy of which had led to the introduction of the Pauli principle in the first place. But it became apparent that the special circumstances under which Bell's inequality could be subjected to experimental test had never been realized in the laboratory. Suddenly, the race was on to perfect an apparatus that could be used to perform the necessary measurements on pairs of correlated quantum particles.

As early as 1946, the physicist John Wheeler, then at Princeton University, had proposed studies on correlated photons produced by electron–positron annihilation. Experiments on annihilation were done in 1949 by Chien-Shiung Wu and Irving Shaknov, and these had confirmed that the photons so produced were correlated, but they had not directly probed the consequences of this entanglement. However, the polarization correlations of two photons emitted in rapid succession (in a 'cascade') from an excited state of the calcium atom proved to be the most accessible to experiment and initially closest to the ideal. Carl A. Kocher and Eugene D. Commins at the University of California at Berkeley used this source in 1966 in a study of correlations between the linear polarization states of the photons, although they did not explicitly set out to test Bell's inequality.

The real repercussions of Bell's 1966 papers were felt through the work of a small group of theoreticians and experimentalists who had read the papers and had become obsessed with the problem that they posed. Abner Shimony had studied philosophy under Rudolph Carnap and had had Eugene Wigner as his Ph.D. adviser. In seeking a career in which he could combine his interests in both physics and philosophy, he had given up a tenured position at the prestigious Massachusetts Institute of Technology to take an appointment at Boston University. In 1968, he took on Michael Horne as a Ph.D. student and gave him Bell's papers, challenging him to see if he could devise a practical experimental test of quantum non-locality. Horne eventually settled on an adaptation of the Kocher–Commins experiment and, together with Shimony, drafted a paper setting out their proposal for a meeting of the American Physical Society. They also lined up an experimentalist, Richard Holt at Harvard, to put their design into practice. They missed the deadline for submission of papers to the meeting and discovered that they had actually been pipped to the post by John Clauser at Columbia University, who had also been strongly influenced by Bell's papers and had been working independently on much the same experimental design. Shimony contacted Clauser, and all four—Clauser, Horne, Shimony, and Holt—combined their efforts to produce the generalization of Bell's inequality suitable for non-ideal cases, as described in Chapter 8.

Clauser moved to the University of California at Berkeley to take up a postdoctoral position with Charles Townes, the inventor of the maser. Townes agreed that Clauser should spend time away from his principal research project on radio astronomy, pursuing an experimental test of quantum non-locality. Clauser talked to Commins, and Commins agreed to provide one of his own graduate students, Stuart J. Freedman, to help extend the original Kocher–Commins design. Clauser wrote to de Broglie, Bohm, and Bell, asking if they were aware of any other similar experiments, performed or proposed. They were not. Thus it was that the first direct tests of Bell's inequality were performed in 1972, by Freedman and Clauser, using the calcium-atom source. These experiments produced the violations of Bell's inequality predicted by quantum theory, but because of some further 'auxiliary' assumptions that were necessary in order to extrapolate the data, only a weaker form of the inequality was tested. These auxiliary assumptions left unsatisfactory loopholes for the ardent supporters of local hidden variables to exploit. It could still be argued that the evidence against such hidden variables was only circumstantial.

Other results followed, but the first comprehensive experiments designed specifically to test the general form of Bell's inequality were those performed by Alain Aspect and his colleagues Philippe Grangier, Gérard Roger, and Jean Dalibard, at the Institut d'Optique Théoretique et Appliquée, Université Paris-Sud in Orsay, in 1981 and 1982. These scientists also made use of cascade emission from excited calcium atoms as a source of correlated photons.

Cascade emission

In the lowest energy (so-called 'ground') state of the calcium atom, the outermost $4s$ orbital is filled with two spin-paired electrons. The vector sum of the spin angular momenta of these electrons is therefore zero, and the state is characterized by a total spin quantum number S equal to zero. The spin multiplicity, given by $(2S + 1)$, is 1, and so the state is called a singlet state.

The total angular momentum of the atom is a combination of the intrinsic angular momentum that the electrons possess by virtue of their spins and the angular momentum they possess by virtue of their orbital motion. We can combine these two kinds of angular momentum in different ways. In the first, we determine separately the total spin angular momentum (characterized by the quantum number S) and the total orbital angular momentum (quantum number L), and combine these to give the overall momentum (quantum number $\mathcal{J}$). In the second, we combine the spin and orbital angular momenta of each individual electron (quantum number j) and combine these to give the overall total. The former method is appropriate for atoms with light nuclei, and we will use it here.

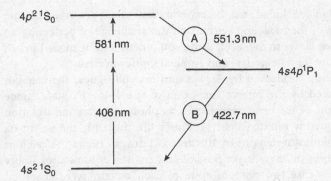

Fig. 9.1 Electronic states of the calcium atom involved in the two-photon cascade emission process.

In fact, for the ground state of the calcium atom, the outermost electrons are both present in a spherically symmetric s orbital and therefore possess no orbital angular momentum: L equals zero, S equals zero, and so $\mathcal{J}$ equals zero. The state is labelled $4s^2\,{}^1S_0$, where the superscript 1 indicates that it is a singlet state, the S indicates that L is equal to zero (S corresponds to $L = 0$, P corresponds to $L = 1$, D corresponds to $L = 2$, etc.), and the subscript 0 indicates that $\mathcal{J} = 0$.

If we use light to excite the ground state of a calcium atom, the photon that is absorbed imparts a quantum of angular momentum to the atom. This extra angular momentum cannot appear as electron spin, since that is fixed by the fact that the electron spin quantum number is constrained to the value $s = \frac{1}{2}$. The angular momentum must therefore appear in the excited electron's orbital motion, and so the value of L must increase by one unit. Promoting one electron from the $4s$ orbital to the $4p$ orbital satisfies this selection rule. If there is no change in the spin orientations of the electrons, the excited state is still a singlet state, $S = 0$, and, since $L = 1$, there is only one possible value for $\mathcal{J}$: $\mathcal{J} = 1$. This excited state is labelled $4s4p\,{}^1P_1$.

Now, imagine that we could somehow excite a second electron (the one left behind in the $4s$ orbital) also into the $4p$ orbital but still maintaining the alignment of the electron spins. The configuration would then be $4p^2$, which can give rise to three different electronic states corresponding to the three different ways of combining the two orbital angular momentum vectors. In one of these states, the orbital angular momentum vectors of the individual electrons cancel, $L = 0$, and since $S = 0$, we have $\mathcal{J} = 0$. This particular doubly excited state is labelled $4p^2\,{}^1S_0$.

If this doubly excited state is produced in the laboratory, it undergoes a rapid cascade emission through the $4s4p\,{}^1P_1$ state to return to the ground state (see Fig. 9.1). Two photons are emitted. Because the quantum number $\mathcal{J}$ changes from $0 \rightarrow 1 \rightarrow 0$ in the cascade, the net angular momentum of the photon pair must be zero. In fact, the photons have wavelengths in the visible region. Photon A, from the $4p^2\,{}^1S_0 \rightarrow 4s4p\,{}^1P_1$ transition, has a wavelength of 551.3 nm (in the green region of the visible spectrum), and photon B, from the $4s4p\,{}^1P_1 \rightarrow 4s^2\,{}^1S_0$ transition, has a wavelength of 422.7 nm (blue).

The Aspect experiments

Alain Aspect had studied the fundamental problems of quantum theory and the EPR argument whilst doing three years' voluntary service in Cameroon. He was also strongly influenced by Bell's papers. He concluded that the experimental tests performed to date in the 1970s had fallen short of the ideal and set himself the challenge of perfecting an apparatus that would get much closer to the ideal, using equipment that he made himself in the basement of the Institute for Theoretical and Applied Optics in Paris.

In the experiments eventually conducted by Aspect and his colleagues, the calcium $4p^2\ {}^1S_0$ state was not produced by the further excitation of the $4s4p\ {}^1P_1$ state, since that would have required light of the same wavelength as photon B, making isolation and detection of the subsequently emitted light very difficult. Instead, the scientists used two high-power lasers, with wavelengths of 406 and 581 nm, to excite the calcium atoms. The very high intensities of lasers make possible otherwise very-low-probability *multi-photon excitation*. In this case two photons, one of each colour, were absorbed simultaneously by a calcium atom to produce the doubly excited state (see Fig. 9.1).

Aspect, Grangier, and Roger actually used a calcium atomic beam. This was produced by passing gaseous calcium from a high-temperature oven through a tiny hole into a vacuum chamber. Subsequent collimation of the atoms entering the sample chamber provided a well-defined beam of atoms with a density of about 3×10^{10} atoms per cubic centimetre in the region where the atomic beam intersected the laser beams. This low density (atmospheric pressure corresponds to about 2×10^{19} molecules per cubic centimetre) ensured that the calcium atoms did not collide with each other or with the walls of the chamber before absorbing and subsequently emitting light. It also removed the possibility that the emitted 422.7-nm light would be reabsorbed by ground state calcium atoms.

Figure 9.2 is a schematic diagram of the apparatus used by Aspect and his colleagues. They monitored light emitted in opposite directions from the atomic beam source, using filters to isolate the green photons (A) on the left and the blue photons (B) on the right. They used two polarization analysers, four photomultipliers, and electronic devices designed to detect and record coincident signals from the photomultipliers. The polarization analysers were actually polarizing cubes, each made by gluing together two prisms with dielectric coatings on those faces in contact. These cubes transmitted light polarized parallel to the plane of incidence (vertical) and reflected light polarized perpendicular to

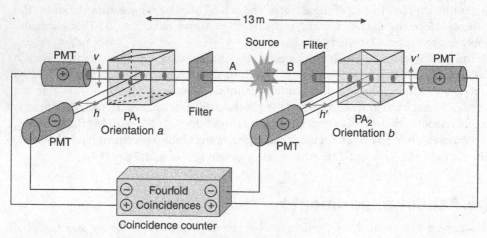

Fig. 9.2 Schematic diagram of the experimental apparatus used by Aspect, Grangier, and Roger (PMT represents a photomultiplier).

this plane (horizontal). Thus, detection of a transmitted photon corresponds in our earlier discussion to a + result, while detection of a reflected photon corresponds to a − result.

The polarizing cubes were neither quite perfectly transmitting for pure vertically polarized light nor perfectly reflecting for pure horizontally polarized light. For light with a wavelength of 551.3 nm, the physicists measured the transmittance of PA$_1$ for vertically polarized light to be 95.0 per cent of the total incident light, and the reflectance of PA$_1$ for horizontally polarized light was similarly found to be 95.0 per cent. They also measured the reflectance of PA$_1$ for vertically polarized light and the transmittance for horizontally polarized light to be around 0.7 per cent, representing a small amount of 'leakage' through the analyser. The equivalent results for light of wavelength 442.7 nm were 93.0 per cent and 0.7 per cent, respectively.

Each polarization analyser was mounted on a platform which allowed it to be rotated about its optical axis. Experiments could therefore be performed for different relative orientations of the two analysers. The analysers were placed about 13 m apart. The electronics were set to look for coincidences in the arrival and detection of the photons A and B within a 20-ns time window. This is large compared with the time taken for the intermediate $4s4p\ ^1P_1$ state to decay (about 5 ns), and so all true coincidences were counted.

Note that to be counted as a coincidence, the photons had to be detected within 20 ns of each other. Any kind of signal passed between the photons, 'informing' photon B of the fate of photon A, for example, must therefore have traveled the 13 m between the analysers and detectors within 20 ns. In fact, it would take about 40 ns for a signal moving at the speed of light to travel this distance. The two analysers were therefore space-like-separated.

Aspect, Grangier, and Roger actually measured coincidence *rates* (coincidences per unit time). After correction for accidental coincidences, the physicists obtained results which varied in the range of 0–40 coincidences per second, depending on the angle between the vertical axes of the polarizers $(b − a)$. They then used these results to derive an experimental measure of the expectation value, $E(a, b)$, for comparison with the theoretical predictions, measuring the expectation value for seven different sets of analyser orientations. The results they obtained are shown in Fig. 9.3. From Chapter 8 and Appendix 18, we know that the quantum theory prediction for $E(a, b)$ is $\cos 2(b − a)$. However, the extent of the correlation observed experimentally was dampened by 'imperfections' in the apparatus. The physicists therefore derived a slightly modified form of the quantum theory prediction that takes these limiting factors into account, including the finite solid angles for detection of the photons (not all photons could be physically 'gathered' into the detection system), the effect of imperfect analyser transmittances and reflectances, the small amount of leakage, and the fact that not all photons incident on the analysers were ultimately detected. The *predictions* of quantum theory, corrected for these instrumental deficiencies, are shown in Fig. 9.3 as the continuous line. As expected, the predictions demonstrate that perfect correlation, $E(a, b) = +1$ when $(b−a) = 0°$, and perfect anticorrelation, $E(a, b) = −1$ when $(b − a) = 90°$, were not quite realized in these experiments. However, it is quite clear that the measured values of $E(a, b)$ agree well with the modified quantum theory prediction, particularly when compared with the straight-line relationship expected from a simple local hidden variable theory (see Fig. 8.4).

Aspect and his colleagues then performed four sets of measurements with analyser orientations $(b − a) = 22\frac{1}{2}°, (d − a) = 67\frac{1}{2}°, (b − c) = −22\frac{1}{2}°$ and $(d − c) = 22\frac{1}{2}°$ (see Chapter 8). This allowed them to test the generalized form of Bell's inequality, derived in

Fig. 9.3 Results of measurements of the expectation value $E(a, b)$ for seven different relative orientations of the polarization analysers. The continuous line represents the quantum theory predictions modified to take account of instrumental factors (see text). Reprinted with permission from Aspect, Alain, Grangier, Philippe, and Roger, Gérard (1982). *Physical Review Letters*, **49**, 91.

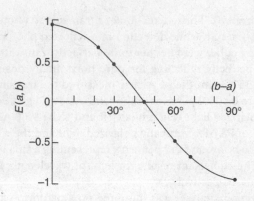

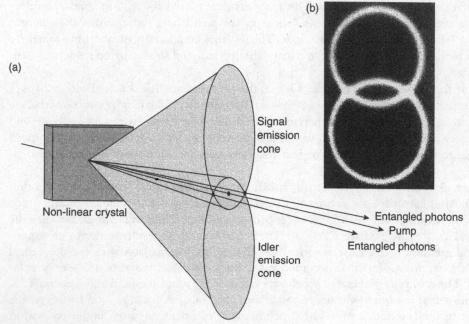

Fig. 9.4 (a) Schematic diagram of the down-converted output from a non-linear crystal pumped by a laser beam. Two spatially separated 'cones' are produced, corresponding to the signal and idler beams. Where these cones intersect, polarization entangled photons are produced. (b) Emission cones, taken through an infrared filter to block out the pump beam. Adapted, with permission, from Kwiat, Paul G., Mattle, Klaus, Weinfurter, Harald, and Zeilinger, Anton, Sergienko, Alexander V., and Shih, Yanhua (1995). *Physical Review Letters*, **75**, 4338.

Appendix 20. For the specific combination of expectation values used in this generalized form of the inequality, quantum theory predicts a value $2\sqrt{2}$, or 2.828, *versus* a local hidden variable prediction of ≤ 2. They obtained the result 2.697 ± 0.015, a violation of Bell's inequality by 83 per cent of the maximum possible predicted by quantum theory. By taking the instrumental limitations into account, the physicists obtained a modified quantum theory prediction for this combination of expectation values to be 2.70 ± 0.05, in excellent agreement with experiment.

Parametric down-conversion

An alternative to cascade emission is available in the form of parametric down-conversion. Certain crystalline materials, such as potassium and lithium niobate and β-barium borate, exhibit small, so-called nonlinear optical effects. Their polarization characteristics vary non-linearly with the strength of the electric field vector. This is usually understood by considering the polarization vector to be described mathematically as a series expansion in terms of the electric field. The first-order term (i.e. the first power in the electric field vector) then describes the normal refractive properties of the crystal. Parametric down-conversion (and its reverse, so-called three-wave mixing) is a second-order effect and becomes important only when the electric field amplitudes are large.

A photon from an intense light source, such as a laser, may be down-converted in these crystals into two orthogonally polarized photons of lower frequency (longer wavelength) in a process in which energy and momentum are conserved. The total energy of the output photons is therefore equal to the energy of the input photon. An input laser beam (often referred to as the 'pump') with a wavelength in the ultraviolet region can be converted into two beams (one generally referred to as a 'signal', the other an 'idler') with wavelengths in the visible and infrared regions.

Different types of down-conversion are possible. Some of these, like the original Aspect experiments, generate pairs of photons entangled by their polarization properties and rely on coincidence counting itself as a method of 'post-selection'—identifying the pairs only *after* detection has occurred. However, there is a form of down-conversion, called type II, in which polarization entangled photon pairs can be generated in two different, controlled, directions. The physics of the non-linear optical conversion in the crystal dictates that the signal and idler photons are emitted into spatially distinct cones, and by careful orientation of the crystal, it is possible to overlap these cones (see Fig. 9.4). The points at which the cones intersect define two directions along which pairs of photons propagate with polarizations that are undetermined (it is not possible to say which photons possess which polarization states prior to measurement) but which are perfectly anti-correlated. Each member of the pair is therefore emitted in a distinct, predictable direction. No photon pairs are 'lost' as a result of being emitted in the 'wrong' directions.

In experimental results reported in 1998 by Christian Kurtsiefer, Markus Oberparleiter, and Harald Weinfurter, coincidence count rates as high as 360,800 per second were recorded, compared with a maximum of 40 coincidences per second in the original Aspect experiments. These physicists used 351.1-nm light from an argon-ion laser as the pump and β-barium borate as the non-linear crystal. Signal and idler photons were generated with the same wavelength, 702.2 nm (twice the pump wavelength) and passed into single mode optical fibres so that they could be separated even further for subsequent analysis using polarization filters and detectors.

These kinds of sources currently provide the highest-quality entangled states, and in the experiments referred to above, correlations corresponding to the generalized form of Bell's inequality were found to be 2.6979 ± 0.0034, a violation of the inequality by 204 standard deviations.

Long-distance entanglement

It is possible to engineer another type of entanglement in a three-level atomic system, such as that shown for the calcium atom in Fig. 9.1. If the highest and lowest states involved in the three-level system have long emission lifetimes (i.e. they may persist for a relatively long time before emitting a photon), then the energy–time uncertainty relation allows us in principle to measure the total energy of the two photons emitted with high precision (remember from Chapter 2 that the greater the uncertainty in time, the lower the uncertainty in energy). If, however, the intermediate state has a much shorter lifetime, then the uncertainty relation denies us precise knowledge of the energy of *each emitted photon individually*. We know the total energy of the photons, but we do not know precisely how much energy is carried by each photon. Conversely, although we do not know the precise time of emission (because of the relatively long lifetime of the highest state) we do know that both photons must be emitted almost simultaneously (because of the short lifetime of the intermediate state).

This situation is analogous to the original EPR experiment which relied on position–momentum entanglement. In this case, the time of emission of a photon pair is uncertain by an amount related to the lifetime of the highest energy state. But the detection of one photon (say photon A) immediately determines the time of emission of photon B, whether or not we choose to detect it. We could just as well set up the experiment to measure the energy of photon A, which would immediately determine the energy of photon B, in exact analogy to the original EPR argument. If we follow the EPR argument and accept their 'reasonable' definition of reality, we would be forced to accept that both the energy and time of emission of photon B are elements of reality and are in principle defined all along. Assuming that they are not defined until a measurement is made implies a collapse of the two-photon state vector.

A hypothetical experiment with energy–time entangled photons was developed by J. D. Franson in 1989. He showed that by introducing interferometers in the paths of photons A and B in front of their respective detectors, it is possible to observe interference effects in the detection coincidence rates as the path lengths of the interferometers (and hence the relative phases of the two photons) are varied. This is equivalent to introducing polarization analysers in the experiments with polarization entangled photons, and Franson was able to demonstrate that quantum theory predicts a sinusoidal oscillation between perfect correlation and anti-correlation as the phase difference is varied. As we saw in Chapter 8, local hidden variable theories do not predict such striking interference.

Franson judged the experiment to be 'difficult, but feasible'.[1] Ten years later, Wolfgang Tittel, J. Brendel, Nicolas Gisin, and H. Zbinden from the Group of Applied Physics at the University of Geneva reported the results of experiments demonstrating a violation of the generalized form of Bell's inequality with energy–time entangled photons detected at observer stations located in Bellevue and Bernex, two small Swiss villages outside Geneva *some 10.9 km apart!*

Instead of relying on atomic emission states, as Franson had originally proposed, the Geneva physicists used type-I parametric down-conversion in a potassium niobate crystal to generate two photons each with a wavelength of 1310 nm, suitable for transmission along installed telecommunications fibres. In this case, the long *coherence time* of the

---

[1] Franson, J. D. (1989). *Physical Review Letters*, **62**, 2208.

pump laser (related to its bandwidth, or degree of monochromaticity) prevents precise knowledge of the time of emission of the two down-converted photons from the crystal. However, the very short coherence time of the emitted photons effectively guarantees a high degree of simultaneity of emission, analogous to Franson's original scheme.

The physicists generated the entangled pairs at a telecommunications station in Geneva and sent one of the photons down an installed fibre-optic cable to Bellevue, 4.5 km north of Geneva, and sent the other to Bernex, 7.3 km southwest of Geneva and 10.9 km from Bellevue. Observation stations were set up in both villages, consisting of all-fibre interferometers and photon-counting detectors. The photons were detected, with detections registered by sending a (classical) light signal back to coincidence electronics located in Geneva. After correcting for accidental coincidences, the physicists obtained correlations corresponding to the generalized form of Bell's inequality of 2.64 ± 0.03, a violation of the inequality by about 21 standard deviations. The Geneva physicists went on to record the results of experiments using three and four interferometers. In all cases, they obtained strong violations of Bell's inequality.

How fast is 'instantaneous'?

These experiments beg some obvious questions. If, for a moment, we assume the collapse or reduction of the state vector to be a real physical phenomenon, then the results of the Geneva experiments clearly indicate that this collapse is transmitted throughout an entire two-particle state vector stretched 10.9 km across at speeds much faster than that of light, as Einstein had always suspected. Can we therefore place a lower limit on how fast this collapse occurs?

The answer is yes. In subsequent experiments by the Geneva group, the fibres connecting the source of entangled photons to the observer stations were aligned so that the time difference in the arrival of the photon pair at their respective detectors was reduced to less than 5 ps. From this, the physicists deduced that the collapse must be occurring at speeds some 20 or 30 million times the speed of light, as viewed from the reference frame of their Geneva laboratory. There are some awkward questions about the most appropriate frame from which to be judging this speed. A change of frame to that from which the cosmic microwave background radiation is seen to be isotropic (uniform in all directions) places a lower limit on the speed of the collapse of 20,000 times the speed of light. Either way, the results appear to violate one of the principal assumptions of Einstein's special theory of relativity.

Testing non-locality without inequalities

All of the experiments described so far have been variants on the EPR experiment using polarization or energy–time entangled photon pairs and which ultimately represent a test of Bell's inequality. However, in 1990, a proposal was put forward for an experiment to test quantum non-locality that does not depend on Bell's inequality at all.

Michael Horne first met Austrian physicist Anton Zeilinger in 1976, and they both met the American Daniel Greenberger 2 years later. They discovered that they all shared an interest in fundamental quantum physics. During one of their many discussions on the subject in the mid-1980s, Greenberger wondered what might result if *three* particles were to become entangled. Horne felt that this was a sufficiently interesting question to warrant further investigation, and Greenberger began to work on it. The results crystallized, and Greenberger, Horne, and Zeilinger (GHZ) outlined the consequences of entanglement of three photons, which have since become known as three-photon GHZ states. Their conclusions were widely, but informally, circulated within the physics community in 1988. They refined their work and published it in the *American Journal of Physics* in 1990, together with Abner Shimony. In that same issue, N. David Mermin at Cornell University published a paper in which he turned the original GHZ idea into a simple experimental concept.

Suppose we can somehow create such three-photon states entangled through their polarization properties which are set as vertical (v) or horizontal (h) in some arbitrary laboratory frame. Further suppose that the initial three-photon state vector is formed from a linear combination of $|v_A\rangle|v_B\rangle|v_C\rangle$ and $|h_A\rangle|h_B\rangle|h_C\rangle$ states, that is, an entangled state in which all three photons are either vertically or horizontally polarized. Here, the subscripts A, B, and C label the three photons. Just as with the two-photon experiments we considered earlier, we can imagine that we can set up polarization analysers to measure either circular polarization (left/right) or linear polarization at some angle to the laboratory axis (v'/h'). We have three detectors, one for each photon. Consider the situation where we measure the v'/h' polarization of photon A and the circular polarization properties of photons B and C. For this particular initial three-photon state vector, quantum theory predicts that whenever photon A is measured to be in a v' state, photons B and C will always be measured in *identical* states of circular polarization, either left–left or right–right (see Appendix 21). Quantum theory also predicts that whenever photon A is measured in an h' state, photons B and C will always be measured in *opposite* states of circular polarization, left–right or right–left. When we permute these combinations, we get the same kind of correlation. For example, if we measure the circular polarization properties of photons A and C and the v'/h' polarization of photon B, we expect that whenever B is observed in a v' state, A and C will be observed in identical states of circular polarization, and when B is observed in a h' state, A and C will be observed in opposite states of circular polarization.

These correlations are relatively easy to explain by expanding the initial three-photon state vector in a basis of the various measurement eigenstates, as shown in Appendix 21. The challenge is to come up with a local hidden variable theory that will explain why these (and only these) combinations of states can be observed. One simple approach is to assume that these correlations are 'built in' to the properties of the three photons the moment they are entangled, through the operation of some hidden variable, λ, which predetermines the measurement outcomes, as described in Chapter 8. We would assume that the photons cannot 'know' in advance what kind of polarization measurement we intend to make on each photon, and so we must therefore assume that the λ parameters are 'two-valued': they carry fixed information that predetermines *both* the circular *and* linear polarization properties of the photons.

As before, we can bring out the nature of the correlation very clearly by ascribing some results to the individual measurements. We can denote detection of left-circularly polarized photons and v' photons as + results, with value +1, and detection of right-circularly polarized photons and h' photons as − results, with value −1. It can be quickly deduced that the correlations between the three photons are such that the total result (the product for all three detected states), is always +1. For example, if photon A is detected in a v' state (+1), this implies that photons B and C are detected in identical states of circular polarization (giving either +1 multiplied by +1 or −1 multiplied by −1). The net result is $+1$. If photon A is detected in a h' state (-1), photons B and C are measured in opposite states of circular polarization (giving $+1$ multiplied by -1 or -1 multiplied by $+1$), with the net result again $+1$. We conclude that this is a feature of the physics of the interactions that give rise to the entangled photons. It is also a feature that any local hidden variable theory must account for.

Now let us consider what happens in our local hidden variable model when we have a situation in which the λ parameter for photon A predisposes detection in a h' state (-1). This means that the λ parameters for photons B and C must be predisposed towards detection as left–right or right–left states. Let us assume that the result is right–left, meaning that the λ parameter for photon B yields a -1 result, and the λ parameter for C yields a $+1$ result. This satisfies the requirements of the correlation and gives a net result of $+1$. But photon A cannot know in advance that it is to be subjected to a linear polarization measurement—we could equally well choose to make the linear polarization measurement on either B or C instead, and the λ parameters must take account of this freedom. If we assume that the λ parameter for photon B is predisposed towards detection in a v' state $(+1)$, this means that A and C must be detected in identical states of circular polarization. We have already assumed above that the λ parameter for C gives a $+1$ result, so we are led to conclude that the λ parameter for A must similarly lead to a $+1$ result if the correlation is to be preserved.

Now we come to photon C. Let us assume that C has a λ parameter predisposed towards detection in a v' state $(+1)$. This means that A and B must be detected in identical states of circular polarization. But now we have a problem. We assumed above that A's λ parameter gives a $+1$ result for circular polarization, and that B's λ parameter gives a -1 result for circular polarization. We cannot now square this with the requirement that both A and B be detected in identical states of circular polarization, as demanded by the correlation. There is no way around this, no matter how hard we try. We must conclude that the particular combination of λ parameters predisposing detection of A, B, and C as h', v', and v' is simply *not admissible in this scheme*. If we were to admit it, we cannot reproduce the correlations predicted by quantum theory for the case where we subject one photon to a linear polarization measurement and the other two to circular polarization measurements.

If we denote the two-valued λ parameter for A as $\lambda_A \binom{-1}{+1}$, where the upper value in brackets determines that A will yield a h' measurement result, and the lower value determines that A will yield a left-circular polarization result, we must acknowledge that the combination $\lambda_A \binom{-1}{+1} \lambda_B \binom{+1}{-1} \lambda_C \binom{+1}{+1}$ is, for some reason, prevented from being formed by the physics of the entanglement process. (We can see this straight away from the values themselves—an upper value of $+1$ in the parentheses for C should imply identical results for the lower values in the parentheses for A and B. In this case it does not.)

There are, in fact, only *four* combinations that are admissible. These are (v', v', v'), (v', h', h'), (h', v', h'), and (h', h', v'), where the detected states are written in order for photons A, B, and C. We can devise a local hidden variable theory for which it is possible to set the λ parameters to prescribe linear polarization states for these four combinations and still meet the correlation requirements for circular polarization.

What if we now set up the experiment to make *only* linear polarization measurements for all three photons? We are left from the above discussion to conclude that only the combinations (v', v', v'), (v', h', h'), (h', v', h'), and (h', h', v') are admissible if the requirements of the earlier correlations are to be satisfied. We would therefore expect to see only these four possible combinations of detected states. We conclude that our local hidden

Table 9.1 Summary of Detected States

If any one of the photons A, B, or C is detected as ...	... this means that the other two photons must be detected in linear polarization states that are ...	
	Local hidden variables	Quantum theory
$|v'\rangle$	The same	Opposite
$|h'\rangle$	Opposite	The same

variables theory says that when photon A, B, or C is measured in a $|v'\rangle$ state, the other two must be detected in *identical* states of linear polarization, both $|v'\rangle$ or both $|h'\rangle$. When photon A, B, or C is detected in a $|h'\rangle$ state, the other two must be detected in *opposite* states of linear polarization, either $|v'\rangle/|h'\rangle$ or $|h'\rangle/|v'\rangle$.

What does quantum theory predict? Remarkably, quantum theory predicts the *exact opposite*. In the experiment in which we make only v'/h' polarization measurements, we can expand the three-photon state vector in this basis of measurement eigenstates, and we conclude that the only visible combinations should be (v', v', h'), (v', h', v'), (h', v', v'), and (h', h', h')—see Appendix 21. Quantum theory says that when we measure one photon in a $|v'\rangle$ state, the other two must be detected in *opposite* states of linear polarization, either $|v'\rangle/|h'\rangle$ or $|h'\rangle/|v'\rangle$. When one photon is detected in a $|h'\rangle$ state, the other two must be detected in *identical* states of linear polarization, both $|v'\rangle$ or both $|h'\rangle$. These contradictory predictions are summarized in Table 9.1.

We are therefore presented with a simple test. Detection of *any* of the combination states predicted by quantum theory in a laboratory experiment effectively rules out all classes of local hidden variables theories[2]—and not an inequality in sight.

The experimental creation of three-photon GHZ states was first reported in 1999 by Dik Bouwmeester, Jian-Wei Pan, Matthew Daniell, Harald Weinfurter, and Anton Zeilinger.[3] These were post-selected triplets formed from two pairs of polarization entangled photons produced by down-converting ultraviolet pulses from a titanium–sapphire laser in a β-barium borate crystal. The fourth photon from the two pairs was used to trigger coincidence detection of the other three. Subsequent direct tests of quantum non-locality using three-photon GHZ states were reported in 2000 by Pan, Bouwmeester, Daniell, Weinfurter, and Zeilinger in the journal *Nature*. They demonstrated a very strong bias for those measurement states predicted by quantum theory, as described above for the situations where the circular polarization properties are measured for two of the photons. Their results for the critical experiment involving only v'/h' polarization measurements are shown in Fig. 9.5. The histogram of measured triple coincidences again shows a clear preference for those states predicted by quantum theory. The small proportions of coincidences predicted by local hidden variables are ascribed to spurious events attributable to unavoidable experimental error.

What if a different initial three-photon state vector is produced? We find that different initial state vectors give rise to different patterns of correlations in the combined states that are predicted, but the fundamental incompatibility between the predictions of local hidden variables theories and quantum theory remains. For example, if the initial state is

---

[2] Subject to closing so-called 'loopholes', as described later in this chapter.
[3] The creation of four-photon GHZ states was subsequently reported by Pan *et al.* in 2001.

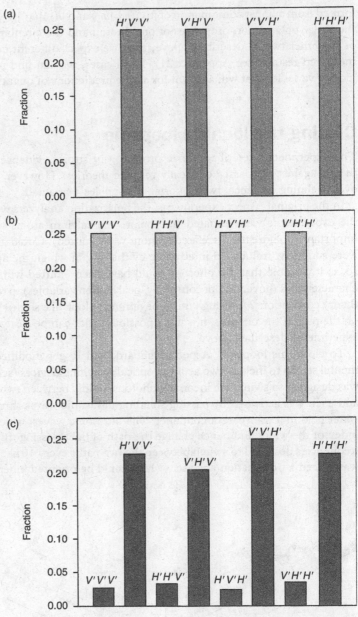

Fig. 9.5 (a) Quantum theory predicts that only the combination states (v', v', h'), (v', h', v'), (h', v', v'), and (h', h', h') will be observed, with equal (25 per cent) probability. (b) In contrast, in order to account fully for correlations arising when one of the three photons is subject to a linear polarization measurement and the other two are subject to circular polarization measurements, local hidden variables theories predict that only the combination states (v', v', v'), (v', h', h'), (h', v', h'), and (h', h', v') are possible. (c) The results show a clear preference for the combination states predicted by quantum theory. Reprinted, with permission, from Pan, Jian-Wei, Bouwmeester, Dik, Daniell, Matthew, Weinfurter, Harald, and Zeilinger, Anton (2000). *Nature*, **403**, 515.

formed from a linear combination of $|v_A\rangle|v_B\rangle|h_C\rangle$ and $|h_A\rangle|h_B\rangle|v_C\rangle$ states, and we measure the linear polarization properties of one photon and the circular polarization properties of the other two as described above, then we expect different correlations between the measured results (see Appendix 21). In all cases, we can find no combination of local hidden variables that will account for all the predictions of quantum theory.

Closing the locality loophole

The experiments described above provide very strong evidence in favour of quantum theory against all classes of locally realistic theories. However, despite this seemingly overwhelming evidence, two important 'loopholes' remained.

In the original Aspect experiments, the polarization analyzers were set in position *before* the experiments were initiated (i.e. before the calcium atoms were excited and, most importantly, before the correlated photons were emitted). Could it not be that the photons were somehow influenced in advance by the way in which the apparatus was set up? If so, is it possible that the photons could have been emitted with just the right physical characteristics (governed, of course, by local hidden variables) to reproduce the quantum-theory correlations? Although this is beginning to look like some kind of grand conspiracy on the part of the photons, it is not a possibility that can be easily excluded by any of the experiments described above.

To close this loophole, Aspect, Dalibard, and Roger modified their original experimental set-up to include two acousto-optical switching devices (see Fig. 9.6). Each device was designed to switch the incoming photons rapidly between two different optical paths, and each was activated by passing standing ultrasonic waves through a small volume of water held in a transparent container. The ultrasonic waves, which change the refractive index of the water and hence change the path of light passing through it, were driven at frequencies designed to switch between the two paths every 10 ns. At the end of each path was placed a polarization analyser (which could be oriented independently of the rest of

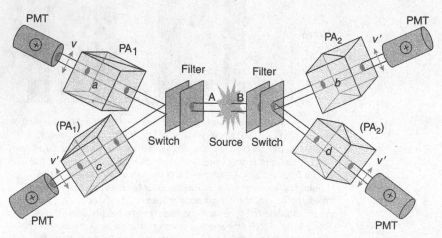

Fig. 9.6 Schematic diagram of the experimental apparatus used by Aspect, Dalibard, and Roger.

the apparatus) and a photomultiplier. The vertical axes of each of the four analysers were oriented in different directions.

By this arrangement, the physicists prevented the photons from 'knowing' in advance along which optical path they would be travelling, and hence through which analyser they would eventually pass. The end result was equivalent to changing the relative orientations of the two analysers *while the photons were in flight*. Any communication between the photons regarding the way in which the apparatus was set up was therefore restricted to the moment of measurement, in principle requiring faster-than-light signalling between the photons to establish the correlation.

The switching arrangement shown in Fig. 9.6 was difficult to operate and run successfully. Aspect and his colleagues could not add to this difficulty by trying to detect photons both transmitted and reflected by the polarization analysers (such an experiment would have required 8 photomultipliers and the necessary coincidence detection). The physicists could therefore only detect those photons *transmitted* by the analysers: they could observe only + results. Fortunately, a version of Bell's inequality had been derived by John F. Clauser, Michael A. Horne, Abner Shimony, and Richard A. Holt in 1969 for just this kind of experiment (see Appendix 22—in fact, many of the experiments described earlier actually tested Bell's inequality in this form).

All the quantities needed to obtain an experimental measure of the same combination of expectation values, but now using switched optical paths, were determined by Aspect and his colleagues. For polarization analyser orientations $a = 0°, b = 22\frac{1}{2}°, c = 45°$, and $d = 67\frac{1}{2}°$, they obtained the result 2.404 ± 0.080, once again in clear violation of the generalized form of Bell's inequality. Taking account of inefficiencies in the polarization analysers and the finite solid angles for detection allowed them to obtain a modified quantum theory prediction of 2.448, in excellent agreement with experiment.

For the purist, however, the locality loophole was still not completely closed by these experiments. The standing ultrasonic waves used to drive the acousto-optical switches did not provide completely random switching, although the two switches were driven at different frequencies. Closing this loophole completely took a little while longer, but in December 1998, Gregor Weihs, Thomas Jennewein, Christoph Simon, Harald Weinfurter, and Anton Zeilinger reported the results of experiments in which the switching was completely random, driven by a physical random number generator.

These physicists used type-II parametric down-conversion to create polarization-entangled photons, each with a wavelength of 702.2 nm, which were spatially separated and directed through fibre optics some 360 m across the science campus at the University of Innsbruck in Austria. The orientations of the polarization analysers at each observer station were set by electro-optic modulators. The modulators were in turn driven by the output of photomultipliers set up to detect photons from a light-emitting diode (LED) passed through a beamsplitter. Photons from the LED were detected by one or other of two photomultipliers and assigned a label '0' or '1'.[4] The physicists estimated that random switching of the analysers occurred within a maximum time of about 100 ns, considerably shorter than the 1.3 μs required to send a signal 360 m across the campus. They obtained a result for the generalized form of Bell's inequality of 2.73 ± 0.02, a violation of 30 standard

---

[4] In fact, this arrangement makes use of the random collapse or reduction of the state vector of the photons from the LED interacting with a beamsplitter as a source of truly random numbers.

deviations and, at 88 per cent of maximum predicted by quantum theory, a stronger violation of the inequality than had been originally reported by Aspect, Dalibard, and Roger.

Closing the efficiency loophole

The numbers of photon pairs actually detected in these experiments are considerably smaller than the numbers actually generated. The principal reasons for this lie both in the method of generation of the entangled states and in the detection system. Entangled states generated by cascade emission sources suffer from the problem that there is no direct correlation between the directions in which the two photons are emitted. The schematic diagram given in Figs 9.2 and 9.6 idealize the situation in that they show photon A only moving to the left (from the reader's perspective) and photon B only moving to the right. In truth, the photons are emitted in all directions, and there is no correlation between the direction of A and the direction of B for any given pair. Only those A photons that happen to have been emitted in the left direction pass through the filter, enter the polarization analyser PA_1, and are detected. Only those B photons that happen to have been emitted in the right direction pass through the second filter, enter the polarization analyser PA_2, and are detected. The coincidence counter seeks to ensure that only photons detected within a given time window (and which can therefore be traced back to the same emission 'event') are counted. Large numbers of A and B photons may be detected outside this time window, implying that they do not originate from the same emission events and are therefore not counted as coincidences. Add to this the general inefficiencies associated with the measuring devices as described above (transmission and reflection inefficiencies, leakages, and general detector inefficiencies) and it quickly becomes clear why only a few per cent of photon pairs created actually contribute to the measurements. Using type-II parametric down-conversion as a source of entangled photons instead of cascade emission certainly eliminates inefficiencies associated with emission directions, but instrumental inefficiencies remain. In the experiments to close the locality loophole performed by Weihs, Jennewein, Simon, Weinfurter, and Zeilinger, the detection efficiency was still only 5 per cent.

Why does this represent a loophole? Because if only a very small proportion of the pairs that are generated are ultimately detected, we are required to assume that the detected pairs represent a true statistical sample (called a 'fair sample') of the total number of created pairs. Recall from our discussion of the Duhem–Quine thesis in Chapter 6 that, when faced with contradictory experimental data, we have to acknowledge that at least one of the group of hypotheses under test is unacceptable and must be changed, but the experiment does not tell us which one. If, because of practical and instrumental reasons, we need to make a number of auxiliary hypotheses in addition to the central hypothesis that the quantum particles under study are Einstein-separable and therefore locally real, then only a weaker form of Bell's inequality is tested. If the predictions of our local hidden-variable theory are not borne out by experiment, we have every right to question the auxiliary hypotheses and so devise a more devious hidden-variables theory that leaves the central hypothesis intact. In the case of the efficiency loophole, we could choose to reject the assumption that the small proportion of photon pairs detected represents a fair sample of the total and argue instead that the experiments are biased in favour of those photon pairs that deliver results in accordance with the quantum-theory predictions. We would have to suppose that, whilst the sub-ensemble of detected photon pairs violates the generalized Bell's inequality, the total ensemble does not. A local hidden-variables theory which, because of data rejection, predicts the same measurement outcomes as quantum theory was first devised by Philip Pearle in 1970. In a more recent model, Nicolas Gisin and B. Gisin described a local hidden-variables theory in which the variables themselves determined the efficiency of the detectors. The theory explained the measured (quantum) correlations whilst at the same time remaining true to Bell's inequality.

Avoiding the fair sample assumption demands an efficiency of pair detection to pair generation of about 83 per cent. This kind of efficiency is likely to prove extremely difficult in any experiment that depends on the generation and detection of photon pairs. There is an alternative approach, however. In February 2001, physicists from the National Institute of Standards and Technology in Boulder, Colorado and the Department of Physics at the University of Michigan published the results of experiments they had conducted on entangled states created using massive, positively charged *ions*.

These were beryllium ions caught in an ion trap. The physicists created coherent superpositions of excited beryllium ion states with different total angular momentum quantum numbers using Raman transitions driven at a wavelength of 313 nm (ultraviolet) by two lasers orientated perpendicular to one another. The excited states are labelled for convenience as $|\uparrow\rangle$ and $|\downarrow\rangle$. Once created, the superposition was 'collapsed' using further pulses of laser light lasting about 400 ns. By carefully controlling the trap strength, the phase of the laser, and the phase of the radio-frequency synthesizer used to fix the frequency of the 'collapse' transition, it was possible to measure the relative frequency of collapse into different measurement eigenstates at different values of the total phase angle, exactly analogous to the measurement of polarization states using different orientations of the polarization analysers.

The possible measurement outcomes in this experiment were $|\psi_{++}\rangle = |\uparrow_A\rangle|\uparrow_B\rangle$, $|\psi_{+-}\rangle = |\uparrow_A\rangle|\downarrow_B\rangle$, $|\psi_{-+}\rangle = |\downarrow_A\rangle|\uparrow_B\rangle$ and $|\psi_{--}\rangle = |\downarrow_A\rangle|\downarrow_B\rangle$ for all possible differences in the total phase angle. Here, A and B label the two ions. A further laser beam was used to detect the individual ion states. Detection photons are preferentially scattered by the $|\downarrow\rangle$ states (which are therefore referred to as 'bright'), with little light scattered by the $|\uparrow\rangle$ states (referred to as 'dark'). The physicists determined the correlations required to test the generalized form of Bell's inequality by measuring the number of ion pairs giving $|\psi_{++}\rangle$ results (both ions dark), $|\psi_{+-}\rangle$ *and* $|\psi_{-+}\rangle$ (one ion bright, one dark, without discriminating between the ions), and $|\psi_{--}\rangle$ (both ions bright). For the set of phase angles required to produce the maximum possible violation of the generalized form of Bell's inequality, they obtained the result 2.25 ± 0.03, a violation of the inequality by more than 8 standard deviations.

The most significant difference between these experiments and earlier studies based on entangled photons is that the fair sampling hypothesis is *not* required: *all* the entangled ion pairs created were detected and contributed to the measured results. Commenting on these results, Philippe Grangier stated that he regards the efficiency loophole to be closed by these experiments. He wrote[5]:

Closing both [the locality and efficiency] loopholes in the same experiment remains a challenge for the future, and would lead to a full, logically consistent rejection of any local realistic hypothesis. Even so, the overall agreement with quantum mechanics seen in all the experimental tests

[5] Grangier, Philippe (2001). *Nature*, **409**, 774–775.

of Bell's inequalities is already outstanding. Moreover, each time a parameter is changed that was considered to be crucial (for example, using time-varying measurements, or increasing the detection efficiency), the experiments show that these changes have no consequence: the results continue to agree with quantum-mechanical principles. This appears rather compelling evidence to me that quantum mechanics is right, and cannot be reconciled with classical mechanics.

To salvage local hidden variables from these experimental results, we would need to invoke a very grand conspiracy indeed, one that somehow exploits both locality and efficiency loopholes *at the same time*, but not independently. This immediately brings to mind another of Einstein's famous quotes (made in a rather different context): 'The Lord is subtle, but he is not malicious.'[6]

In any case, it would seem to be only a matter of time before the experimental difficulties associated with an 'ultimate' test of quantum non-locality are overcome; one that closes both the locality and efficiency loopholes simultaneously.

So, where does all this leave local reality? Physicists who have commented on these results, including David Bohm and John Bell who rejected the Copenhagen view, have largely accepted that these experimental tests create great difficulties for theories which feature a local reality. The message seems to be this: reality is non-local, so get used to it. We are obliged to give up local reality and accept that there can be some kind of 'spooky action at a distance', perhaps involving some kind of strange communication between distant parts of the universe at speeds faster than that of light. As Einstein noted, this appears to conflict with the postulates of special relativity.

Although the independent reality advocated by the realist does not have to be a local reality, it is clear that the experiments described here leave the realist with a lot of explaining to do. An observer changing the orientation of a polarizer *does* appear somehow to affect the behaviour of a distant photon, no matter how distant it is. Whatever the nature of reality, it cannot be as simple as we might have thought at first.

Do these experiments necessarily represent the end of this story as far as experimental physics is concerned? Commenting on the Aspect experiments in 1985, Bell thought not[7]:

It is a very important experiment, and perhaps it marks the point where one should stop and think for a time, but I certainly hope it is not the end. I think that the probing of what quantum mechanics means must continue, and in fact it will continue, whether we agree or not that it is worth while, because many people are sufficiently fascinated and perturbed by this that it will go on.

---

[6] The phrase 'Raffiniert ist der Herr Gott. Aber Boshaft ist Er Nicht' is carved in stone above the fireplace in a room in Fine Hall, Princeton University, in memory of Einstein.

[7] Bell, J. S. in Davies, P. C. W. and Brown, J. R. (1986). *The ghost in the atom*. Cambridge University Press, Cambridge, p. 52.

10

Complementarity and entanglement

The results of the experiments described in the last chapter all point fairly unambiguously to a reality that is decidedly non-local in nature, at least in the quantum domain. Of course, these are results that come as no surprise to those willing to accept one of the most fundamental principles of the Copenhagen interpretation—that quantum phenomena are describable only in terms of the dual classical physical concepts of waves and particles. The wave properties of quantum entities lend them inherently non-local characteristics, and it is only when we forcibly 'collapse' the wave function by making a measurement that we are taken aback by what appears to be contradictory or counter-intuitive behaviour.

All our attention so far has been focused on the properties and behaviour of entangled quantum particles. The experiments described in Chapter 9 were all rather esoteric, involving complicated apparatus and somewhat complicated analyses. They seem to take the interested spectator a long way from what might appear to be the heart of the matter: wave–particle duality. After all, it was Bohr's insistence on the complementary nature of wave and particle properties that became one of the foundation stones of the Copenhagen interpretation. The experiments demonstrate in a roundabout way that this complementarity creates a direct conflict between quantum theory and local reality. Is there a less roundabout way of showing this?

In the last chapter, we saw that closing the locality loophole involved switching between different analyser orientations while the emitted photons were in flight. The choice between the nature of the measurement was therefore delayed with respect to the transitions that originally created the photons. Is it possible to make this a delayed choice between measuring devices of a more fundamental nature?

For example, in our discussion on quantum measurement in Chapter 5, we imagined the situation in which a single photon passes through a double-slit apparatus to impinge on a piece of photographic film. We know that if we allow a sufficient number of photons individually to pass through the slits, one at a time, an interference pattern will be built up. This observation suggests that the passage of each photon is governed by wave interference, so that it has a greater probability of being detected (producing a spot on the film) in the region of a bright fringe (see Fig. 2.1). It would seem that the photon literally passes through both slits simultaneously and interferes with itself. As we noted earlier, the sceptical physicist who places a detector over one of the slits in order to show that the photon passes through one or the other does indeed prove their point—the photon is detected, or not detected, at one slit. But then the interference pattern can no longer be observed.

Advocates of local hidden variables could argue that the photon is somehow affected by the way we choose to set up our measuring device. It thus adopts a certain set of physical characteristics (owing to the existence of hidden variables) if the apparatus is set up to show particle-like behaviour, and adopts a different set of characteristics if the apparatus is set up to show wave interference. However, if we can design an apparatus that allows us to choose between these totally different kinds of measuring device, we could delay our choice until the photon was (according to a local hidden variable theory) 'committed' to showing one type of behaviour. We suppose that the photon cannot change its 'mind' *after* it has passed through the slits, when it discovers what kind of measurement is being made.

Delayed choice

In 1978, the physicist John Wheeler proposed just such a delayed-choice experiment, which is in effect a modified version of the double-slit apparatus described above. This experiment was performed in the laboratories of two independent groups of researchers: Carroll O. Alley, Oleg G. Jakubowicz, and William C. Wickes from the University of Maryland and T. Hellmuth, Herbert Walther, and Arthur G. Zajonc from the University of Munich. Both groups used a similar experimental approach, a somewhat simplified version of which is described below.

The basic apparatus employed by Alley and colleagues is shown schematically in Fig. 10.1. A pulse of light from a laser was passed through a beamsplitter which, like a half-silvered mirror, transmitted half the intensity of the incident light and reflected the other half. The split light beams followed two paths, indicated as A and B in Fig. 10.1. Fully reflecting mirrors were used to bring the two beams back into coincidence inside a triangular prism called a Köster prism.

The recombined beams show wave-interference effects. Viewed in terms of a wave picture, the relative phases of the waves (positions of the peaks and troughs) at the point where the beams recombine in the Köster prism determine whether they show constructive interference (peak coincides with peak) or destructive interference (peak coincides with trough). The relative phases of the waves could be adjusted simply by changing the length of one of the paths. In Fig. 10.1, a 'phase-shifter' is shown in path A.

In fact, the inner axis of the triangular prism represents another beam-splitting surface, arranged to provide another 90° phase difference between light reflected from it and transmitted through it. Light reflected from this surface was detected by photomultiplier 1, and transmitted light was detected by photomultiplier 2. Thus, if the light waves entering the prism from paths A and B were already out of phase by 90° as a result of the different lengths of the paths, then the light reflected from the beam-splitting surface was 180° out of phase (peak coincident with trough), giving destructive interference. No light was detected by photomultiplier 1. However, light transmitted through the prism was shifted back into phase (peak coincident with peak), giving constructive interference. This light was detected by photomultiplier 2. The advantage of this arrangement was that interference effects were readily observed by the simple fact that all the light was detected by only one

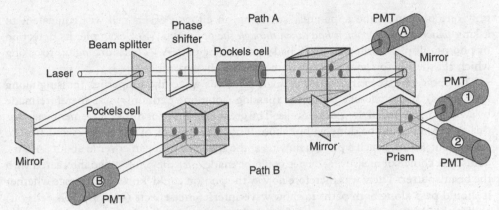

Fig. 10.1 Schematic diagram of the experimental apparatus used by Alley and colleagues to perform delayed-choice measurements. A laser beam is split along two paths—A and B—by a beamsplitter. Path A contains a phase-shifter, Pockels cell, and polarizing cube, and path B contains a Pockels cell and polarizing cube. The beams are recombined in the prism. With the Pockels cells switched off, the phase shifter can be used to bring the beams to coincidence in the prism, such that interference between them is maximized, with the result that photons are detected only by photomultiplier 2. Reducing the light intensity in the apparatus so that, on average, only one photon passes through at a time, either Pockels cell can be switched on *after* the photon has passed through the beamsplitter. This has the effect of discovering which path the photon has taken.

photomultiplier (in this case photomultiplier 2). Blocking one of the paths, and thereby preventing the possibility of interference, resulted in equal light intensities reaching these photomultipliers.

Performing the experiment with the laser light intensity reduced, so that only one photon passed through the apparatus at a time, resulted in the expected detection of the photons only by photomultiplier 2. In this arrangement, the photon behaved as though it had passed along both paths simultaneously, interfering with itself inside the triangular prism, in exact analogy with the double-slit experiment.

The researchers also inserted two optical devices called Pockels cells, one in each path. Without going into too many details, a Pockels cell consists of a crystal across which a small voltage is applied. The applied electric field induces birefringence in the crystal—in effect, it becomes a polarization rotator. If the conditions are right, vertically polarized light passing through a birefringent crystal can emerge horizontally polarized. A permanent polarizing filter was used in conjunction with each Pockels cell to reflect any horizontally polarized light out of the path and into a photomultiplier. Photomultiplier A monitored light reflected out of path A, and photomultiplier B monitored light reflected out of path B.

If both Pockels cells were switched off (no voltage applied), the vertically polarized light passed down both paths undisturbed and recombined in the triangular prism to show interference effects. If either Pockels cell was switched on, the vertically polarized light passing through the active cell became horizontally polarized and was deflected out of its path and detected, preventing the observation of interference effects. Thus, with only one photon in the apparatus, switching on either Pockels cell was equivalent to asking *which path the photon had taken through the apparatus.* (For example, its detection by photomultiplier A showed that it had passed along path A.) This is analogous to asking which slit the photon goes through in the double-slit experiment.

The choice between measuring a single photon's wave-like properties (passing along both paths) or particle-like properties (passing along one path only) was therefore made by switching on one of the Pockels cells. The great advantage of this arrangement was that this switching could be done within about 9 ns. The lengths of the paths A and B were each about 4.3 m, which a photon moving at the speed of light can cover in about 14.5 ns. Thus, the choice of measuring device could be made *after* the photon had interacted with the beamsplitter. There was therefore no way the photon could 'know' in advance whether it should pass along both paths to show wave-interference effects (both Pockels cells off) or if it should pass along only one of the paths to show localized particle-like properties (one Pockels cell on).

Both groups of researchers reported results in agreement with the expectations of quantum theory. Within the limitations set by the instruments, with one of the Pockels cells switched on, photons were indeed detected in one or other of the two paths, and no interference could be observed. With both Pockels cells off, photons were detected only by photomultiplier 2, indicative of wave-interference effects. Of course, according to the Copenhagen interpretation, the wave function of the photon develops along both paths. If one of the Pockels cells is switched on, the detection of a photon directed out of either path collapses the wave function instantaneously, and we infer that the photon was localized in one or other of the two paths.

Wheeler's 'Great Smoky Dragon'

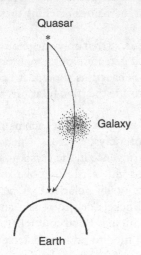

Quasar

Galaxy

Earth

Fig. 10.2 The gravitational lens effect offers a means of performing the delayed-choice experiment on a cosmological scale.

John Wheeler has described this behaviour in a particularly picturesque way. Like the photon entering the delayed-choice apparatus, Wheeler's 'Great Smoky Dragon' has a sharply defined tail. Our knowledge of the tail therefore seems complete and unambiguous. The point at which the photon is detected—the mouth of the Dragon—is similarly sharply defined. However, the middle of the Dragon is a fog of uncertainty: '. . . in between we have no right to speak about what is present'.

Wheeler has also suggested that the delayed-choice experiment can be performed on a cosmological scale, by making use of the gravitational lens effect. Two close-lying quasi-stellar objects (quasars), labelled 0957 + 561A, B are believed to be one and the same quasar. One image is formed by light emitted directly towards earth from the quasar. A second, virtual, image is produced by light emitted from the quasar which would normally pass by the earth but which is bent back by an intervening galaxy (this is the gravitational lens effect—see Fig. 10.2). The light reaching earth from the quasar can therefore travel via two paths. If we choose to combine the light from these paths, we can, in principle, obtain interference effects.

We seem to have the power to decide by what route (or routes) any given photon emitted from the quasar will travel to earth billions of years *after* it set out on its journey. Wheeler wrote[1]:

in a loose way of speaking, we decide what the photon *shall have done* after it has *already* done it. In actuality it is wrong to talk of the 'route' of the photon. For a proper way of speaking we recall once more that it makes no sense to talk of the phenomenon until it has been brought to a close by an irreversible act of amplification: No elementary phenomenon is a phenomenon until it is a registered (observed) phenomenon.

The possibility of actually observing interference by bringing together light from the same quasar that has travelled along two different paths was believed to be impossible, leaving the cosmological version of the delayed-choice experiment firmly in the realms of thought. The problem lies with the coherence time of the light emitted. The photons might set off from the quasi-stellar source in phase (with their peaks and troughs 'in step'), but over the kinds of distances they would need to travel to earth and the time it would take to do this, any phase relationship would be expected to be lost. The difference between the two paths that can be taken is considerably longer (in time) than the coherence time of the light emitted, effectively ruling out any hope of demonstrating interference between them and hence allowing the opportunity to carry out a delayed-choice experiment. More recently, however, another form of gravitational lensing has been discovered which offers the possibility of smaller path differences and better prospects for experiment. We may yet see the results of delayed-choice experiments conducted on photons that set off from their source literally millions of years ago.

Watching the electrons

Look back at the description of the double-slit experiment in Chapter 1 and, in particular, the results of the electron interference experiment shown in Fig. 2.1. Perhaps there is something about these results that continues to nag in the backs of our minds. Wave–particle duality is manifested by the appearance of bright spots on the photographic film, showing where individual particles have been detected, but grouped into alternate bright and dark bands characteristic of wave interference.

But wait a moment. It is clear that we can only *detect* particles, whether through the chemical processes occurring in a photographic emulsion or through the physical processes occurring in a photomultiplier or similar device. We understand that this is so because these detection processes require that the initial interaction between object and measuring device involves a whole quantum particle which cannot be subdivided. Thus, a single electron or photon interacts with an ion in the photographic emulsion, initiating a chain of chemical reactions which ultimately results in the precipitation of a large number of silver atoms. This initial interaction appears to localize the particle: it reacts with this particular ion at this particular place on the film.

[1] Wheeler, J. A. (1981) In *The American Philosophical Society and The Royal Society: papers read at a meeting, June 5.* American Philosophical Society, Philadelphia, PA. Reproduced in Wheeler, J. A. and Zurek, W. H. (eds.) (1983). *Quantum theory and measurement.* Princeton University Press, Princeton, NJ.

The evidence for the quantum particle's wave-like properties derives from the *pattern* in which a large number of individual particles are detected. According to the Copenhagen interpretation, this pattern arises because the wave function or state vector of each quantum particle has greater amplitude in some regions of the film compared with others, owing to interference effects generated by its passage through the two slits. Before it is detected, a quantum particle is 'everywhere' on the film, but it has a greater probability of interacting with an ion in those regions of the film where the modulus-square of the state vector is large. These regions become bright fringes.

In his *Lectures on physics*, Richard Feynman introduced a hypothetical variant of the two-slit experiment using electrons. By introducing a light source behind the two slits, he proposed to observe through which slit a particular electron had passed. This is directly analogous to Einstein's thought experiment involving careful control of the transfer of momentum between the electron and a plate containing a single slit, used as a means of identifying the trajectory that the electron will subsequently follow through a second plate with two slits (see Chapter 7). In Feynman's thought experiment, light scattered by the electron as it emerges from one or other of the two slits reveals through which slit it has passed (see Fig. 10.3). Of course, Feynman concludes that the very act of trying to gain 'which way' information prevents us from observing interference effects. And, as in Bohr's initial defence of quantum theory in the face of Einstein's early challenges, he cites the 'clumsiness' of our measurement combined with the uncertainty principle as the mechanism by which we are prevented from observing simultaneous particle-like ('which-way') and wave-like (interference) behaviour. In this case, it is the sizeable scale of the interaction between an electron and a photon which knocks the electron 'off course', destroying the phase relationship and washing out the interference fringes.

Feynman wrote[2]:

If an apparatus is capable of determining which hole the electron goes through, it *cannot* be so delicate that it does not disturb the pattern in an essential way. No one has ever found (or even thought of) a way around the uncertainty principle. So we must assume that it describes a basic characteristic of nature ... if a way to 'beat' the uncertainty principle were ever discovered, quantum mechanics would give inconsistent results and would have to be discarded as a valid theory of nature.

But we saw in Chapter 7 that Bohr was eventually forced to abandon the 'disturbance' argument as a means of defending the Copenhagen interpretation against EPR's 'bolt from the blue'. He had to do this for the simple reason that such a defence requires an almost classical realist conception of the measurement interaction, consistent with EPR's 'reasonable' definition of reality and virtually admitting that the probabilistic nature of quantum theory originates in the statistics of clumsy interactions, the nature of which we remain ignorant. Bohr had to deny Einstein's realism and therefore shifted his defence to a much more subtle (and somewhat vague) argument to the effect that the complementary wave–particle nature of quantum entities essentially *precludes* simultaneous observation of wave-like and particle-like behaviour. This incompatibility derives from the limits placed on our ability to acquire deeper knowledge in the quantum domain. In essence, we impose the classical concepts of waves and particles in designing our measuring apparatus, and the incompatibility of these classical concepts does the rest. Though Bohr was

[2] Feynman, Richard P., Leighton, Robert B., and Sands, Matthew (1965). *The Feynman lectures on physics*. Volume III. Addison-Wesley, Reading, MA, pp. 1–9.

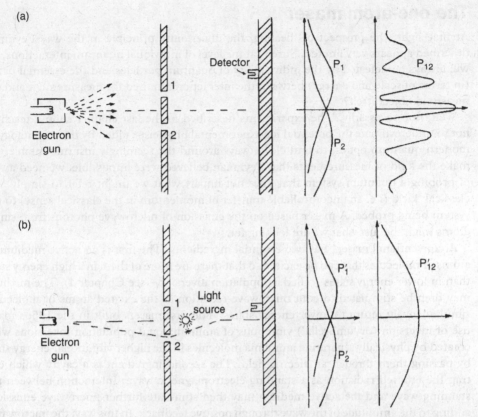

Fig. 10.3 (a) An electron gun provides a source of electrons which pass through a plate with two slits or holes and strike a second plate on which a movable detector is mounted. As the detector is moved up and down the plate, electrons are detected in a pattern which conforms to the predicted interference pattern, P_{12}. With either slit blocked, the resulting pattern is that of diffraction through the single, open slit, P_1 or P_2. (b) A light source is now placed behind the plate with two slits. As electrons pass through, they scatter photons from the light source and can therefore be revealed to have passed through either slit 1 or slit 2. However, the pattern of detection at the second plate is now that of a simple sum of the diffraction patterns from the two slits, $P'_{12} = P'_1 + P'_2$: interference has disappeared. Reprinted with permission from Feynman, Richard P., Leighton, Robert B., and Sands, Matthew (1965). *The Feynman lectures on physics*. Volume III. Addison-Wesley, Reading, MA, figs. 1-3 and 1-4.

adamant that this had to be the way the quantum world works, he had no evidence for this assertion.

From Bohr's shifting position, we can infer that *complementarity* prevents us from watching the electrons in the way that Feynman envisaged, *not* the uncertainty principle. It suggests that if we could ever find a way of 'beating' the uncertainty principle, the results of any such experiments would still remain consistent with quantum theory's predictions. Complementarity would ensure this.

The one-atom maser

At first sight, the prospect of 'beating' the uncertainty principle in the way Feynman described appears very limited. Surely, at the level of individual quantum interactions, we will always be defeated by the indivisibility of quantum particles and the essential parity (in terms of scale and energy) between the interaction required for measurement and the measured object itself?

Well, not necessarily. The experiments described in the last chapter should teach us not to underestimate the potential for experimental physicists, aided by the technology of modern quantum optics, to find clever ways around the seemingly insurmountable. To make the kind of measurements that Feynman believed were impossible, we need a way of probing a quantum system that does not impart what we might refer to simply as a classical 'kick' (i.e. an uncontrollable transfer of momentum in the classical sense) to the system being probed. A maser based on the emission of microwave photons from single atoms might be just what we are looking for.

A conventional maser[3] has two essential ingredients. The first is an active medium of atoms or molecules that can be excited so that there are more of them in a high energy state than in lower energy states (called a population inversion—see Chapter 4). The medium may then be stimulated to emit microwave radiation as the excited atoms or molecules simultaneously return to lower energy states. The first masers built in the 1950s made use of inversion (or 'umbrella') vibrations of ammonia, and population inversions were created by physically separating ammonia molecules in the higher vibrational-energy state by passing them through an electric field. The second ingredient is a cavity which can trap the emitted radiation as a standing electromagnetic wave. Interaction between the standing wave and the active medium may then stimulate further microwave emission, adding to the amplitude of the wave through positive feedback. In this way, the microwaves are amplified.

In the mid-1980s, Herbert Walther and his colleagues at the Max Planck Institute for Quantum Optics and the Universities of Munich and Wuppertal developed a so-called micromaser based on emission from single atoms of rubidium. A beam of rubidium atoms is first passed through a device which selects atoms with a narrow range of velocities before passing through the beam of an ultraviolet laser and into a micromaser cavity. Narrowing the velocity range allows control over the amount of time each atom spends inside the cavity. The laser excites the more abundant isotope ^{85}Rb to an atomic state labelled $63p_{3/2}$, in which the excited electron is removed so far from its initial orbit that the state behaves as though the atom consists of a central point charge with a single orbiting electron (and is consequently called a *Rydberg state*). The cavity is made of pure niobium cooled to below its superconducting temperature using liquid helium, isolating the interior from any stray electromagnetic fields and considerably reducing the number of thermal 'black body' photons emitted from the inside surface of the cavity.

The first excited atom to pass through the cavity spontaneously emits a microwave photon as it falls to the lower energy $61d_{3/2}$ state. A standing electromagnetic wave is established, which stimulates subsequent atoms to emit the same wavelength radiation as they pass in turn through the cavity. Having passed through the cavity, the ^{85}Rb atoms

---

[3] Maser stands for microwave amplification by the stimulated emission of radiation. Masers were the forerunners to lasers.

present in the upper $63p_{3/2}$ state are ionized and detected. Maser operation is registered by observing a *depletion* in the number of atoms present in the upper state (and hence a depletion in the number of ions produced). With about 1750 atoms passing through the cavity per second, each taking $50\mu s$, there are on average 0.09 atoms in the cavity at any one time, implying that more than 90 per cent of the interactions between the atoms and the cavity are single-atom events. Nevertheless, a standing wave corresponding to approximately 100 microwave photons can be established in the cavity.[4]

The nature of the microwave emission is such that no classical 'kick' is imparted to a rubidium atom as it passes through the cavity. And yet the very existence of microwave photons inside the cavity is a telltale sign that a rubidium atom did indeed pass through it.

Which way did it go (again)?

It is a relatively short step from the real properties of micromaser cavities to a thought experiment which uses these in an attempt to gain 'which way' information whilst at the same time observing interference. Just such a thought experiment was proposed in 1991 by Marlan Scully, at the Center for Advanced Studies at the University of New Mexico and Berthold-Georg Englert and Herbert Walther at the Max Planck Institute for Quantum Optics and the University of Munich.

Instead of electrons, we make use of the quantum properties of the rubidium atoms themselves to provide a source of two-slit interference resulting from the overlap of their atomic wave functions or state vectors. Instead of Feynman's light source, we deploy a micromaser cavity in front of each slit (see Fig. 10.4). Initially, the micromaser cavities are 'switched off'. Rubidium atoms are then fired one by one through this apparatus and presumably pass through one or other of the two cavities and hence through one or other of the two slits. Their passage through one or other of the slits can then be registered by the fact that one of the cavities now contains a microwave photon and is 'switched on'. We can therefore gain 'which way' information simply by looking to see which of the two cavities has been switched on by passage of a rubidium atom through it. The atoms suffer no classical 'kick' as a result of emitting microwave photons, so they continue with the spatial components of their state vectors intact, passing through the slits and eventually impinging on a detector, which allows us to observe interference fringes. We seem to have been able to discover 'which way' information, demonstrating the rubidium atom's particle-like properties, without preventing us from observing interference effects characteristic of wave-like behaviour.

What would Bohr have made of this challenge? In their proposal, Scully, Englert and Walther observed that[5]:

On the other hand, Bohr would not have been distressed by the outcome of these considerations, as wave-like (interference) phenomenon is lost as soon as one is able to tell which path the atom traversed. Quantum mechanics contains a built-in safeguard such that the loss of coherence in measurements on quantum systems can always be traced to correlations between the measuring apparatus and the system being observed.

[4] Rempe, G., Scully, M. O., and Walther, H. (1991). *Physica Scripta*, **T34**, 5–13.

[5] Scully, Marlan O., Englert, Berthold-Georg, and Walther, Herbert (1991). *Nature*, **351**, 111.

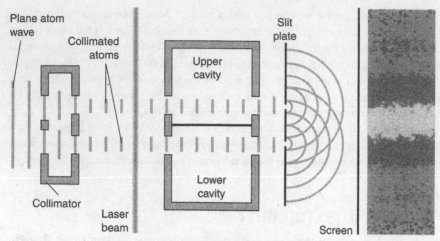

Fig. 10.4 Scully, Englert, and Walther's scheme for measuring 'which way' information using micromaser cavities. Rubidium atoms are formed into two collimated beams, excited to the upper $63p_{3/2}$ maser state and passed through two micromaser cavities, one atom at a time. Whilst in the cavity, an excited atom spontaneously emits a microwave photon, leaving behind a 'telltale' signal from which it is possible to determine which cavity it passed through, and therefore which slit it passed through on the way to the screen. Reprinted with permission from Englert, Berthold-Georg, Scully, Marlan O., and Walther, Herbert (1994). *Scientific American*, **271**, 86.

The safeguard these physicists refer to arises when the entire system described above is considered from the perspective of quantum theory. It might be true that the rubidium atoms can pass through the micromaser cavities with the spatial components of their state vectors intact (thereby 'beating' the uncertainty principle) but, in leaving behind a 'telltale' microwave photon, the quantum states of the rubidium atoms have become inexorably entangled with the quantum states of the microwave photons used to detect their presence, in a way exactly analogous to the 'infinite regress' implied in the paradox of Schrödinger's cat (see Chapter 7 and Appendix 15). This logic is described in more detail in Appendix 23.

What this analysis suggests is that the orthogonality of the two possible micromaser detection states—which describe the situation in which either the first or second cavity is switched 'on'—effectively destroys the interference terms. If we force the atom to reveal which slit it went through, quantum theory appears to deny us the possibility of observing interference effects through a mechanism that has nothing whatsoever to do with the 'clumsiness' of the measurement. If wave–particle duality is the fundamental characteristic lying at the heart of quantum theory, then complementarity—not the uncertainty principle—is the mechanism by which it operates. And, perhaps for the first time, we have here a mathematical basis from which to understand how complementarity actually *works*.

To be sure, these are thought experiments, and, since the publication of the proposal in 1991, there has been some lively correspondence in the pages of the journal *Nature* and elsewhere, with some physicists criticizing this analysis and arguing for the supremacy of uncertainty over complementarity. Some have argued that it is in fact impossible for a rubidium atom to emit a microwave photon without experiencing some kind of 'kick'—in this case, a quantum-mechanical momentum transfer rather than a clumsy classical 'kick'. Scully and his colleagues have defended their original position. To this debate, we must add a number of reports of experiments based on the original ideas of Scully, Englert, and Walther but which do not use micromasers as the 'which way' detectors. Experiments with light scattered from two trapped atoms were reported in 1993 by U. Eichmann, J. C. Bergquist, J. J. Bollinger, J. M. Gilligan, W. M. Itano, and D. J. Wineland at the National Institute of Standards and Technology in Boulder, Colorado and M. G. Raizen at the University of Texas in Austin in which 'which way' information could be derived from the polarization of the scattered photons. In 1995, Thomas Herzog, Paul Kwiat, Harald Wienfurter, and Anton Zeilinger at the University of Innsbruck in Austria used photon pairs generated by type-I parametric down-conversion to perform experiments in which 'which way' information was available from polarization measurements (more on these experiments below). Interference could be observed if 'which way' measurements were not performed but disappeared if they were. In 1998, S. Durr, T. Nonn, and G. Rempe of the University of Konstanz in Germany reported the results of experiments using an atom interferometer in which a microwave field was used to store 'which way' information in internal states of rubidium atoms. All these reports argued for an absence of the kind of classical 'kick' (and the more subtle quantum momentum transfer) required to destroy interference through the uncertainty principle. Yet, in all cases, the interference pattern was lost as soon as 'which way' information was gained.

But what if we don't look?

It would be tempting to conclude from these experiments that the detection of 'which way' states characteristic of particle-like behaviour collapses the wave function at that point, rendering impossible the further detection of interference characteristic of wave-like behaviour. But this is not quite how Scully, Englert, and Walther's original thought experiment was set up. They deliberately formed *two* atomic beams and placed the micro-maser cavities in *front* of the two slits, not behind them. If it is accepted that the spatial component of the atomic state vector is not disturbed by the emission of a photon inside one of the cavities, then, aside from entanglement of the state vector with the photon states of the cavities, as described above, it is difficult to see how von Neumann's projection postulate can contribute to our understanding of the process. We would have to suppose that, although the two atomic beams are spatially separated by passing through the collimator, as illustrated in Fig. 10.4, atoms within these beams are still described by a single state vector (after all, the two beams are formed from one 'plane wave' of atoms, and the spatially separated components must remain in phase if interference is to be observed). Then, indeed, we would suppose that detection of a rubidium atom in one or other of the two cavities collapses the wave function at this point.

But now here is a thought. What if we choose not to look to see which cavity has been switched on? By choosing not to gain 'which way' information, can we expect the interference pattern to be restored? What if we wait until the atom has traversed the apparatus and has been detected and *then* choose whether or not we want to look to see which way it had gone? Can we switch the interference pattern on and off by choosing whether to look or not *after* the atom has been detected? If we can, does this imply that we can somehow 'uncollapse' the wave function?

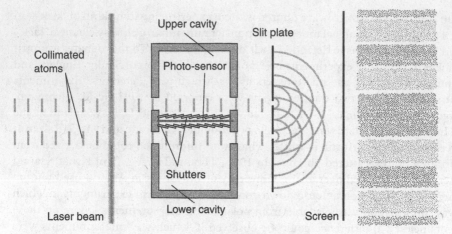

Fig. 10.5 The micromaser cavities are now isolated from each other using shutters. If these shutters are opened after the rubidium atom has emitted a photon and been subsequently detected on the screen (but before we look to see which cavity the photon is in), the 'which way' information is erased. Reprinted with permission from Englert, Berthold-Georg, Scully, Marlan O., and Walther, Herbert (1994). *Scientific American*, **271**, 92.

Discussions of this kind of quantum 'weirdness' have helped sell many a popular science book. The answers to these questions are, however, very subtle. It may, perhaps, be surprising to learn that according to Scully, Englert, and Walther's original analysis of this hypothetical quantum 'eraser' experiment, the interference pattern does indeed come back if we choose not to look, but the mechanism by which it returns is quite sophisticated.

Scully and his colleagues proposed a further thought experiment in which they imagined that the two micromaser cavities are isolated from each other by a central photodetector which is, in turn, isolated from the contents of each cavity by a shutter (see Fig. 10.5). With the shutters open, the two cavities form a single large cavity with a photodetector placed along its central axis. If a single atom passes through this arrangement with the shutters closed, we know it will emit a microwave photon, but we cannot tell in advance in which cavity the photon will be emitted. The correct quantum-mechanical description of the result is therefore two 'partial waves', one in each cavity, representing a 50:50 probability of finding the photon in one or the other. We allow the atom to continue through the apparatus until it is detected. If we now open the shutters, these partial waves combine to produce a different kind of correlation. They may combine constructively at the point where we have placed the photodetector, resulting in detection of a single photon in the larger cavity. Alternatively, they may combine destructively at the point of the photodetector, with the result that the photon goes undetected. The probability of detecting the photon (and thereby 'erasing' it) and not detecting it is again 50:50.

Of course, by opening the shutters, we have denied ourselves the possibility of gaining 'which way' information. The interference pattern is predicted to return, but to see it, we must correlate detection of the atom with detection (or non-detection) of the photon. Assume we code each spot recorded by detection of an atom which is correlated with detection of a photon using a red colour, and we code detection of an atom which is correlated with non-detection of the photon using a blue colour. If we then look at the

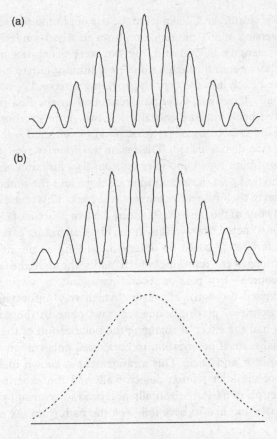

Fig. 10.6 By monitoring only rubidium atoms that are detected coincidently with detection of a microwave photon (i.e. red spots as described in the text), we recover an interference pattern that looks something like (a). Monitoring atoms detected coincidentally with non-detection of a photon (i.e. blue spots) yields an interference pattern given by (b). If we do not discriminate between red and blue spots, we get the sum of interference fringes and anti-fringes, which yields what looks like a scatter pattern, (c).

resulting pattern of spots through different coloured filters, we would see an interference pattern formed by the red spots and a second interference pattern formed by the blue spots but with the latter shifted from the former by 180°. If we describe the pattern formed by the red spots as interference fringes, then the pattern formed by the blue spots would be described as 'anti-fringes', where the peak of a fringe coincides with the trough of an anti-fringe (see Fig. 10.6). If we do not look at the pattern through coloured filters and therefore do not differentiate between red and blue, we cannot see any interference pattern at all (in fact, we get a scatter pattern identical with that associated with 'which way' detection). The mathematical basis for this explanation is given in Appendix 24.

Scully's pizza

These ideas are fascinating, and quantum theory again seems to have all the answers. But can these ideas be transformed into real experiments? Recognizing the enormity of the challenge to experimentalists, Marlan Scully upped the stakes by offering a free pizza to whoever could provide a convincing demonstration of a practical quantum 'eraser'.[6]

[6] Yam, Philip (1996). *Scientific American*, January, 30.

In 1995, Herzog, Kwiat, Weinfurter, and Zeilinger made use of photon pairs formed by type-I parametric down-conversion of 351-nm radiation from an argon-ion laser. 'Signal' and 'idler' photons with wavelengths of 633 and 789 nm were generated by a direct pass of the argon-ion pump laser beam through a non-linear lithium iodate crystal. The pump beam was then reflected back into the crystal, providing a second opportunity to create another signal–idler pair. The signal–idler pair generated by the first pass of the pump beam was also reflected back into the crystal using two more mirrors, with the end result that the photon pairs created by direct (first pass) or reflected (second pass) pump beams could no longer be distinguished. The result was interference, detectable as 'fringes' in the count rates of both signal and idler photons as a function of the phase difference accumulated by the first pass signal and idler photons and the pump photons in propagating from the crystal to their respective mirrors and back. This phase difference could be varied by moving any one of the mirrors. In effect, the two sources of signal–idler pairs (first pass and second pass) acted like two slits in the classical Young experiment.

With this arrangement, interference could be observed, so long as no attempt was made to identify which photons were following which paths through the apparatus, and hence identifying their source—first pass or second pass (and, by analogy, identifying which 'slit' they had passed through). However, 'which way' information could be simply obtained by, for example, placing a quarter wave plate in the path of the first-pass idler photons. This had the effect of changing the polarization of the first-pass idler photons from their initial vertical polarization to horizontal polarization once they had made their way to the mirror and back. This arrangement is shown in Fig. 10.7. Placing a polarizer in front of the idler photon detector allowed the experimenters to identify the source of the idler photons. A horizontally polarized idler meant that it had to have come from the first pass and had to have followed the path from the non-linear

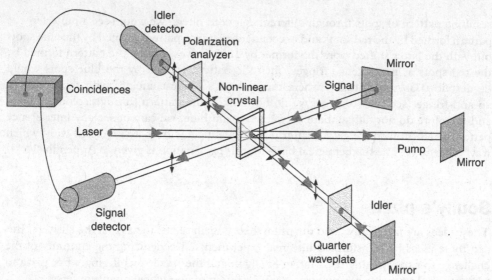

Fig. 10.7 Experimental arrangement used by Herzog, Kwiat, Weinfurter, and Zeilinger to perform 'which way' versus interference measurements on two pairs of photons produced by type-I parametric down-conversion.

crystal to the mirror and back through the crystal before being detected. A vertically polarized idler had to have come from the second pass and had to have propagated from the crystal directly to the detector. Obtaining 'which way' information for the idler photons implied equivalent 'which way' information for the signal photons, too. The end result was that interference in both signal and idler count rates disappeared (see Fig. 10.8(a)).

This 'which way' information could be erased simply by rotating the polarizer to an angle of 45°, preventing the possibility of learning if the idler photons were horizontally or vertically polarized. This arrangement was sufficient to restore interference in the idler count rate, but not the signal count rate (see Fig. 10.8(b)). Measuring coincidences in the detection of signal and idler photons within a 5-ns time window ensured that the 'which way' and interference effects were being recorded for *single* photons.

The physicists also reported the results of experiments which get even closer to the quantum 'eraser' thought experiment originally proposed by Scully, Englert, and Walther. Placing a second quarter wave plate in the path of the first-pass signal photons and a polarizing beamsplitter in front of the signal photon detector allowed the possibility of *regaining* 'which way' information for the *idler* photons. As before, detection of a horizontally polarized signal photon implied that this was a first-pass signal, and the coincident idler was therefore a first-pass idler. Detection of a vertically polarized signal photon implied a second-pass signal and hence a second-pass idler. The interference in the idler count rate duly disappeared (see Fig. 10.8(c)). And, as expected, no interference could be seen in the coincidence rates between idler and horizontally polarized signal photons or idler and vertically polarized signal photons (coincidences I and II in Fig. 10.8(c)).

But now the same game could be played again. Rotating the polarizing beam splitter through 45° prevented the possibility of gaining 'which way' information. This time, however, interference was not restored in the count rates of either signal or idler photons. But interference *did* return if detection of an idler photon was *correlated* with detection of a signal photon at either +45° or −45°, analogous to correlation with red or blue spots in the original thought experiment (see Fig. 10.8(d)). This interference was revealed in the coincidence count rates, which formed fringes (+45°) and anti-fringes (−45°). If the signal photon polarization was not distinguished, no interference could be seen.

History does not record if Herzog, Kwiat, Weinfurter, and Zeilinger claimed Scully's pizza.

The mathematical analysis of these experiments as outlined in Appendices 23 and 24 make it clear that there is a direct relationship between tests of complementarity and tests of quantum non-locality. Again, we should not, perhaps, be too surprised by this. Interference effects are the direct manifestation of non-local behaviour. These effects can be 'encoded' in the mathematical structure of quantum entanglement—in this case, entanglement of the states responsible for interference with the states to be used to detect 'which way' information. These states cannot be disentangled without forcing the system to reveal one type of behaviour or another. They cannot be disentangled to reveal both types of behaviour simultaneously.

Though still a subject for debate, a consensus is building that complementarity—and hence quantum non-locality and entanglement—is the mechanism responsible for the

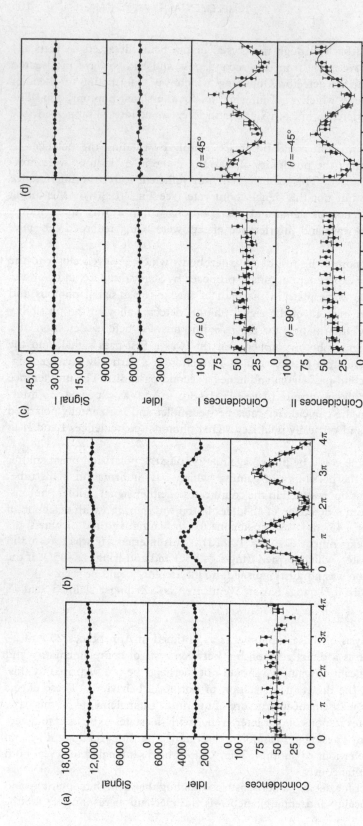

Fig. 10.8 Results of the experiments reported by Herzog, Kwiat, Weinfurter, and Zeilinger. (a) With the arrangement as shown in Fig. 10.7, the detection of 'which way' information for the idler photons destroyed interference in the observed signal, idler, and coincidence count rates. (b) By reorienting the polarization analyser to an angle of 45°, the 'which way' information was erased, and interference in the idler and coincidence count rates was restored. (c) To the arrangement in Fig. 10.7 was added a quarter wave plate and polarizing beamsplitter in the signal photon beams (see text). Detection of the polarization of the signal photons restored the possibility of obtaining 'which way' information for the idler photons, with the result that interference for the idler photons was again destroyed. (d) Reorienting the polarizing beamsplitter to an angle of 45° restored interference, but only in the *coincidence* count rates. Coincidences between idler photons and signal photons polarized at +45° produced interference fringes, as shown (Coincidences I), whereas coincidences with signal photons polarized at −45° produced anti-fringes (Coincidences II). Adapted with permission from Herzog, Thomas J., Kwiat, Paul G., Weinfurter, Harald, and Zeilinger, Anton (1995). *Physical Review Letters*, **75**, 3034, figs. 2 and 3.

mutual exclusivity in the dual wave–particle nature of quantum entities, what Richard Feynman described as the 'central mystery' at the heart of quantum theory.

Superluminal communications?

Experimental physicists are delightfully resilient individuals. All the experiments described so far appear to point unambiguously to a description of nature that is decidedly non-local at the quantum level, and therefore profoundly strange. These results may have set the blood racing among theoreticians and philosophers (as we will see in Part V) but experimentalists are generally more pragmatic folk. They do not necessarily need to understand what quantum non-locality and entanglement mean in order to seek some practical applications.

Whether or not we accept that correlated photons are objectively real entities which exist independently of our instruments, the results of all these experiments suggest an interesting possibility. Can we exploit the communication that seems to take place between distant photons to send faster-than-light messages, perhaps at speeds of tens of thousands or tens of millions times that of light? To answer this question, we need to devise a simple procedure by which information might be communicated between two distant observers, and then see if such a procedure works in principle.

The feature of the physics of the correlated photons that we must try to exploit is the instantaneous realization of a specific polarization state for photon B at the moment that photon A is detected to be in a specific polarization state. Consider the AMAZING Super-luminal Communications System^{TM}, manufactured and marketed by the AMAZING Company of Reading, England. This system has three parts, a transmitter, a receiver, and a 'line' provided by a central source of correlated photons emitted continuously in opposite directions, at regular intervals of, say, 10 ns. The photons that make up a pair are timed to arrive coincidentally at the transmitter and receiver.

The transmitter is built around an electro-optical switch, which switches the incoming A photons between two different optical paths. One path leads to a polarizing filter oriented with its axis of maximum transmission in the vertical direction, and the other leads to a polarizing filter oriented horizontally. The transmitter electronics recognizes detection of a photon through the vertical polarizer as a '1' and detection of a photon through the horizontal polarizer as a '0'. The receiver, located on the moon, has only one polarization filter, oriented vertically.

Now suppose that Alice (A), located in Reading, wishes to inform her friend Bob (B) on the moon that the local temperature is currently 19 C. This number can be encoded as a binary number (in fact, the binary form of 19 is 10011). We plug this binary number into the electronic system that controls the electro-optical switch. When the system wants to send a 1, the next incoming A photon is switched through to the vertical polarizer. Its detection forces photon B into a vertical polarization state (because of the correlation between the photons—$E(a, a) = +1$), which passes through the vertical polarizer in the receiver and is detected. This signal is recognized by the receiver electronics as a 1, and the digit has therefore been transmitted instantaneously (or, at least, at speeds of tens of thousands or tens of millions times that of light).

When the transmitter wants to send a 0, the next incoming A photon is switched through to the horizontal polarizer. Its detection forces the next photon B into a horizontal polarization state, which is blocked by the polarizer in the receiver. Since the receiver expects the next photon within 10 ns of the previous one, it recognizes non-detection as a 0.

This process continues until all the binary digits have been sent. Bob decodes the binary number and learns that the temperature at Alice's location is 19 C. The information takes about 50 ns to transmit, compared with the 1.3 s or so that it takes a conventional signal to travel the 240,000 miles from the earth to the moon. This represents a time saving of a factor of about 30 million. In fact, this factor is unlimited, since the communication is in principle instantaneous, and we can place the transmitter and receiver an arbitrarily long distance apart (although we may have to wait a while for the 'line' to be established).

Of course, if this scheme had any chance of working whatsoever, it would have been patented years ago. It does *not* work because an A photon passed through to the vertical polarizer is not automatically forced into a state of vertical polarization. According to quantum theory, it has equal probabilities for vertical or horizontal polarization, and we have no means of predicting in advance what the polarization will be. Thus, simply switching the A photon through to the vertical polarizer does not guarantee that photon B will be forced into a state of vertical polarization. In fact, despite switching between either path in the transmitter, there is still an unpredictable 50:50 chance that photon B will be transmitted or blocked by the polarizer in the receiver. No message can be sent unless Alice informs Bob of the results she obtained by means of a conventional 'classical' communication channel, which restricts the speed of communication to that of light. Of course, if she has to resort to a conventional communication, she might just as well use this to inform Bob of the temperature. (The AMAZING Company of Reading, England, recently filed for intellectual bankruptcy.)

Actually, it has been argued that our inability to exploit the apparent faster-than-light signaling between distant correlated photons to send meaningful information allows quantum theory and special relativity peacefully to coexist. Special relativity is founded on the postulate that the speed of light represents the ultimate speed of transmission of any *conventional* information. Perhaps it does not matter that the speed of transmission of influences within an entangled pair may be tens of thousands or tens of millions times the speed of light if we cannot use this to send any useful information. Whatever the nature of the communication between distant correlated photons, it is certainly *not* conventional.

Qubits and quantum computing

The information transmitted in the hypothetical example described above is reduced to its most basic level—the 'bit'—which can have values of either 0 or 1, and which forms the basis of all computing. Like classical information bits, the two-state basis sets $|v\rangle/|h\rangle, |v'\rangle/|h'\rangle$ and $|L\rangle/|R\rangle$ all appear to be composed of 'polar opposites', vertical versus horizontal, left versus right, hinting at a very powerful analogy with classical information theory. The quantum version of the classical bit is the quantum bit or *qubit*. A qubit is a quantum system comprising just two orthonormal states, $|0\rangle$ and $|1\rangle$, which form a basis set in 'qubit space'. The most significant difference between classical bits and qubits is that we can form coherent superpositions of $|0\rangle$ and $|1\rangle$, which has some interesting consequences for processing *quantum information*.

The different possible coherent superpositions of the qubit states form the surface of a *qubit sphere*, shown in Fig. 10.9. In this diagram, the basis states $|0\rangle$ and $|1\rangle$ lie at the 'north' and 'south' poles of the sphere (emphasizing that they are indeed 'polar opposites'), and

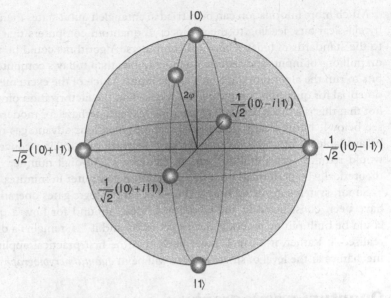

Fig. 10.9 A qubit sphere. Coherent superpositions of the generic quantum states $|0\rangle$ and $|1\rangle$ lie on the surface of the sphere, and incoherent superpositions lie closer to the centre. All states represented as 'polar opposites' on the sphere form an orthonormal basis in the qubit space. After Tittel, Wolfgang and Weihs, Gregor (2001). *Quantum information and computation*, **1**, 3. Reproduced with permission of the authors.

coherent superpositions of these states in which the expansion coefficients are equal lie around the equator. Superpositions in which the expansion coefficients are not equal lie on a geodesic between the north and south poles at a polar angle given by the ratio of the coefficients.

If we substitute the states $|v\rangle$ and $|h\rangle$ for $|0\rangle$ and $|1\rangle$, we can form a linear superposition that corresponds to the left-circular polarization state $|L\rangle$ and another corresponding to the right-circular polarization state $|R\rangle$, as shown. Superpositions corresponding to $|v'\rangle$ and $|h'\rangle$ lie on the geodesic indicated in Fig. 10.9 with a polar angle 2φ. The figure makes it clear that there are many more possible coherent superpositions of the qubit states which can be written as superpositions of $|v\rangle$ and $|h\rangle$ or of $|L\rangle$ and $|R\rangle$. What makes all these different possible combinations coherent is the *phase relationship* between the two qubit states—they form a coherent superposition if the ratio of the expansion coefficients is $e^{i\varphi} (= \cos\varphi + i\sin\varphi)$, which becomes i when 2φ is 90° and $-i$ when 2φ is 270°. If this phase relationship is not maintained, the states formed from the superposition no longer lie on the surface of the sphere but fall closer to its centre, and completely incoherent superpositions (with no phase relationship between the states) lie at the centre of the sphere, corresponding to completely depolarized light. Although the qubits referred to above are all envisaged as photon polarization states, there is in principle no restriction on the types of states that can be thought of as qubits. Quantum states (often based on photons) with differences in spatial directions, time, and frequency can all be used to form qubit systems.

Much more information can be carried in entangled qubit states than can ever be carried by classical bits, leading to the prospect of quantum computers that are vastly superior to the standards of today's classical computers. Algorithms could be run simultaneously on millions of inputs and require no more qubits than today's computers require classical bits to run the algorithm once on a single input. Much of the excitement surrounding the potential for quantum computing has focused on the factorization of very large numbers, not that this is in itself particularly exciting (except, perhaps for modern cryptographers—see below), but it clearly demonstrates the tremendous advantages of speed that could be available. It has been estimated that a network of a million conventional computers would require over a million years to factorize a 250-digit number. Yet this feat could theoretically be performed on a single quantum computer in minutes.[7]

So far, systems involving only a couple of simple logic gates operating on a few qubits have been constructed in the laboratory. Ideas abound for how a quantum computer could be built but no practical device yet exists and it '...remains a dream waiting to be realised.'[8] Rather, it is much more likely that the first practical application of quantum mechanics at the level of single entities will be in *quantum cryptography*.

Quantum cryptography

History is littered with 'unbreakable' codes that proved to be all-too-breakable, often with deadly consequences for those sending or receiving secret information. But information can be encoded in entangled qubits, and because it is impossible to access this information without destroying the entanglement, quantum systems based on entangled pair states provide a potentially unassailable mechanism for secure transmission. Suppose that Alice wishes to send secret information to her colleague, Bob. She first sets up a central source of entangled photon pairs and sends one photon (photon A) to herself and the other (photon B) to Bob, perhaps by passing the photons down fibre-optic cables. Both Alice and Bob then randomly choose the basis in which they measure the photon polarizations, $|v\rangle/|h\rangle$, $|v'\rangle/|h'\rangle$ or $|L\rangle/|R\rangle$. After they have both detected a large number of photons, they publicly compare events where they successfully detected a photon and report the basis in which they made this detection, without revealing the result they obtained (i.e. without saying if they recorded a 0 or 1). They both know that for those events in which they both successfully detected a photon *and* used the same basis, the results will be correlated—they will have either both recorded a 0 or both recorded a 1. The resulting bit sequence, known only to Alice and Bob, can then be used as an *encryption key*. (Note that the need for comparison of the events by conventional communication channels again implies that no useful information can be transmitted at speeds faster than that of light.) Alice and Bob can now happily send more conventional coded messages (using a 'one-time pad', for example), secure in the knowledge that only they share the key.

To test for the presence of an eavesdropper (traditionally referred to as 'Eve'), Alice and Bob can perform tests of Bell's inequality. If the photons arrive unperturbed, correlations between the photons will exceed Bell's inequality, and it can be concluded that nobody has been eavesdropping, since this would have required a measurement, and a measurement would have destroyed the entanglement and noticeably reduced the correlation.

---

[7] See Deutsch, David (1997). *The fabric of reality*. Penguin Books, London, pp. 200, 216.

[8] Milburn, Gerald J. (1998). *The Feynman processor*. Perseus Books, Cambridge, MA, p. 171.

If Eve were somehow to gain full knowledge of all the polarization states of the photons transmitted to Alice and Bob (necessary to acquire the encryption key), then the correlations would no longer violate Bell's inequality. Eve's presence will have been detected, and Alice and Bob simply discard the key as unsafe.

Quantum cryptography was first proposed in the early 1980s by Stephen Weisner (then at Columbia University), Charles Bennett at IBM and Gilles Brassard at the University of Montréal. Several single photon protocols (not requiring entangled pair states) have been proposed, and many experiments have been performed to demonstrate the feasibility of secure transmission of information using single quantum entities. The protocol outlined above is named for Arthur Ekert at Oxford University who, in 1991, independently proposed a system of quantum cryptography based on entangled pair states. Any experiment involving a rapid change of polarization measurement basis, such as those used to close the locality loophole described in Chapter 9, are in effect practical realizations of the Ekert protocol.

Modern 'classical' cryptographic systems are based on *asymmetric keys*. Alice sends a message encrypted using a key that Bob has made public (also known as a *public key*). Bob deciphers Alice's message using a second, private, decryption key. The system is based on mathematical structures called modular functions, and security is guaranteed because the (private) decryption key is formed from randomly chosen prime numbers, and the (public) encryption key is formed by multiplying the prime numbers together. If the numbers are large enough, Eve cannot get the decryption key by factorizing the public encryption key, as this requires enormous computer power and lots of time. This system, termed RSA after the mathematicians Ronald Rivest, Adi Shamir, and Leonard Adelman who came up with the idea of using modular functions to create an asymmetric cipher,[9] is particularly suited to the protection of electronic communications and transactions. Interestingly, quantum computers using an algorithm developed by Peter Shor at AT&T Bell Laboratories in New Jersey would quickly render the RSA system vulnerable. If quantum computers are ever developed to the point where they pose a real threat to the RSA system, quantum cryptography could provide an instant solution.

Quantum teleportation

A third area of active research in the practical application of entanglement is *quantum teleportation*. Suppose Alice chooses to entangle her photon A with another photon in an unknown polarization state (which we will call photon X). She could do this, for example, by carefully bringing the two photons together in a Köster prism, causing them to interfere, and so causing them to lose their individual identity. She now detects the photons using photomultipliers placed either side of the prism, similar to the arrangement shown for photomultipliers 1 and 2 in Fig. 10.1. We know that this will yield one of four possible outcomes resulting from interference of the photons. These involve the detection of both photons by one or other photomultiplier, detection of one photon by each photomultiplier, or no detection at all. If the photons are detected one by each photomultiplier, the effect of this measurement is to project photon B into the polarization state previously carried

---

[9] In fact, public-key cryptography was first developed at the United Kingdom's Government Communications Headquarters in Cheltenham.

by photon X, which is still unknown (see Appendix 25). In effect, photon B replicates the state of photon X and 'becomes' photon X. X has therefore been 'teleported' from Alice to Bob. Alice must communicate the result she observed to Bob if he is to recognize his photon B as being a teleported photon. In fact, if Alice communicates all the results she obtains, event by event, Bob can replicate the state of photon X either by doing nothing or by rotating its plane of polarization in a way that depends on the specific results Alice obtained. By this means, all of the photons sent to Bob can be converted into teleported photons. (And, we note again, a classical communication which can occur at speeds no faster than that of light is required for meaningful information to be carried from Alice to Bob.)

Is this really teleportation, of the kind that fans of the science-fiction television series *Star Trek* are only too familiar? Although teleporting one photon is quite different from teleporting the whole of Captain James T. Kirk, aspects of the physics involved confirm that this is exactly what it is. Photons are indistinguishable quantum particles characterized only by their quantum states (up/down, left/right, vertical/horizontal). If we are able to replicate exactly the state of a photon at position A in the state of another photon at position B, then to all intents and purposes, we have sent the photon from A to B. Note that we have not created a *copy* of the initial photon (X was destroyed in the measurements described above), as this is actually forbidden in quantum physics.[10] Indeed, if we could entangle the (unknown) states of every quantum particle that makes up James T. Kirk with one of a pair of entangled quantum particles, make the necessary measurements (thereby destroying Kirk in his current location), and communicate the results (to Scotty, in the transporter room), Scotty could in principle use this information to replicate Kirk. We should not let our lively imaginations be distracted by the fact that this is certainly impossible with present-day or near-future technology, and may be forever impossible for classical objects such as people.

The possibility of teleporting a photon was first raised by Charles Bennett, Gilles Brassard, Claude Crépeau, Richard Jozsa, Asher Peres, and William Wooters in a paper published in *Physical Review Letters* in 1993. Experimental demonstrations of one-photon teleportation followed some 5 years later. Most recently, a team at the Australian National University in Canberra publicly announced teleportation of a laser beam (consisting of billions of photons) about 1 m across the laboratory.[11]

Was Einstein wrong?

Non-locality and entanglement are now established as unassailable experimental facts, as the very active research on quantum information demonstrates. Where does this leave Einstein's realism? It is certainly a strange irony of the history of science that Einstein, having laid the foundations of quantum theory through his revolutionary vision, should have become one of the theory's most determined critics. When he launched his attack on the theory in 1935 with his charge of incompleteness, he could not have possibly anticipated the work of Bell, 30 years later. At the time they were made, the arguments between Bohr and Einstein were purely academic arguments between two eminent physicists with

---

[10] This is the quantum 'no cloning' principle, first articulated in 1982 by William Wooters and Wojciech Zurek.

[11] See *The Times* (2002) 18 June, 3. Physicists at Geneva University and the Danish University of Aarhus have since succeeded in teleporting qubits over distances of 2 km.

different philosophies: realism versus anti-realism. Bell's theorem changed all that. The arguments became sharply focused on practical matters that could be put to the test in the laboratory. If, like the great majority of the physics community, we are prepared to accept that the experiments described in this chapter and in Chapter 9 have been correctly interpreted, we must also accept that Einstein's charge of incompleteness is unsubstantiated, at least in the *spirit* in which that charge was made in 1935.

How would Einstein have reacted to these results? Of course, any answer to such a question is bound to be subjective. However, it seems reasonable to suppose that he would have accepted the results (and their interpretation in terms of non-local behaviour) at face value. Not for him a relentless striving to find more loopholes through which some sense of local reality might be preserved. It is also reasonable to suppose that he would not have been persuaded by these results to change his position regarding the interpretation of quantum theory. While accepting that the results are essentially correct, I suspect he would have still maintained that their interpretation contains 'a certain unreasonableness'. He might have marvelled at the unexpected subtlety of nature, but his conviction that 'God does not play dice' was an unshakeable foundation on which he built his personal philosophy.

So, was Einstein wrong? In the sense that the EPR paper argued in favour of an objective reality for each quantum particle in an entangled pair independent of the other and of the measuring device, the answer must be yes. But if we take a wider view and ask instead if Einstein was wrong to hold to the realist's belief that the physics of the universe should be objective and deterministic, we must acknowledge that we cannot answer such a question. It is in the nature of theoretical science that there can be no such thing as certainty. A theory is only 'true' for as long as the majority of the scientific community maintain a consensus view that the theory is the one best able to explain the observations. And the story of quantum theory is not over yet.

Was Bohr right?

I feel sure that Bohr would have been delighted by the results of these experiments, if not particularly surprised by them. They appear to be a powerful vindication of complementarity and graphically demonstrate the central, crucial role of the measuring device. Perhaps Bohr would have been quick to point out that the methods used to predict the results of the experiments on entangled quantum particles and on complementary wave–particle behaviour are actually based on some of the simplest of experimental observations with polarized light. Observations such as those that led to Malus's law formed the basis of our derivations of the projection amplitudes given in Appendix 12 and which were used in our analysis of the experiments in Chapter 9. Seen in this light, quantum theory is no more than a useful means of interrelating different experimental arrangements, allowing us to take the results from one to predict the outcome of another. We cannot go beyond this because, according to Bohr's anti-realist philosophy, we have reached the limit of what is knowable. The questions we ask of nature must always be expressed in terms of some kind of macroscopic experimental arrangement.

Does this mean that the tests of Bell's inequality, delayed-choice, and quantum eraser experiments prove that the Copenhagen interpretation is the only possible interpretation of quantum theory? I do not think so. We should recall here the arguments made in Chapter 6. The Copenhagen interpretation insists that, in quantum theory, we have reached the limit of what we can know. Despite the fact that this interpretation emerges unscathed from the experimental tests described in this chapter and in Chapter 9, there are some physicists who argue that it offers nothing by way of explanation. The non-locality and indeterminism of the quantum world create tremendous difficulties of interpretation, which the Copenhagen view dismisses with a metaphorical shrug of the shoulders. The non-locality, some would argue, is an apparent feature of properties that are themselves not directly observable and which therefore have no basis in reality, so why worry? For other physicists, however, this is not good enough. We will see in Part V that, while some of the suggested alternative interpretations seem bizarre, they are in principle no less bizarre than the Copenhagen interpretation itself.

Readers inclined to a less metaphysical outlook might ponder the merits of such alternatives. Why bother to seek strange new theories when a much tried and tested theory is already available? Surely, any alternative will be so contrived and artificial that it will be worthless compared with the simple elegance of quantum theory? But look once more at the postulates of quantum theory described in Chapter 4. What could be more contrived and artificial than the quantum state vector? Where is the justification for postulate 1, apart from the fact that it yields a theory that works? What about the problems of quantum measurement highlighted by the paradox of Schrödinger's cat? How are we meant to account for the 'collapse' of the wave function? If these questions cause you to stop and think, and perhaps reveal a hint of doubt in your mind, then you will see why some physicists continue to argue that the Copenhagen interpretation cannot be the answer.

Bohr himself once said that: ' . . . anyone who is not shocked by quantum theory has not understood it'.

Part V

Alternatives

11

Pilot waves, potentials, and propensities

So, where do we go from here? It is apparent from the last two chapters that nature denies us the easy way out. The clear experimental demonstration of quantum entanglement rules out the idea of local hidden variables, one of the simpler solutions to the conceptual problems of quantum theory. For the time being at least, we are forced to live with the consequences of entanglement, non-locality, and 'spooky' action at a distance. This might make some of us uncomfortable, but there is no solace in the Copenhagen interpretation, which does not provide us with any answers and is still regarded by many to be no inter-pretation at all. And, lest we forget, there remain the problems of interpreting the meaning of quantum probability and the collapse of the wave function. Even those who adhere to the Copenhagen view tend to put aside or disregard the conceptual difficulties that these issues raise as they analyse the results from the latest particle-accelerator experiment. This is not a very satisfactory situation.

In Part V, we will survey some of the alternatives to the Copenhagen interpretation that have been put forward in the years since quantum theory was first developed. Although these alternatives are quite different from one another and from the original theory, we will find that they possess a common thread. In every case, they attempt to eliminate the conceptual problems of quantum probability or the collapse of the wave function by introducing some additional feature into the theory. This is at least consistent with Einstein's belief that quantum theory is somehow incomplete. Such features are designed to bring back determinism and causality, or to break the infinite regress of the quantum measurement process, as illustrated by the paradox of Schrödinger's cat, or both. The fact that some rational scientists are prepared to go to such lengths to obtain an aesthetically or metaphysically more appealing version of the theory demonstrates the extent of the discomfort they experience with the 'Kopenhagener Geist'.

Of course, if they are to work effectively, none of these alternatives should make predict-ions that differ from those of orthodox quantum theory for any experiment yet performed. In fact, few, if any, even hint at the possibility that they could be subjected to stringent test through experiment. For many scientists, who have been brought up to regard observation and experiment as the keys to unlocking the mysteries of the physical world, a theory that cannot be tested is of no practical value. However, we should perhaps recall that our ability to perform precise measurements on the world is a relatively new development in the history of man's search for understanding. Without the kind of speculative thinking that the positivists dismiss as non-scientific, there could have been no science (and, for that matter, no positivism) in the first place.

We now recognize that our problems start with the phenomenon of entanglement and are commonly manifested in quantum wave-like diffraction and interference effects. Maybe we are reasonably relaxed about these effects in the context of light. Until the advent of quantum theory, we were content with a description of the properties and behaviour of light that is based principally on the concept of electromagnetic waves. We obviously cannot ignore Planck, Einstein's light-quantum hypothesis, and the rise of the photon, but photons are peculiar quantum entities that are thought to have zero rest mass and are constrained to travel at one fixed speed. We are, perhaps, more comfortable with the idea of interference effects involving particles that have no mass. We experience greater discomfort when we force ourselves to face up to the reality of diffraction and interference involving 'bits' of matter—particles with mass. Almost by its very nature, mass is a property that we intuitively think of in localized terms. Look back once again at the results of the electron-interference experiment shown in Fig. 2.1. This picture *still* nags. Each spot is the signature of an indivisible quantum particle that possesses a precisely defined mass. If we imagine the experiment in which the electron density is reduced to the point where there is, on average, only one electron passing through the apparatus at a time, then we are obliged to accept that a particle possessing mass exhibits non-local behaviour. We conclude that mass, which we instinctively think of as a property of localized entities, is something that can be somehow 'distributed' over an extended region of space. When the electron interacts with other entities that make up some kind of measuring device, that same mass is somehow dragged from all the regions of space where there was a non-zero probability of finding the electron and converges on the space where it is eventually to be found. The electron is 'here', and a bright spot appears on a piece of photographic film.

Is this a realistic picture? Can we really 'spread' an electron through space just like we would spread so much butter on a slice of bread? If it is and we can, then we must accept that we have no mechanism readily to hand to explain *how* the electron instantaneously 'converges' at the point of detection, how a *substance* extended over a potentially vast region of space collapses to one location. Perhaps we would wish to argue that electrons themselves are, like photons, 'peculiar' quantum entities that behave, on balance, more like waves than particles. But we know by now that quantum theory will not let us draw such simplistic boundaries. Indeed, diffraction and interference effects have been demonstrated with particles with much larger masses. In fact, diffraction of a beam of large molecules consisting of 60 carbon atoms has been observed in the laboratory. These are the so-called 'soccer ball' C_{60} or buckminsterfullerene molecules, spherical structures of pure carbon shaped like tiny soccer balls.[1]

---

[1] See Arndt, Markus, Nairz, Olaf, Voss-Andreae, Julian, Keller, Claudia, van der Zouw, Gerbrand, and Zeilinger, Anton (1999). *Nature*, **401**, 680–682. The analogies and puns are endless. The ratio of the diameter of a C_{60} molecule and the spacing of the silicon nitride grating used to observe the diffraction pattern is comparable with the ratio of the diameter of a conventional soccer ball and the width of a goal (according to FIFA standards), giving a potentially whole new meaning to the term 'bend it like Beckham'.

De Broglie's pilot waves

As it appears that the only things we can detect are particles, Einstein's suggestion that the particles are real entities that follow precisely defined trajectories is very persuasive. The experimental evidence summarized in Chapters 9 and 10 encourages us to dismiss the possibility that any such trajectories are determined by local hidden variables, but are there any other ways in which the particles' motions might be predetermined?

In 1926, Louis de Broglie proposed an alternative to Born's probabilistic interpretation of the wave function. Suppose, he said, that quantum entities like electrons and photons are independently *real* particles, represented as singularities (point particles) moving in a *real* field. In addition, a second field is admitted which has the same (statistical) significance as the wave function in Schrödinger's wave mechanics and is open to the same probabilistic interpretation. This is different again from Schrödinger's wave mechanics, which attempted to explain everything in terms of waves only. De Broglie suggested that the equations of quantum mechanics admit a *double solution*: a continuous wave field, which has statistical significance, and a second field, containing a point-like solution corresponding to a localized particle, representing the quantum of the field. The velocity of the particle is linked to the phase of the wave function by a relation that has analogies in classical physics, and the particle is to all intents and purposes a classical particle. However, the continuous wave field can be diffracted and can exhibit interference effects.

De Broglie's double-solution interpretation was subsequently simplified to a construction involving a point particle moving in a continuous wave field. This is de Broglie's *pilot wave* interpretation: the motion of the classical particle follows the wave field, and the particle is more likely to follow a path in which the amplitude of the wave field is large. Thus, the square of the amplitude of the wave field is still related to the probability of 'finding' the particle, but this is now because the real particle, which is always localized, has a forced preference for regions of space in which the wave amplitude is large.

In terms of the double-slit experiment, we can imagine that the wave field interferes with itself as it passes through the slits, producing a pattern of bands of alternating large and small amplitudes on the photographic film. As a particle moves in the field, it is *guided* by the field amplitude and therefore has a greater probability of arriving at the film in a region which we will come to recognize as a bright fringe when a sufficient number of particles has been detected. The particle is not prevented from following a trajectory that leads to it being detected in the region of a dark fringe, but this is much less probable because the amplitude of the field along such a path is small. Unlike Bohr's complementarity, which offers us a choice between waves *or* particles depending on the nature of the measuring device, de Broglie's pilot wave interpretation suggests that reality is composed of waves *and* particles.

De Broglie published his ideas in 1926 and 1927, and presented them at the fifth Solvay Conference in October 1927. Einstein commented that he thought de Broglie was searching in the 'right direction', although he stopped short of giving his full backing to the approach de Broglie had taken. During this early period in the development of quantum theory, its interpretation had not yet crystallized: no single interpretation was dominant, and the air was full of all manner of concepts. It had been Einstein's remark connecting the wave function with a 'ghost field' that had led Born to develop his probabilistic interpretation in the first place, although de Broglie's proposal that such a field is physically real differs completely from Born's view that the wave function in some way represents only our 'state of knowledge' of the quantum particle. As we saw in Chapter 8, Einstein had himself searched in this direction in May 1927, only to reject the approach as 'too cheap'.

Schrödinger was still intent at this stage on establishing a premier position for his wave mechanics, and, since Einstein's support was equivocal, strong noises of dissent from Hendrik Kramers and Wolfgang Pauli and Heisenberg's elaboration of the uncertainty principle were enough to bury de Broglie's approach for over 20 years. De Broglie's further discussions with members of the Copenhagen school (notably Pauli) raised doubts in his own mind about the validity of his theory, and by early 1928, he had all but abandoned it. He did not include it in a course on wave mechanics he taught at the Faculté des Sciences in Paris later that year. In fact, de Broglie became a convert to the Copenhagen view.

It is important to note that the pilot-wave theory is a hidden-variable theory. The hidden variable is not the pilot wave itself—that is already adequately revealed in the properties and behaviour of the wave function of quantum theory. It is actually the particle *positions* that are hidden. Now we know from the results of the experiments described in the last two chapters that two correlated quantum particles cannot be locally real, so the pilot-wave idea can be sustained only if we acknowledge that the objectively real particles are subject to non-local 'spooky' influences at a distance. The pilot-wave theory reintroduces the classical concept of causality—classical particles are directed along classical traject-ories dictated by extended wave fields. But change the wave field simply by reorienting a polarization analyser whilst two entangled particles are in mid-trajectory, and the physical states of both the particles are obliged to respond instantaneously. It seems that we cannot have it both ways: either quantum theory is already complete or we must introduce non-local hidden variables which, in turn, appear to make the theory incompatible with special relativity. Either way, it is clear that Einstein was not very satisfied.

Quantum potentials

The American physicist David Bohm was initially an advocate of Bohr's notion of complementarity and made it the interpretational basis for his book *Quantum theory*, first published in 1951. However, his presentation of the EPR thought experiment set him thinking, and he eventually became dissatisfied with the Copenhagen view. Strongly encouraged and influenced by Einstein, Bohm sparked off a renewal of interest in the question of hidden variables through the two papers he published on this subject in 1952 in the journal *Physical Review*. Bohm's hidden variable theory has much in common with de Broglie's pilot-wave idea and, in fact, has been described as the pilot-wave theory car-ried to its logical conclusion. For this reason, Bohm's redevelopment is often referred to as the *de Broglie–Bohm theory*. Bohm continued to develop and refine this theory, despite the general indifference of the majority of the physics community at the time.

At issue was the Copenhagen school's outright *denial* that individual quantum systems could be described objectively. Bohm wrote[2]:

The usual interpretation of the quantum theory is self-consistent, but it involves an assumption that cannot be tested experimentally, *viz.*, that the most complete possible specification of an individual

[2] Bohm, David (1952). *Physical Review*, **85**, 166–193.

system is in terms of a wave function that determines only probable results of actual measurement processes. The only way of investigating the truth of this assumption is by trying to find some other interpretation of the quantum theory in terms of at present 'hidden' variables . . . the mere possibility of such an interpretation proves that it is not necessary for us to give up a precise, rational, and objective description of individual systems at a quantum level of accuracy.

Bohm was not specifically seeking a 'new' theory or a return to simple classical physics. Rather, he acknowledged that quantum theory was constructed on a set of assumptions of which the most important, the 'completeness' postulate (postulate 1—see Chapter 4) is not subject to experimental test. The Copenhagen interpretation is founded on this completeness postulate, and, rather than accept it at face value as the Copenhagen school demanded, Bohm wanted to explore the *possibility* that other descriptions and hence other interpretations are conceivable *in principle*.

The development of Bohm's theory involves a reinterpretation of Schrödinger's wave function as representing an objectively real field. Schrödinger had also started with this interpretation but had jumped to the (entirely logical) conclusion that his wave equation implied an exclusively wave-like description for quantum systems. As de Broglie had done, Bohm resisted this logical interpretation and reworked Schrödinger's wave equation into a form resembling a fundamental dynamical equation in classical physics that is actually a statement of Newton's second law of motion, and which therefore is much more closely associated with a particle interpretation (details are provided in Appendix 26). Bohm simply assumed that the wave function of the field can be written in a form containing real amplitude and phase functions. In itself, the assumption of a specific form for the wave function represents no radical departure from conventional quantum theory. However, following de Broglie, Bohm now assumed the existence of a real particle, following a real trajectory through space, its motion embedded in the field and tied to or 'guided' by the phase function by the imposition of a 'guidance condition'. Every particle in every field therefore possesses a precisely defined position *and* a momentum, and follows trajectories determined by their respective phase functions. The equation of motion is then found to depend not only on the classical potential energy (usually denoted V) but also on a second, so-called *quantum potential* (denoted U).

The quantum potential is intrinsically non-classical and is alone responsible for the introduction of quantum effects in an otherwise classical description. Take out the quantum potential or allow it to tend to zero, and the equations of the de Broglie–Bohm version of quantum theory revert to the classical equations of Newtonian mechanics (in the form of something called the Hamilton–Jacobi formalism). True to its nature, the quantum potential has some peculiar properties. Obviously, it is separate and distinct from the classical potential V, and therefore can exert effects on the particle in regions of space where V disappears.[3] But the wave function is a solution of the Schrödinger wave equation that is dependent on V, and, as the quantum potential U is determined by the wave function, there is therefore a connection between both types of potentials. This connection is a subtle

---

[3] The notion of influences being carried at the quantum level in places where a classical potential does not exist was to provide the foundation for another of Bohm's discoveries, made in 1959 with his student Yakir Aharanov whilst at Bristol University. This is the Aharanov–Bohm effect—a physical influence on the phase of an electron passing close to but completely screened from a region of high electromagnetic field strength. Although it can be demonstrated that the electromagnetic field is zero in the region of the electron, the electron nevertheless responds to it.

one. Bohm found that U depends only on the *mathematical form* of the wave function, not its amplitude. This means that the effect of U can be large, even in regions of space where V is negligible, and the amplitude of the wave function is small. This contrasts markedly with the effects exerted by classical potentials (such as a Newtonian gravitational potential), which tend to fall off with distance. A particle moving in a region of space in which no classical potential is present can still be influenced therefore by the quantum potential, and some of the cherished notions of classical physics—such as straight-line motion in the absence of a (classical) force—must be abandoned. In the de Broglie–Bohm theory, the classical potential is able to exert some decidedly non-classical influences, *mediated* by the wave function and the quantum potential. This must be so, as the quantum potential has to be capable of accounting for the effects of diffraction and interference.

The particle position and its trajectory are 'defined' at all times during its motion, and it is therefore not necessary in principle to resort to probabilities. When we consider a large number of particles (an ensemble) all describable in terms of the same wave function, the above reasoning can still be applied; that is, there is nothing in principle preventing us from following the trajectories of each particle. However, in practice, we do not usually have access to a complete specification of all the particle initial conditions, and, just as in Boltzmann's statistical mechanics, we calculate probabilities as a *practical* necessity.

This contrasts strongly with the notion of quantum probability. In conventional quantum theory, the wave function or state vector is really a calculation tool for probabilities interpreted as the relative frequencies of possible outcomes of repeated measurements on an ensemble of identically prepared systems. These outcomes are not determined until a measurement is made. In the de Broglie–Bohm theory, the particle motions are predetermined, and we calculate probabilities because we are ignorant of the initial conditions of all the particles in the ensemble. These probabilities refer to individual states of individual particles—their positions and their trajectories—not measurement outcomes. Measurement therefore has no 'magical' role: the measurements merely tell us the actual states or positions of the particles or their actual trajectories through an apparatus, which are determined all along.

The probability is still related to the amplitude of the wave function, but this does not mean that the wave function has only a statistical significance. On the contrary, it is assumed that the wave function has a strong physical significance—it also determines the shape of the quantum potential.

A causal explanation of quantum phenomena

In the early 1990s, two detailed accounts of the de Broglie–Bohm theory appeared in print: *The undivided universe*, by David Bohm and Basil Hiley, and the *Quantum theory of motion*, by Peter Holland. Holland argued cogently that the theory represents a return to a picture of the world much more in tune with the basic instincts of most physicists. He wrote[4]:

This is the unspoken contradiction at the heart of quantum physics: physicists do want to find out 'how nature is' and feel they are doing this with quantum mechanics, yet the official view which

---

[4] Holland, Peter R. (1993). *The quantum theory of motion.* Cambridge University Press, Cambridge, pp. 9, 17.

most workers claim to follow rules out the attempt as meaningless! . . . [The de Broglie–Bohm theory showed that one] *could* analyse the causes of individual atomic events in terms of an intuitively clear and precisely definable conceptual model which ascribed reality to processes independently of acts of observation, *and* reproduce all the empirical predictions of quantum mechanics . . . It is thus very much a 'physicists' theory' and indeed puts on a consistent footing the way in which many scientists instinctively think about the world anyway.

But the notions of indeterminism, quantum uncertainty, and probability are so deeply embedded in the conventional interpretation that it seems hard to understand how these can be overturned through some simple mathematical tricks—adopting a specific form for the wave function and tying the particle to the wave function through the guidance condition. And yet, this is in fact all it takes.

The resurgence of interest in the pilot-wave approach created by Bohm's 1952 papers was gradual, but to a certain extent, it was also irresistible. This has led to a substantial elaboration of the de Broglie–Bohm theory not only in terms of its mathematical formalism, its concepts, and their properties, but also in terms of its use as a tool for providing an entirely causal explanation of quantum phenomena. The interpretation of Heisenberg's uncertainty principle, for example, reverts to a 'disturbance' picture. The particle position and momentum are in principle precisely defined, but measurement interactions have significant impacts on the wave function (and hence the quantum potential), and their non-commutation is a manifestation of the mutual exclusivity of the different types of measurement required. The measuring device remains central in so far as the whole device (including components that may be placed far from detectors) determines both the classical and quantum potentials, and it is impossible to make measurements on a quantum system without disturbing it in an essential way. There is no conflict here with Bohr's contention that the measuring device has a fundamental role which cannot be ignored. In the de Broglie–Bohm theory, changing the measuring device (which might amount to no more than changing the orientation of a polarizing filter) changes the wave function and hence the quantum potential: all future trajectories of quantum particles passing through the apparatus are thus determined by the new experimental arrangement. The Bohr of pre-1935 would have been satisfied with this account and its implications for his notion of complementarity.

Seeing is believing, it is said. It is one thing to describe how quantum phenomena can be given a causal explanation, but it is quite another to show it, even in pictures. Bohm's 1952 papers rekindled some interest and eventually led de Broglie in the mid-1950s to draw together some like-minded physicists to explore the theory's implications. Writing in the foreword to Bohm's 1957 book *Causality and chance in modern physics*, de Broglie suggested that: 'One can, it seems to me, hope that these efforts will be fruitful and will help to rescue quantum physics from the cul-de-sac where it is at the moment.' But it was the publication, in 1979, of pictures of the quantum potential corresponding to electrons passing through a two-slit apparatus, together with their trajectories, that encouraged more physicists to sit up and take note.

In the context of the de Broglie–Bohm theory, what happens in such a situation is straightforward. Electrons pass, one at a time if the beam intensity is low enough, through *one slit or the other*. Each electron is accompanied by its own wave, described in terms of a wave function that is a solution of the Schrödinger wave equation for this particular experimental arrangement. It is the *wave* that passes through both slits, recombining to produce a resultant wave with peaks corresponding to constructive interference and troughs corresponding to destructive interference. The resulting quantum potential was calculated for a specific set of experimental parameters by C. Philippidis, C. Dewdney, and Basil Hiley, and is reproduced in Fig. 11.1(a) and (b).

Fig. 11.1(a) shows the quantum potential as viewed from the screen looking back towards the slits, and Fig. 11.1(b) shows the same picture viewed from in front of the slits looking towards the screen. The potential peaks in the regions of the slits and evolves within a short distance into a complex series of oscillations, building to a large, central peak. Further from the slits, the potential decays into a more uniform structure consisting of alternating plateaus and troughs. A single electron passing through one of the slits is initially 'diffracted' by the shape of the quantum potential lying immediately beyond the slit, but its subsequent motion is affected by the large peaks lying between the slits. The electron is constrained to move on the same 'side' of the two-slit apparatus as the slit it initially passed through and is pushed along into the series of plateaus and troughs. It might be imagined that the electron eventually becomes trapped in one of the troughs in the quantum potential, but this is not the case. The troughs exert a relatively strong force on the electron but are not of sufficient magnitude (or 'depth') in relation to the electron's kinetic energy to trap it. In fact, the effect of the troughs is to cause the electron to zigzag as it crosses from plateau to trough to plateau. At any one instant, the electron spends more time crossing the plateaus. Positioning the screen a certain fixed distance beyond the plate with the two slits effectively captures a 'snapshot' of this dynamical motion. The plateaus become bright fringes, and the troughs become dark fringes. The actual trajectories computed by running electrons with certain initial conditions over the quantum potential are shown in Fig. 11.2. This figure shows that the individual trajectories diverge immediately beyond each slit, but they do not cross each other. The effect of the large peaks in the potential lying between the two slits is to prevent trajectories from crossing into the other 'side' of the apparatus. The zigzag motion reflecting the traversal of the troughs is also clearly seen. A single electron will follow *one* of the trajectories depicted in Fig. 11.2. It passes through one or other of the two slits, follows one of the predetermined trajectories, and is detected as a bright spot on the screen—refer back to Fig. 2.1(a). As more electrons pass through the apparatus, the variation in their initial conditions means that they individually pass through different slits and follow different trajectories. The end result is a pattern of spots on the screen, which reflects the clumping of the various trajectories that the electrons followed through the apparatus—see Fig. 2.1(d).

The quantum potential is the medium through which influences on distant parts of a correlated quantum system are transmitted. The measurement of some property (such as vertical polarization or spin orientation) of one of a pair of correlated particles changes the quantum potential in a non-local manner, so that the other particle takes on the required properties without the need for a collapse of the wave function. In an analysis of the de Broglie–Bohm theoretical description of the EPR experiment with correlated electron spins, particle entanglement is reflected in the structure of the quantum potential. Measurement of the spin of one particle causes the quantum potential to exert a 'torque' on the spin of the other, such that the spins of both particles are correlated. To preserve the relationship between de Broglie–Bohm theory and the postulates of special relativity, it is again necessary to note that such apparently instantaneous transmission of 'guidance information' cannot be exploited to send information of a more practical and useful sort. Although such influences might be transmitted at speeds faster than that of light, they

(a)

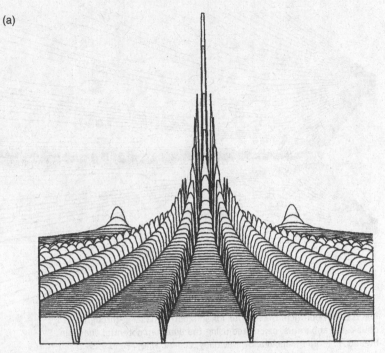

(b)

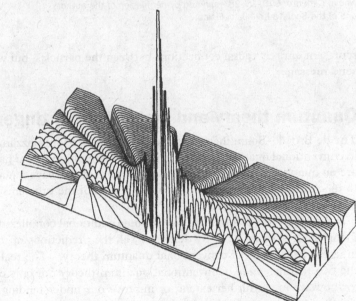

Fig. 11.1 Theoretical calculation of the shape of the quantum potential for an electron passing through a double-slit apparatus. (a) Potential as seen from the screen looking back towards the slits. (b) Potential as seen from above and in front of the slits. From Philippidis, C., Dewdney, C., and Hiley, B. (1979). *Nuovo Cimento*, **52B**, 15–28, reprinted by permission of the authors and of the Società Italiana di Fisica.

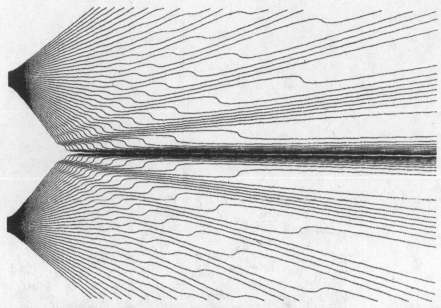

Fig. 11.2 Theoretical trajectories for an electron passing through a double-slit apparatus, calculated using the quantum potential shown in Fig. 11.1. From Philippidis, C., Dewdney, C., and Hiley, B. (1979). *Nuovo Cimento*, **52B**, 15–28, reprinted by permission of the authors and of the Società Italiana di Fisica.

represent entirely causal connections between the particles, but we cannot exploit this to send messages.

Quantum theory and historical contingency

The de Broglie–Bohm theory resolves many of the more puzzling conundrums inherent in conventional quantum theory. The assumption of waves *and* particles means that there are no quantum jumps, and there is no collapse of the wave function. Probability reverts to its classical interpretation. These resolutions of the problems of quantum theory are bought at a price, however.

It can be argued that for the price of some additional complexity, the de Broglie–Bohm theory adds nothing new. By definition, all the predictions of the theory should be no different from those of conventional quantum theory.[5] The real difference lies in what the two theories *mean*. Conventional quantum theory not only offers us no suggestions for how we might further extend or improve our understanding of the quantum world but denies that such understanding is possible in principle. In contrast, de Broglie–Bohm theory offers us an interpretation of phenomena at the quantum level, which draws heavily on our classical preconceptions, suitably conditioned by the properties of the quantum

---

[5] There are some potential exceptions, however. For example, de Broglie–Bohm theory may have something different to say about the nature of time at the quantum level, specifically the time taken for quantum particles to 'tunnel' through a classical potential barrier. See Holland, Peter R. (1993). *The quantum theory of motion*. Cambridge University Press, Cambridge, pp. 211–215.

potential. It offers us trajectories that are 'really there', though we can not see them, and therefore offers us the prospect of extending and improving our understanding through exercising our ingenuity: finding new ways to probe nature that might one day reveal the presence of something that is currently hidden.

There are many practical objections to the de Broglie–Bohm theory. For example, the wave in de Broglie–Bohm theory can exert a strong influence on the particle through the form of the quantum potential, but there is no reciprocal reaction of the particle on the wave, seemingly at odds with classical mechanics in the form of Newton's third law of motion. However, the theory was originally developed simply to demonstrate that a causal (not necessarily entirely classical) interpretation of quantum phenomena is possible in principle. De Broglie–Bohm theory is non-relativistic, some serious problems arise if we attempt to assign properties other than position to the 'Bohm particle', such as charge, and the theory is not without its own measurement problems. But solutions to these problems at least appear accessible compared with the apparent inaccessibility of solutions to the philosophical conundrums associated with conventional quantum theory.

The de Broglie–Bohm theory is also criticized for being non-local and counter-intuitive, which has always seemed to me to be an odd objection. It might be pointed out that the purpose of the de Broglie–Bohm theory was not somehow to restore an entirely classical description to the quantum domain. The experiments described in Chapters 9 and 10 demonstrate reasonably unambiguously that the quantum world is inherently non-local, and we are therefore required to attempt to make some sense of the 'spooky' action at a distance that this implies. Physicists who object to the de Broglie–Bohm theory because it is somehow not sufficiently classical are really saying that they believe that conventional quantum theory is the only *quantum* description and, as such, is a complete description. It would seem that any attempt to restore aspects of a classical interpretation must be all or nothing (meaning that it must be local)—in their view, there can be no 'halfway house'. The de Broglie–Bohm theory is just such a halfway house: it is the consequence of accepting the basic non-locality of the quantum world whilst also accepting that the conventional theoretical description of this world is incomplete.

Today, the de Broglie–Bohm theory retains a small, dedicated following within the community of concerned physicists and philosophers but remains firmly outside the mainstream of quantum physics and features in few textbooks on the subject. It is in all respects equivalent to conventional quantum theory yet allows a profoundly different interpretation of events occurring at the quantum level, one which is much more closely aligned with the physicists' intuitive perceptions of the way in which the world works. Modern textbooks are filled with a theoretical account of quantum phenomena that is, quite frankly, baffling to most students until they learn to get used to it. Why is it this version of events that appears rather than an account which retains the basic notions of causality and determinism? It can, and has, been argued that the reason is a simple one of historical contingency.[6] This is a potentially disturbing argument for anyone with an idealistic view of how science progresses. Disturbing because the choice between equivalent, competing rivals for one of the most important fundamental

---

[6] See Cushing, James T. (1994). *Quantum mechanics: historical contingency and the Copenhagen hegemony.* University of Chicago Press, Chicago.

physical theories might have been driven simply by the *order in which things happened* rather than more compelling arguments based on notions of truth or explanatory power.

The implicate order

The quantum potential effectively interconnects every region of space into an inseparable whole. This aspect of 'wholeness' is central to the de Broglie–Bohm theory, as indeed it is to the Copenhagen interpretation. Our day-to-day use of the quantum theory depends on our ability to factorize the wave function into more manageable parts (e.g. in an approximation routinely applied in chemical spectroscopy, a molecular wave function is factorized into separate electronic, vibrational, rotational, and translational parts). Under some circumstances, the wave function, and hence the quantum potential, can be factorized into a discrete set of subunits of the whole. However, when we come to deal with experiments on pairs of correlated quantum particles, we should not be surprised if the wave function cannot be factorized in this way. The non-local connections between distant parts of a quantum system are determined by the wave function of the whole system. In one sense, the de Broglie–Bohm theory takes a 'top-down' approach: the whole has much greater significance than the sum of its parts and, indeed, determines the behaviour and properties of the parts. Contrast this with the 'bottom-up' approach of classical physics, in which the behaviour and properties of the parts is assumed to determine the behaviour of the whole.

During the 1960s and 1970s, Bohm delved more deeply into the whole question of *order* in the universe. He developed a new approach to understanding the quantum world and its relationship with the classical world which contains, and yet transcends, Bohr's notion of complementarity. Bohm has described one early influence as follows[7]:

I saw a programme on BBC television showing a device in which an ink drop was spread out through a cylinder of glycerine and then brought back together again, to be reconstituted essentially as it was before. This immediately struck me as very relevant to the question of order, since, when the ink drop was spread out, it still had a 'hidden' (i.e. non-manifest) order that was revealed when it was reconstituted. On the other hand, in our usual language, we would say that the ink was in a state of 'disorder' when it was diffused through the glycerine. This led me to see that new notions of order must be involved here.

Bohm reasoned that the order (the localized ink drop) becomes *enfolded* as it is diffused through the glycerine. However, the information content of the system is not lost as a result of this enfoldment: the order simply becomes an implicit or *implicate* order. The ink drop is reconstituted in a process of *unfoldment*, in which the implicate order becomes, once again, an explicate order that we can readily perceive.

Bohm went further in his book *Wholeness and the implicate order*, first published in 1980. He recognized that the equations of quantum theory describe a similar enfoldment and unfoldment of the wave function. We understand and interpret classical physics in terms of the behaviour of material particles moving through space. The order of the classical world is therefore enfolded and unfolded through this fundamental motion. In Bohm's

[7] Bohm D., in Hiley, B. J. and Peat, F. D. (eds.) (1987). *Quantum implications*. Routledge & Kegan Paul, London, p. 40.

quantum world, the acts of enfoldment and unfoldment are themselves fundamental. Thus, all the features of the physical world which we can perceive and which we can subject to experiment (the explicate order) are realizations of potentialities contained in the implicate order. The implicate order not only contains these potentialities but also determines which will be realized. He wrote:

the implicate order provided an image, a kind of metaphor, for intuitively understanding the implication of wholeness which is the most important new feature of the quantum theory.

With one very important exception, the implicate order represents a kind of ultimate hidden variable—a deeper reality which is revealed to us through the unfoldment of the wave function.

Bohm extended and adapted his original hidden variable theory, guided by the holistic approach afforded by his theory of the implicate order. By modifying the equations of quantum-field theory, he has done away with the need to invoke the existence of independent, objectively real particles. Instead, particle-like behaviour results from the convergence of waves at particular points in space. The waves repeatedly spread out and reconverge, producing 'average' particle-like properties, corresponding to the constant enfoldment and unfoldment of the wave function. This 'breathing' motion is governed by a super quantum potential, related to the wave function of the whole universe:

We have a universal process of constant creation and annihilation, determined through the super quantum potential so as to give rise to a world of form and structure in which all manifest features are only relatively constant, recurrent and stable aspects of this whole.

Pure metaphysics? Certainly. But Bohm has done nothing more than adopt a particular philosophical position in deriving his own cosmology. As we have seen, analysis of the Copenhagen interpretation reveals that it too is really nothing more than a different philosophical position. The difference between these two is that the philosophy of the Copenhagen school is made 'scientific' through the use of the (entirely arbitrary) postulates of quantum theory. Bohm has argued that the reason orthodox quantum theory is derived from these postulates rather than postulates based on an implicate order or similar construction is merely a matter of historical precedent.

Popper's propensities

Karl Popper was one of the twentieth century's most influential philosophers of science. His position on quantum theory is easily summarized: he was a realist. While not agreeing in total with all the ideas advanced by Einstein and Schrödinger, it is clear from his writings that he stood in direct opposition to the Copenhagen interpretation, and in particular to the positivism of the young Heisenberg. Although Popper interacted with various members of the Vienna Circle (particularly Rudolph Carnap), he did not share the Circle's philosophical outlook. Inspired instead by the Polish philosopher Alfred Tarski, Popper was motivated by a desire to search for *objective* truth, a motivation that he held in common with Carnap, although their methods differed considerably.

The last century witnessed an important debate between philosophers and scientists concerning the nature of probability. In 1959, Popper published details of his own *propensity* interpretation of probability which, he argued, has profound implications for quantum probabilities. This interpretation is best illustrated by reference to a simple example, and we will use here an example used extensively by Popper himself.

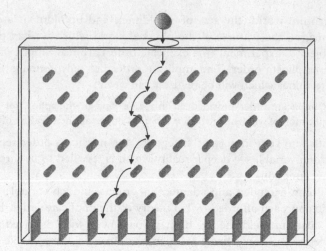

Fig. 11.3 Popper's pinboard. The marble passes through the pinboard in a sequence of left or right jumps. The propensity for the marble to exit the grid at a particular point is determined by the properties of the marble *and* the grid as a whole.

The grid shown in Fig. 11.3 represents an array of metal pins embedded in a wooden board. The grid is enclosed in a box with a transparent front, so that we can watch what happens when a small marble, selected so that it just fits between any two adjacent pins, is dropped into the grid at its centre, as shown. On striking a pin, the marble may move either to the left or to the right. The path followed by the marble is then determined by the sequence of random left versus right jumps as it hits successive pins. We measure the position at the bottom of the grid at which the marble comes to rest.

Repeated measurements made with one marble (or with a 'beam' of identical marbles) allow us to determine the frequencies with which the individual marbles exit at specific places on the grid. These we can turn into statistical probabilities in the usual way. If the marble(s) always enter the grid at the same point, and if the pins are identical, we would expect a uniform distribution of probabilities, with a maximum around the centre and thinning out towards the extreme left and right. The shape of the distribution simply reflects the fact that the probability of a sequence of jumps in which there are about as many left jumps as there are right is greater than the probability of obtaining a sequence in which the marble jumps predominantly to the left or right.

Popper has argued that each probability is determined by the *propensity* of the system as a whole to produce a specific result. This propensity is a property of the marble *and* the 'apparatus' (the pinboard). Change the apparatus, perhaps by removing one of the pins, and the propensities of the system—and hence the probabilities of obtaining specific results—change instantaneously, even though the paths of individual marbles may not take them anywhere near the region of the missing pin.

According to Popper, reality is composed of particles only. The wave function of quantum theory is a purely statistical function, representing the propensities of the particles to produce particular results for a particular experimental arrangement. Change the arrangement (by changing the orientation of a polarizing filter or by closing a slit), and the propensities of the system—and hence the probability distribution or wave function—changes instantaneously. For Popper, the Heisenberg uncertainty relations are merely relations representing the scattering of objectively real particles.

Popper's interpretation does have an intuitive appeal. The collapse of the wave function represents not a physical change in the quantum system but rather a change in the state of our knowledge of it. Going back to the pinboard, before the 'measurement', the marble can exit from the grid at any position, and the probabilities for each are determined by the propensities inherent in the system. During the measurement, the marble is observed to exit from one position only. Of course, the probabilities themselves have not changed, as is readily shown by repeating the measurement with another marble, but the *system* has changed. We can define a new set of probabilities for the new system: the probability for the marble to be found at its observed point of exit being unity and all others being zero.

This last point can be made clear with the aid of another example drawn from the quantum world. Imagine a photon impinging on a half-silvered mirror. Suppose that the probability that the photon is transmitted through the mirror is equal to the probability that it is reflected and, for simplicity, we set these equal to $\frac{1}{2}$. These probabilities are related to the propensities for the system (photon plus mirror and detectors). We would say that the probability of detecting a transmitted photon a relative to the system before the measurement b, given by $p(a, b)$, is equal to $\frac{1}{2}$. Similarly, the probability of detecting a reflected photon, 'not a' or $-a$, relative to the system b, $p(-a, b)$ is also equal to $\frac{1}{2}$. Now, suppose that we detect a transmitted photon. According to Popper, the system (photon plus mirror and detectors) has now changed completely, and it is necessary to define two new probabilities relative to the new system. Because a transmitted photon has been detected, these probabilities are $p(a, a) = 1$ and $p(-a, a) = 0$. The original probabilities $p(a, b)$ and $p(-a, b)$ have *not* changed, since they refer to the system *before* the measurement was made. These probabilities apply whenever the experiment is repeated.

It is only in what Popper calls the 'great quantum muddle' that $p(a, a)$ is identified with $p(a, b)$ and $p(-a, a)$ with $p(-a, b)$ and the process referred to as the collapse of the wave function. He writes[8]:

No *action* is exerted upon the [wave function], neither an action at a distance nor any other action. For $p(a, b)$ is the propensity of the state of the photon relative to the original experimental conditions...the reduction of the [wave function] clearly has nothing to do with quantum theory: it is a trivial feature of probability theory that, whatever a may be, $p(a, a) = 1$ and (in general) $p(-a, a) = 0$.

Popper's common-sense approach is very seductive, the argument is compelling, and it all seems very straightforward. However, the propensity interpretation runs into some difficulties when we attempt to use it to explain the wave-like behaviour of quantum entities. In particular, the only way to explain wave interference effects is to suggest, as Popper does, that the propensities themselves are real and can somehow interfere. Popper concludes that this interference is evidence that the propensities are physically real rather than simply mathematical devices used to relate the experimental arrangement to a set of probabilities. He thus writes of particles and their associated 'propensity waves' or 'propensity fields'. In a lecture to the World Congress of Philosophy in 1988, he stated[9]:

Propensities, like Newtonian attractive forces, are invisible and, like them, they can act: they are actual, they are real. We are therefore compelled to attribute a kind of reality to mere possibilities, especially to weighted possibilities, and especially to those that are as yet unrealized and whose fate

[8] Popper, K. R. (1982). *Quantum theory and the schism in physics*. Unwin Hyman, London, p. 77.
[9] Popper, K. R. (1990). *A world of propensities*. Thoemmes, Bristol, p. 18.

will only be decided in the course of time, and perhaps only in the distant future . . . The world is no longer a causal machine—it can now be seen as a world of propensities, as an unfolding process of realizing possibilities and of unfolding new possibilities.

This is clearly taking us back towards de Broglie's pilot-wave idea, and indeed, Popper has noted that: 'As to the pilot waves of de Broglie, they can, I suggest, be best interpreted as waves of propensities'.

As noted above, the pilot-wave theory is a non-local hidden-variable theory. Popper's outlook was decidedly realist and classical, and he was initially reluctant to accept the non-locality implied in his propensity interpretation, agreeing with Einstein that the idea of superluminal influences passing between two distant correlated quantum particles 'has nothing to recommend it'. However, Popper's views changed as the experimental results became increasingly difficult to explain in terms of any locally realistic theory. If it is accepted that there can be non-local, superluminal influences transmitted via the propensity field, then Popper's emphasis on the need for a holistic approach suggests that there is little to choose between Popper's propensities and Bohm's concept of an implicate order.

12

An irreversible act

If we are not persuaded of the approach to recovering causality in quantum physics through the de Broglie–Bohm theory (perhaps because, like Einstein, we think this route 'too cheap') or Popper's reinterpretation of the meaning of probability, we must seek alternative solutions to the problems of quantum probability and the collapse of the wave function. Even if we choose to accept what quantum probability tells us concerning the fundamental indeterminism and acausality of the quantum world, we are still left to trouble over quantum measurement.

This is, perhaps, the greatest source of discomfort that scientists experience with the Copenhagen interpretation of quantum theory. Given some initial set of conditions, the equations of quantum theory describe the future time evolution of a wave function or state vector in a way that is quite deterministic. The state vector moves through Hilbert space in a manner completely analogous to a classical wave moving through Euclidean space. If we are able to calculate a map of the amplitude of the state vector in Hilbert space, we can use quantum theory to tell us what this map should look like at some later time.

However, when we come to consider a measurement, the Copenhagen interpretation requires us to set aside these elegant deterministic equations and reach for a completely different tool. These equations do not allow us to compute the probabilities for the state vector to be projected into one of a set of measurement eigenstates—this must be done in a separate step. The measurement eigenstates are determined at the whim of the observer, but which result will be obtained with any one quantum particle is quite indeterminate. And we learn from Schrödinger's cat that quantum theory has nothing whatsoever to say about where in the measurement process this projection or collapse of the wave function is supposed to take place.

It is true that most scientists are primarily concerned about the deterministic part of quantum theory in that they are interested in using it to picture how fundamental particles, atoms, or molecules behave in the absence of an interfering observer. For example, molecular quantum theory can provide striking pictures of molecular electronic orbitals which we can use to understand chemical structure, bonding, and spectroscopy. Little thought is given to what these pictures might mean in the context of a measurement—it is enough for us to use them to imagine how molecules *are*, independently of ourselves and our instruments. But our information is derived from measurements. It is derived from processes in which the deterministic equations of motion do not apply. It is derived from processes that present us with profound conceptual difficulties. The search for solutions to the quantum measurement problem has produced some spectacularly bizarre suggestions. We will consider some of these here and in the next two chapters.

The arrow of time

It is our general experience that, apart from in a few science-fiction novels, time flows only one way: forwards. Why? The equations of both classical and quantum mechanics appear quite indifferent to the direction in which time flows. With the possible exception of the collapse of the wave function (which we will discuss at length below), replacing t by $-t$ in the equations of classical or quantum mechanics makes no difference to the validity or applicability of the equations. When we abandon the idea of an absolute time, as special relativity demands, the equations do not even recognize a 'now' distinguishable from the past or future. But our perceptions are quite different: the flow of time is an extremely important part of our conscious experience.

Imagine a collision between two atoms (Fig. 12.1). The atoms come together, collide with each other, and move apart in different directions with different velocities. Run this picture backwards in time, and we see nothing out of the ordinary: the atoms come together, collide, and move apart. The physics of the exact time-reverse of the collision is no different in principle from the physics viewed in forward time.

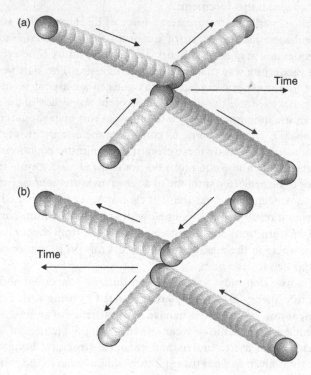

Fig. 12.1 (a) Collision between two atoms (pictured here as rigid spheres) seen in forward time. (b) Time reversal of (a), in which the momenta of the atoms are exactly reversed. The collision in (b) looks no more unusual than that in (a).

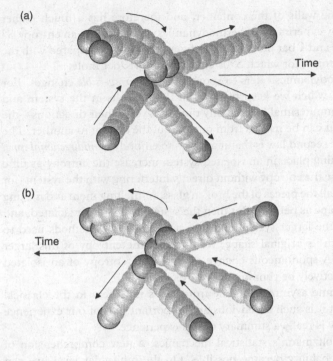

Fig. 12.2 (a) Collision between an atom and a diatomic molecule in which the molecule is dissociated into two atoms, seen in forward time. The apparent curvature in the trajectory of the diatomic molecule is intended to convey the impression that the molecule is rotating. (b) Time reversal of (a), in which the three atoms come together, and a diatomic molecule forms. Now, the time-reversed collision looks 'odd' in the sense that it seems a most unlikely event.

Now imagine a collision between an atom and a diatomic molecule (Fig. 12.2). This time, we suppose that the collision is so violent that it smashes the molecule into two atomic fragments. All three atoms move apart in directions and with velocities that are themselves determined by the initial conditions. Again, the equations are indifferent to this collision run in reverse: bring together the three atoms with exactly the opposite momenta, and the molecule will re-form. However, we now sense that this process looks 'wrong' when run in reverse, or at least looks very unlikely. In reality, such processes do occur but in separate steps: two atoms may come together to form a diatomic molecule, often with some excess energy which makes it unstable and therefore short-lived. The molecule will break apart again unless it collides with another atom, molecule, or the walls of its container, which serves to remove the excess energy and thereby stabilize it.

The picture involving a simultaneous three-body collision looks wrong because it is our general experience that a system does not spontaneously transform from a more complicated to a less complicated state (a broken glass spontaneously reassembling itself, for example). The important difference between the time-reversed collisions shown in Figs 12.1 and 12.2 is that in Fig. 12.2, the number of degrees of freedom is larger—we need more position and velocity coordinates to describe what is going on. This tendency for the physical world always to transform (or disperse) into something more complicated is embodied in the second law of thermodynamics, which can be stated as follows:

For a spontaneous change, the entropy of an isolated system always increases.

Students of science are usually taught to understand entropy as a measure of the 'disorder' in a system. Thus, a crystal lattice has a very ordered structure, with its constituent atoms or molecules arranged in a regular array, and it therefore has a low entropy. However, a gas consists of atoms or molecules moving randomly through space, colliding with each other and with the walls of the container, and therefore has a much higher entropy. This is confirmed by experimental thermodynamics: diamond has an entropy $S°$ of 2.4 J/K per mole at 298 K and 1 bar pressure. This figure should be compared with the entropy of gaseous carbon atoms, for which $S°$ is equal to 158 J/K per mole.

The second law of thermodynamics refers to spontaneous or *irreversible* changes. For a reversible change—one in which we keep track of all the motions in the system and can apply (in principle) an infinitesimal force at any time to reverse their directions—the entropy does not increase but can be moved from one part of the system to another. The most important aspect of the second law is that it appears to embody a *unidirectional time*. All spontaneous changes taking place in an isolated system increase the entropy as time increases. We cannot decrease the entropy without directly interfering with the system (for example, collecting together all the pieces of the broken glass, re-melting them and making a new glass with the same shape as before). But then the system is no longer isolated, and when we come to consider the larger system and take account of the methods used to re-melt the glass and re-form its original shape, we find that the entropy of this larger system will have increased. A spontaneous change in which the entropy of an isolated system decreased would effectively be running backwards in time.

Exactly where does this time asymmetry come from? It is not there in the classical equations of motion, and yet it is such an obvious and important part of our experience of the world. The second law is really a summary of this experience.

With the emergence of Boltzmann's statistical mechanics, a new comprehension of entropy as a measure of probability became possible. On the molecular level, we can now understand the second law in terms of the spontaneous transition of a system from a less probable to a more probable state. A gas expands into a vacuum and evolves in time towards the most probable state in which the density of its constituent atoms or molecules is uniform. We call this most probable state the equilibrium state of the gas. However, this talk of probabilities introduces a rather interesting possibility. A spontaneous transition from a more probable to a less probable state (decreasing entropy) is not disallowed by statistical mechanics—it merely has a very low probability of occurring. Thus, the spontaneous aggregation of all the air molecules into one corner of a room is not impossible, just very improbable. The theory seems to suggest that if we wait long enough (admittedly, much longer than the present age of the universe), such improbable spontaneous entropy-reducing changes will eventually occur. Some scientists (including Einstein) concluded from this that irreversible change is an illusion: an apparently irreversible process will be reversed if we have the patience to wait. We will return to this argument below.

If spontaneous change must always be associated with increasing disorder, how do we explain the highly ordered structures (such as galaxies and living things) that have evolved in the universe? Some answers are being supplied by the theory of *chaos*, which describes how amazingly ordered structures can be formed in systems far from equilibrium.

Time asymmetry and quantum measurement

What does all this have to do with quantum measurement? To see how arguments about spontaneous changes and the second law fit into the picture, it is necessary to delve a little into quantum statistical mechanics. It is neither desirable nor necessary for us to go too deeply into this subject in this book. Instead, we will draw on some useful concepts, basic observations, and some ideas that have been presented in greater detail elsewhere (see the bibliography for some excellent references on this subject).

Quantum statistical mechanics is essentially a statistical theory concerned with ensembles of quantum particles. Consider an ensemble of N quantum particles all present in a quantum state denoted by the state vector $|\psi\rangle$. Such an ensemble is said to be in a *pure state*. The state vector of each particle in the ensemble can be expressed as a superposition of the eigenstates of the operator corresponding to some measuring device. Suppose there are n of these eigenstates; $|\psi_1\rangle, |\psi_2\rangle, |\psi_3\rangle, \ldots, |\psi_n\rangle$. We know from our discussion in Chapter 5 that the probability for any particle in the ensemble to be projected into a particular measurement eigenstate is given by the modulus-squared of the corresponding projection amplitude (or the coefficient in the expansion). After the measurement has taken place, each particle in the ensemble will have been projected into one, and only one, of the possible measurement eigenstates. The quantum state of the ensemble is now a *mixture*, the number of particles present in a particular eigenstate being proportional to the modulus-squares of the projection amplitudes. This is just another way of looking at the problem of the collapse of the wave function. The act of quantum measurement transforms a pure state into a mixture. John von Neumann showed that this transformation is associated in quantum statistical mechanics with an increase in entropy. Thus, irreversibility or time asymmetry appears as an intrinsic feature of quantum measurement.

This is all fine as far as it goes, but the problem remains that the equations of motion derived from the time-dependent Schrödinger equation do not allow such a transformation. If a quantum system starts as a pure state, it will evolve in time as a pure state according to the equations of motion. This is because, in mathematical terms, the action of the time evolution operator in transforming a state vector at some time t into the same state vector at some later time t' is equivalent in many ways to a simple change of coordinates. Abrupt, irreversible transformation into a mixture of states appears to be possible in quantum measurement only through the collapse of the wave function.

From being to becoming

The Nobel prize-winning physical chemist Ilya Prigogine has argued that we are dealing here with two different types of physics. He identified a physics of *being*, associated with the reversible, time-symmetric equations of classical and quantum mechanics, and a physics of *becoming*, associated with irreversible, time-asymmetric processes which increase the entropy of an isolated system. He rejects the argument that irreversibility is an illusion or approximation introduced by us, the observers, on a completely reversible world. Instead, he advocates a 'new complementarity' between dynamical (time-symmetric) and thermodynamic (time-asymmetric) descriptions. This he does in an entirely formal way by defining an explicit microscopic *operator* for entropy and showing that it does not commute with the operator governing the time-symmetric dynamical evolution of a quantum system.

According to Prigogine, introducing a microscopic entropy operator has certain consequences for the equations describing the dynamics of quantum systems. Specifically, he shows that the equations now consist of two parts—a reversible, time-symmetric part equivalent to the usual description of quantum state dynamics and a new irreversible, time-asymmetric part equivalent to an 'entropy generator'. Prigogine's approach is *not* to attempt to derive the second law from the dynamics of quantum particles but to assume its validity and then seek ways to introduce it *alongside* the dynamics. In his book *From being to becoming*, published in 1980, he wrote[1]:

The classical order was: particles first, the second law later—being before becoming! It is possible that this is no longer so when we come to the level of elementary particles and that here we must *first* introduce the second law before being able to define the entities.

It is interesting to note that Prigogine's approach parallels that of Boltzmann a century earlier. Boltzmann attempted to find a molecular mechanism that would ensure that a non-equilibrium distribution of molecular velocities in a gas would evolve in time to a Maxwell (equilibrium) distribution. The result was a dynamical equation that contains both reversible and irreversible parts, the latter providing an entropy increase independently of the exact nature of the interactions between the molecules. Like Prigogine, Boltzmann could not derive this equation from classical dynamics—he just had to assume it.

Prigogine concludes his book with the observation that:

The basis of the vision of classical physics was the conviction that the future is determined by the present, and therefore a careful study of the present permits the unveiling of the future. At no time, however, was this more than a theoretical possibility.

Indeed, one of the most important lessons to be learned from the new theory of chaos is that, even in classical mechanics, our ability to predict the future behaviour of a dynamical system depends crucially on our knowing exactly its initial conditions. The smallest differences between one set of initial conditions and another can lead to very large differences in the subsequent behaviour, and it is becoming increasingly apparent that in complex systems, we simply cannot know the initial conditions precisely enough. This is not because of any technical limitation on our ability to determine the initial conditions; it is a reflection of the fact that predicting the future would require *infinitely* precise knowledge of these conditions.

Prigogine again:

Theoretical reversibility arises from the use of idealisations in classical or quantum mechanics that go beyond the possibilities of measurement performed with any finite precision. The irreversibility that we observe is a feature of theories that take proper account of the nature and limitation of observation.

In other words, it is reversibility, not irreversibility, which is an illusion: a construction we use to simplify theoretical physics and chemistry.

Decoherence

The Copenhagen interpretation of quantum theory leaves unanswered the question of just where the collapse of the wave function is meant to take place. John Bell wrote of the 'shifty split' between measured object and perceiving subject[2]:

What exactly qualifies some physical systems to play the role of 'measurer'? Was the wave function of the world waiting to jump for thousands of years until a single-celled living creature appeared? Or did it have to wait a little longer, for some better qualified system... with a Ph.D.?

---

[1] Prigogine, Ilya (1980). *From being to becoming*. W. H. Freeman, San Francisco, CA.
[2] Bell, J. S. (1990). *Physics World*, **3**, 33.

Prigogine attempts to account for the irreversibility of quantum measurement by incorporating an 'entropy generator' into the formalism, in effect building the second law of thermodynamics directly into the structure of quantum theory. There is an alternative approach, however, which, to a certain extent, addresses Bell's 'shifty split' and makes a formal distinction between the microscopic world of individual quantum entities and the macroscopic world of measuring devices without adding anything more to the quantum formalism.

Bohr recognized the importance of the 'irreversible act' of measurement linking the macroscopic world of measuring devices and the microscopic world of quantum particles. Some years later, John Wheeler wrote about an 'irreversible act of amplification' (see Chapter 10). The truth of the matter is that we gain information about the microscopic world only when we can amplify elementary quantum events like the absorption of photons, and turn them into perceptible macroscopic signals, such as the deflection of a pointer on a scale. If we are prepared to accept the concept of a quantum superposition operating at the microscopic level, maybe we should merely note the simple fact that we never appear to see such superpositions in the macroscopic world. This would appear to be a question of scale: phenomena that are commonplace at the level of individual quanta (photons, electrons, atoms) disappear in bulk matter (pointers and cats). If we could somehow account for this transformation from microscopic to macroscopic using the existing formalism of quantum theory, it would not be necessary to add anything further to it. Indeed, perhaps this process of bridging between the microworld and the macroworld is a logical place to find the collapse of the wave function, or some alternative. Schrödinger's cat might then be spared at least the discomfort of being both dead and alive, because the act of amplification associated with registering a radioactive emission by the Geiger counter settles the matter before a superposition of macroscopic states can be generated.

Perhaps the answer lies in the realization that quantum systems that we can prepare in superposition states are relatively simple and have a limited number of degrees of freedom, whereas the apparatus used to convert and amplify information gathered as a result of some measurement is highly complex, composed of many electrons, atoms, and molecules, with many degrees of freedom. The physicist Dieter Zeh was the first to note that the interaction of a state vector with a measuring apparatus and its 'environment' can lead to rapid, irreversible decoupling or 'dephasing' of the components in a superposition in such a way that interference terms are destroyed, and we are consequently prevented from observing interference in macroscopic objects. He wrote[3]:

Each state will now produce macroscopically correlated states: different images on the retina, different events in the brain, and different reactions of the observer. The different components represent two completely decoupled worlds. This decoupling describes exactly the "reduction of the wave function." As the "other" component cannot be observed any more, it serves only to save the consistency of quantum theory.

This decoupling or dephasing of the state vector is accompanied by a strong coupling to the innumerable states of the apparatus and its environment, and is now commonly referred to as *decoherence*. Its analogue in classical physics is the dissipation of energy through frictional or damping effects, although we require decoherence to work on a

[3] Zeh, H. D. (1970). *Foundations of Physics*, **1**, 69–76.

much faster timescale than these classical phenomena. Indeed, in the expanding literature of studies of decoherence, it is being investigated as a possible fundamental basis for the laws of thermodynamics, the irreversible increase in entropy, and the arrow of time.

Viewed this way, decoherence provides a kind of 'quantum censorship', eliminating the awkward interference terms predicted by quantum theory long before they can have any embarrassing macroscopically observable consequences. For example, in the paradox of Schrödinger's cat, we suppose that a coherent quantum state vector is produced, corresponding to the linear superposition of the state vector of a decayed atom and the state vector of an intact atom. The theory of decoherence focuses on the properties of the *density matrix*, which contains all the probability information available for a specified quantum system (see Appendix 9). The density matrix for Schrödinger's cat state contains elements corresponding to the probability of observing a decayed atom, the probability of observing an intact atom, and two interference terms involving both decayed and intact atoms. This is illustrated in Fig. 12.3(a) in which elements of the density matrix appear as peaks in a three-dimensional map of probability. All four components that emerge from the superposition are equally probable, and the density matrix has four equal peaks. The

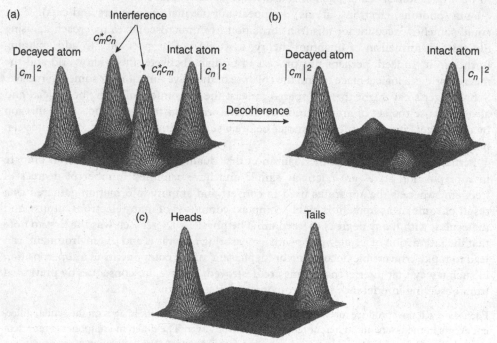

Fig. 12.3 (a) Density matrix of a quantum-mechanical system corresponding to a linear superposition of a decayed and intact radioactive atom. This map of probability shows four peaks corresponding to the (equal) probabilities of obtaining a decayed atom, an intact atom, and 'cross-terms' corresponding to interference. The elements of the matrix are also shown. (b) As a result of the decoherence effect, the interference has been suppressed, and the resulting density matrix closely approximates that of classical probability calculus, illustrated in (c) by reference to tossing a coin. Adapted with permission from Tegmark, Max and Wheeler, John Archibald (2001). *Scientific American*, **284**, p. 73.

paradox arises when we convert this density matrix into one involving the cat as, according to the time-dependent Schrödinger equation, this conversion is linear, and we end up with finite probabilities for observing rather disturbing interferences between a live cat and a dead cat. However, according to the theory of decoherence, the elements in the density matrix which correspond to the interference terms decay to zero extremely rapidly as the atom interacts with the apparatus and its environment. This is shown in Fig. 12.3(b), in which the peaks of probability corresponding to the interference terms have now been suppressed, and the density matrix has an appearance closely resembling that of classical probability calculus, illustrated in Fig. 12.3(c) using the example of tossing an (unbiased) coin.

The coherence of a suitably prepared state vector turns out to be extremely fragile—interactions with a few photons or atoms can quickly result in a loss of phase correlation and turn an overtly quantum system into something that quickly takes on the appearance of a classical system. In this case, however, the collapse of the wave function is not an instantaneous projection into a specific measurement eigenstate, as postulated in the conventional interpretation of quantum theory; it is rather a physical process that takes a finite time.

To illustrate this last point, we can consider the process of localizing a particle in space. We tend to observe that classical objects always appear to be localized and do not exhibit interference (they are *either* 'here' *or* 'there'), whereas quantum objects can be described in terms of delocalized state vectors and do exhibit interference. The transition from quantum, delocalized objects to classical, localized objects can be related to a *decoherence factor*, $e^{-t/\tau}$, where τ is the *decoherence time*. The decoherence time is related to the size of the object under study and the number of interacting particles in its environment. The smaller the decoherence time, the faster the state vector loses coherence and evolves, together with the apparatus and its environment, into a classical object. A large molecule with a radius of about 10^{-6} cm moving through the air has a decoherence time estimated to be of the order of 10^{-30} s,[4] meaning that the molecule is localized within an unimaginably short time and behaves to all intents and purposes as a classical object. Removing the air and observing the molecule in a laboratory vacuum increases the estimated decoherence time to one-hundredth of a femtosecond (10^{-17} s), which is becoming almost long enough to be imaginable. Placing the molecule in intergalactic space, where it is exposed only to interactions with the cosmic microwave background radiation, increases the estimated decoherence time to 10^{12} s, meaning that the molecule may persist in a delocalized state for a little under 32,000 years. In this environment, the molecule may show decidedly quantum effects. In contrast, a dust particle with a radius of a thousandth of a centimetre—a thousand times larger than the molecule—has a decoherence time of a microsecond (10^{-6} s) in intergalactic space. The dust particle therefore persists as a largely classical particle, even where interactions with the environment are at their lowest possible level.

Clearly, for quantum entities such as photons, electrons, and individual atoms, the decoherence times will be longer in all the different environments considered. But when large numbers of interacting particles are involved, as in the interaction of the quantum entities with a macroscopic measuring apparatus and its environment, the decoherence

[4] These estimates are taken from Omnès, Roland (1994). *The interpretation of quantum mechanics*. Princeton University Press Princeton, NJ. The original calculations were reported in Joos, E. and Zeh, H. D. (1985). *Zeitschrift für physik*, **B59**, 223–243.

time becomes extremely short, and for all practical purposes, localization can be assumed to be essentially instantaneous. An electron passing through a double-slit apparatus has a long decoherence time and so persists as a delocalized object. Its state vector passes through both slits and interferes. The resulting pattern of alternating large and small amplitudes in the state vector (the interference fringes) would continue to persist were if not for the fact that we place a piece of photographic film in the way, and this serves to couple the state vector to the large number of states of the molecules in the photographic emulsion and thence to a larger environment. The decoherence time is dramatically reduced, the electron is localized in a time so incredibly short it appears instantaneous, and we find that the electron is now either 'here' or 'there'.

The kinds of timescales over which decoherence is expected to occur for any meaningful example of a quantum system interacting with a classical measuring device suggest that it will be impossible to catch the system in the act of losing coherence. There are, however, systems of scale intermediate between microscopic and macroscopic (and which are therefore sometimes called *mesoscopic*). Such mesoscopic systems are characterized by decoherence times measured in microsecond to millisecond timescales and, therefore, offer the prospect of observing decoherence in real time. Such systems involve entangled states of trapped ions, which have been described as the laboratory equivalents of Schrödinger cat states.[5]

As we might expect, decoherence theory does beg some further questions. The interference terms in a coherent quantum system are suppressed when the number of interacting particles increases to large values. If coherence is really so fragile, how is it, then, that we can routinely observe coherent superpositions of many photons, allowing us to see interference effects at macroscopic levels (i.e. using 'ordinary' light)? The answer is that the interactions occurring in an electromagnetic field involving large numbers of photons are primarily photon–photon interactions. Such interactions do exist, but only as high-order corrections in quantum electrodynamics, and are therefore extremely weak. To a first approximation, photons do not interact with themselves at all and therefore do not represent a significant source of decoherence in an intense electromagnetic field. Coherence survives, and interference at a macroscopic level can be observed.

There is a further question. If the effect of decoherence is to eliminate certain elements of the density matrix, why is it that only elements containing the potentially embarrassing interference terms are chosen? This amounts to the existence of a 'privileged basis'—the density matrix becomes diagonal (cf. Fig. 12.3) only when written in the basis of actual (i.e. observed) measurement outcomes, such as position. Quantum theory assigns equal validity to any one of a number of different bases (including superpositions), but many of these are inappropriate choices because they are removed from our direct experience. The appropriate choices are privileged for the very reason that they accord with our (classical) experience. We make measurements and record that a particle was 'here' rather than 'there'. The positions 'here' and 'there' become the privileged basis in which to write the density matrix, and we find that decoherence prevents us from observing any interference effects in which the particle is both 'here' and 'there'. In decoherence theory, this is explained by arguing that the measurement outcomes are precisely those elements that are 'robust' to decoherence, meaning that the physical interactions involved in measurement

---

[5] For examples, see Haroche, Serge (1998). *Physics Today*. July, 36–42.

and that produce decoherence are themselves responsible for selecting a 'natural' basis. Referring back to Fig. 12.3(a), the effect of decoherence on this density matrix is literally to *filter out* the non-robust elements involving interference, leaving the robust elements corresponding to the decayed atom and intact atom. These then become the measurement outcomes.

The problem of objectification

Decoherence can potentially explain much of the dramatic difference we observe between the microscopic world of quantum entities and our macroscopic world of direct experience, in a way that is intuitively appealing. It makes connections with classical thermodynamics and, because nothing is added to the basic formalism, opens up the possibility that we might one day be able to understand the physical world using only one theoretical framework—quantum mechanics—rather than two.

To a certain extent, decoherence theory has entered the mainstream of quantum physics and provides an important and essential link between the quantum world and the classical world. There are some things, however, that it does not explain. We might be able to eliminate embarrassing macroscopic interference effects, but we are still left with no mechanism for explaining why a specific measurement should give *this* specific outcome. In the double-slit example discussed above, there is nothing in the theory that tells us why the electron should be found 'here' rather than 'there'. An electron must still be measured to be either spin-up *or* spin-down, or a photon must be measured to be either vertically *or* horizontally polarized. Applying decoherence theory to the problem eliminates any interferences between spin-up and spin-down or vertical and horizontal polarization but still appears to leave us with a description in which both results are equally valid: 'here' *and* 'there', up *and* down, or vertical *and* horizontal. The Copenhagen interpretation simply tells us that this choice is undetermined prior to measurement (all outcomes are potentially equally valid: A *and* B) and determined after measurement (only one outcome is observed: A *or* B). We are left searching for a mechanism that helps us to understand how 'and' becomes 'or'. Decoherence does not provide a mechanism, and to some, it therefore does not properly solve the measurement problem. In John Bell's words[6]:

The idea that elimination of coherence, in one way or another, implies the replacement of 'and' by 'or', is a very common one among solvers of the 'measurement problem'. It has always puzzled me.

The mathematician Roger Penrose makes similar observations[7]:

So, knowing what the density matrix is, does not help us to determine that the cat is actually either alive or dead. In other words, the cat's aliveness or deadness is not contained in the density matrix—we need more . . . What we do not have is a thing which I call **OR** standing for *Objective Reduction*. It is an objective thing—either one thing or the other happens objectively. It is a missing theory. **OR** is a nice acronym because it also stands for 'or', and that is indeed what happens, one **OR** the other.

The theoretician Roland Omnès has called this the problem of 'objectification'. In his book *Understanding quantum mechanics*, Omnès denies that the problem really exists, arguing that the property of uniqueness of a result at the end of a measurement is a matter

[6] Bell, J. S. (1990). *Physics World*, **3**, 33.
[7] Penrose, Roger (2000). *The large, the small and the human mind. Canto edition*. Cambridge University Press, Cambridge, p. 82.

for the relation of the theory to physical reality (assumed by Omnès to be an entirely empirical reality) and therefore a matter of interpretation. Objectification or the uniqueness of reality has always to be added to a theory as an assumption in order to make contact with *any* kind of reality. The fact that classical physics is deterministic does not in itself imply a reality that is unique. The probabilistic basis of quantum theory makes the gap between theory and reality much more obvious, but the gap nevertheless exists in all theoretical descriptions.

Omnès is merely shifting the burden, however. Bell's objections to the contention that decoherence solves the measurement problem are really derived from the loss of mechanism associated with our understanding (or lack thereof) of quantum measurement *and* quantum probability. The difference between Fig. 12.3(b) and 12.3(c) is that we have an intuitive mechanism to explain how a coin tossed in the air will give the result 'heads' or the result 'tails' with a 50 per cent probability for each (see Chapter 5), and we know that Fig. 12.3(c) reflects only our ignorance of the dynamics of the coin in an individual measurement. In contrast, the probabilities in Fig. 12.3(b) have quite a different interpretation. Omnès (and other advocates of decoherence theory) wants to retain quantum probability in its original Born interpretation and uses decoherence to connect the quantum world (with all its superpositions and indeterminism) to the macroscopic world where experience tells us that all the superpositions have disappeared. It should not be surprising that not everyone is satisfied with this approach.

GRW theory

Decoherence theory provides us with a physical basis for understanding the difference between the quantum and the classical worlds, but it does not remove the need for an interpretational framework, such as the Copenhagen interpretation or alternatives to be described in Chapters 13 and 14. Bell has argued that there is another way to avoid the 'shifty split' without running into the problem of objectification.

The Italian physicists G. C. Ghiradi, A. Rimini, and T. Weber (GRW) formulated just such a theory in 1986. This theory has been subsequently refined and extended by Philip Pearle and his colleagues, but we will continue to refer to it here as GRW theory. To the usual non-relativistic, time-symmetric equations of motion, GRW added a non-linear term which subjects the state vector to random, spontaneous localizations in configuration space. Their ambition was primarily to bridge the gap between the dynamics of microscopic and macroscopic systems in a unified theory. To achieve this, they introduced two new constants whose orders of magnitude were chosen so that (1) the theory does not contradict the usual quantum theory predictions for microscopic systems, (2) the dynamical behaviour of a macroscopic system can be derived from its microscopic constituents and is consistent with classical dynamics, and (3) the wave function is collapsed by the act of amplification, leading to well-defined individual macroscopic states of pointers and cats, and so on.

The first of these constants refers to the minimum physical distance between the centres of mass of the components of the superposition required to trigger localization. GRW specified a minimum distance of about 10^{-5} cm. The second of these new constants represents the frequency of spontaneous localizations of the state vector describing the superposition. GRW set this localization frequency to a value of 10^{-16} per second. This implies that the state vector is localized about once every billion years. However, this frequency is sensitively dependent on the *number* of particles involved, such that a complex system with N particles localizes at a frequency of N times 10^{-16} per second. In practical terms, the state vector of a microscopic system consisting of individual or small numbers of quantum entities *never* localizes: it continues to evolve in time according to the time-symmetric equations of motion derived from the time-dependent Schrödinger equation. With these choices for the constants, there is no practical difference between the GRW theory and orthodox quantum theory for microscopic systems. However, for a macroscopic system consisting of, say, 10^{23} particles, the localization frequency becomes 10^7 per second; that is, the state vector of one of the particles is localized within about 100 ns.

We would, perhaps, tend to think of the spontaneous GRW localizations as a necessary part of a measurement process, but localization is driven by interactions with macroscopic systems, and these do not have to be associated with measurement. However, because a measuring device is a large object like a photomultiplier (or a cat), the state vector is collapsed in the very early stages of the measurement process. Bell wrote that in the GRW extension of quantum theory: '[Schrödinger's] cat is not both dead and alive for more than a split second'.

There are obvious parallels with decoherence theory in that the behaviour at microscopic versus macroscopic levels is traced to the number of particles involved. Both theories maintain that microscopic systems never localize or never decohere and behave, therefore, pretty much as orthodox quantum theory describes. Similarly, both theories maintain that escalating the (quantum) interactions to a macroscopic level localizes or decoheres the state vector virtually instantaneously. The difference between the approaches is that the 'collapse' into a single state is taken to be real in GRW theory, whereas decoherence serves only to remove embarrassing interferences, still leaving us with equal probabilities for the 'robust' states.

The GRW theory serves only to sharpen the collapse of the wave function and make it a real and necessary part of the process of amplification. It does not solve the problems associated with quantum entanglement and the need to invoke 'spooky action at a distance' implied by the results of the experiments described in Chapters 9 and 10. The GRW theory would predict that in these experiments, the detection and amplification of any quantum entity automatically collapse the whole (spatially quite delocalized) wave function. The properties of any other, not yet detected, entities change from being possibilities into actualities at the moment this collapse takes place. Bell himself demonstrated that this action at a distance need not imply that messages must be sent between the photons and that, therefore, there is nothing in the GRW theory to contradict the demands of special relativity. In fact, although GRW originally formulated their theory as an extension of non-relativistic quantum mechanics, they have since generalized it to include the effects of special relativity and can apply it to systems containing identical particles. Bell summarized his feelings thus[8]:

I think that Schrödinger could hardly have found very compelling the GRW theory as expounded here—with the arbitrariness of the jump function, and the elusiveness of the new physical constants. But he might have seen in it a hint of something good to come.

[8] Bell, J. S. (1987) In Kilmister, C. W. (ed.) *Schrödinger: centenary celebration of a polymath*. Cambridge University Press, Cambridge, pp. 41–52.

The GRW theory attempts to resolve the problem of the collapse of the wave function by introducing a spontaneous localization of the state vector governed by physical constants that depend on the distance between components in the superposition and on the number of particles involved. The superposition of states that can give rise to interference effects is therefore a key characteristic of the microscopic world of quantum particles but is absent from our macroscopic world of experience. We therefore never see superposition states involving different pointer positions or live and dead cats.

But introducing new physical constants is always less satisfactory than having the solution to the problem emerge 'naturally' from the theory itself especially, as Bell noted, if the constants are 'elusive'.

Penrose and the geometry of space–time

The mathematical fusion of quantum theory and special relativity into quantum field theory was fraught with difficulties. The mathematics tended to produce irritating infinities which were eventually removed through the process of renormalization—a process still regarded by some physicists as rather unsatisfactory. However, these difficulties pale into insignificance compared with those encountered when attempts are made to fuse quantum field theory with general relativity. If this merging of the two most successful of physical theories could ever be accomplished, the result would be a theory of *quantum gravity*. Some progress has been made in recent years on the development of a theoretical account of quantum gravity, and we will examine this in more detail in Chapter 14.

In Einstein's general theory of relativity, the action at a distance implied by the classical (Newtonian) force of gravity is replaced by a curved space–time. The amount of curvature in a particular region of space–time is related to the density of mass and energy present (since $E = mc^2$). The mathematician Roger Penrose has suggested that a linear superposition of quantum states will begin to break down and eventually collapse into a specific eigenstate when a region of significant space–time curvature is entered. Unlike decoherence or the GRW theory, in which the number of particles is the key to the collapse, in Penrose's theory, it is the density of mass-energy which is important.

Gravitational effects are likely to be insignificant at the microscopic level of individual atoms and molecules, and so the state vector describing individual or small numbers of quantum entities is expected to evolve in the usual time-symmetric fashion according to the dynamical equations of quantum theory. Penrose suggests that it is the *difference* between gravitational fields in the space–times of different measurement possibilities which is important, and he estimates a timescale for the collapse of the order of $\hbar$ divided by the energy E required to move the matter involved in the measurement between the space–times corresponding to the different measurement outcomes. Thus, if we are dealing with a quantum superposition subjected to a conventional macroscopic measuring device or a cat, the amount of matter involved is significant, E is large, and the timescale of the collapse is consequently significantly reduced. If the 'measurement' interaction occurs only at the level of a single quantum entity, then the amount of matter involved is considerably smaller, E is very small, the timescale for collapse is considerably extended, and the interaction therefore merely serves to propagate the superposition through the larger quantum system. Penrose has estimated that a quantum superposition involving a speck of water with a radius of a thousandth of a centimetre would collapse within about a millionth of a second. A similar superposition involving a proton would, if undisturbed, take a few million years to collapse. Penrose does not deny the importance of the environment in all of this and suggests that it may be necessary to consider the differences in space–time geometries not just of the immediate objects involved in measurement but also of their environments. His approach can therefore be thought of as a form of decoherence theory in which he has made the initial collapse an explicit feature (an objective reduction) triggered by differences in space–time geometries.

There are, however, drawbacks with this approach, too. To a certain extent, it suffers from the same problem of requiring a 'privileged basis'—there is nothing in the theory that indicates what the preferred states are and therefore what should be regarded as 'unstable' superpositions to be eliminated by the collapse. Penrose believes that he can resolve this problem by identifying the preferred states as stationary solutions of what he calls the *Schrödinger–Newton equation* (the non-relativistic Schrödinger equation wave equation supplemented by a classical, Newtonian gravitational potential). The implication is that the preferred states should emerge 'naturally' as eigenfunctions of the Schrödinger–Newton equation, in much the same way that the quantum numbers emerged naturally in the eigenfunctions of the original Schrödinger equation. Penrose has further proposed an experiment which could determine the validity not only of his own approach to objective reduction but of all approaches that assume a real, physical basis for the collapse.[9] He believes that there is a 'reasonable prospect', in the 'not-too-distant future' of resolving this question experimentally. Together with William Marshall, Christoph Simon, and Dik Bouwmeester, Penrose has subsequently issued a preprint containing proposals for an experiment to create quantum superposition states consisting of some 10^{14} atoms by bouncing a single photon off a tiny mirror mounted on a mechanical oscillator and which are 'within reach of current technology'.[10]

Macroscopic realism

Of course, in the 67 years since Schrödinger first introduced the world to his cat, no one has ever reported seeing a cat in a linear superposition state (at least, not in a reputable scientific journal). Decoherence theory suggests that such interferences are suppressed in a kind of 'quantum censorship' as a result of interactions between the state vector of the microscopic system with the myriad states of the macroscopic measuring device and its environment. The GRW theory suggests that such a thing is impossible because the wave function collapses much earlier in the measurement process, triggered by spontaneous localizations induced by some kind of quantum 'noise'. Penrose believes that such a thing is impossible because the collapse is triggered by differences in the space–time geometries of the two measurement outcomes long before the cat is (or is not) killed.

What all these prescriptions do in their different ways is to draw a sharp distinction between the microscopic world of quantum entities and the macroscopic world of directly perceivable objects and attempt to provide a reason why we never see superpositions of macroscopic states. They are saying that there exists a microscopic, quantum realism describable in terms of conventional quantum theory with all its superpositions and

[9] See Penrose, Roger (2000). *The large, the small and the human mind. Canto edition.* Cambridge University Press, Cambridge, pp. 196–200.

[10] Marshall, William, Simon, Christoph, Penrose, Roger, and Bouwmeester, Dik (2002). quant-ph/0210001 v1, 30 September.

interference, and there exists a separate macroscopic classical realism in which objects maintain their distinctness and never interfere. The physicist Anthony Leggett has identified the GRW and Penrose theories as substantiating a hypothesis of *macroscopic realism*: they introduce additional features into conventional quantum theory to make the collapse of the wave function objective and explicit, and thereby deny that superpositions of macroscopically distinct states occur in nature. Because the line between microscopic and macroscopic is not all that clear, Leggett offers an approach for determining the extent of 'distinctness' between macroscopic objects based on the difference of one or more 'extensive' physical properties of the objects in question (such as total charge, magnetic moment, position, momentum, etc.) and the level of 'disconnectivity' between them. This latter parameter is not well defined in a quantitative sense but can be considered to be represented by the number of correlated particles that it would be necessary to measure in order to distinguish an entangled system of two N-particle systems from a classical system with no entanglement. For example, if the systems consist of just one particle each (as in the original EPR thought experiment), then their disconnectivity is correspondingly small. If, however, the systems consist of many particles (dust grains, water specks, golf balls, or cats), then the disconnectivity is correspondingly large. Both the extensive difference and disconnectivity measures are required because it is clearly possible to establish superpositions in systems with significant extensive difference (in an interferometer, say) where the disconnectivity is low.

The assertion of macroscopic realism, then, is that it is not possible to generate a superposition of distinct macroscopic states defined to have significant extensive difference and disconnectivity. But is this true?

Superpositions of distinct macroscopic states

There have been an increasing number of experiments performed in recent years which set out to establish if it is possible to observe quantum behaviour in ever-larger objects. The observation of diffraction effects in a beam of C_{60} molecules was mentioned in Chapter 11. But it is the observation of quantum interference effects in superconducting devices that provides the strongest challenge to the notion of macroscopic realism.

As described in Chapter 4, electrons are fermions and obey the Pauli exclusion principle, but when considered as though they are a single entity, two spin-paired electrons have no net spin and can, under the right conditions, collectively form a boson. Like other bosons (such as photons), these pairs of electrons can 'condense' into a single quantum state. When a large number of pairs so condense in a superconductor, the result is a macroscopic quantum state extending over large distances (i.e. several centimetres). In this condensed state, which lies lower in energy than the normal conduction band of the superconducting material, the electrons experience no resistance. The attraction between the electrons is mediated by lattice vibrations and so is very weak and easily overcome by thermal motion (hence the need for very low temperatures). The distance between each electron in a pair is consequently quite large, and so many such pairs overlap within the metal lattice. The state vectors of the pairs likewise overlap, and their peaks and troughs line up just like light waves in a laser beam. The result can be a macroscopic number of electrons ($\sim 10^{20}$) moving through a metal lattice with their individual state vectors locked in phase.

The attentions of theoretical and experimental physicists have focused on the properties of *superconducting rings*. Imagine that an external magnetic field is applied to a metal ring, which is then cooled to its superconducting temperature. The current which flows in the surface of the ring forces the magnetic field to flow outside the body of the material. The total field is just the sum of the applied field and the field induced by the current flowing in the surface of the ring. If the applied field is removed, the current continues to circulate (because the electrons feel no resistance), and an amount of magnetic flux is 'trapped'. According to the quantum theory of superconductivity, this trapped flux is quantized: only integer multiples of the so-called superconducting flux quantum (given by $h/2e$, where e is the electron charge) are allowed. These different flux states therefore represent the quantum states of an object of macroscopic dimensions—such superconducting rings are usually about a centimetre or so in diameter. The existence of these states has been confirmed by experiment.

In a superconducting ring of uniform thickness, the quantized magnetic flux states do not interact. The quantum state of the ring can be changed only by warming it up, changing the applied external field, and then cooling it down to its superconducting temperature again. However, the mixing of the flux states becomes possible if the ring contains a Josephson junction. This is essentially a small region of the ring in which an insulator has been inserted but which is narrow enough to allow *quantum tunnelling* of pairs of electrons from one side to the other. A variety of quantum interference effects then become possible, and the rings are generally called superconducting quantum interference devices (SQUIDs). These devices are incredibly sensitive, and are used to measure magnetic field strengths in a variety of medical applications. The smallest change in magnetic flux that can be detected in one second using a typical SQUID is about 10^{-32} joules, corresponding roughly to the energy required to raise a single electron 1 mm in the earth's gravitational field. More sensitive devices approach the limits imposed by Heisenberg's uncertainty principle.[11]

Interestingly, this sensitivity is a feature of the Josephson junction, not the ring. This means that a macroscopic variable (such as a measurable flux of electrons around the ring) can be controlled by a microscopic amount of energy in a manner that has little to do with the physical size of the ring itself. The question then becomes: Is it possible to create superpositions of distinct macroscopic states of such devices—for example, states in which large numbers of electrons are flowing in *opposite* directions around the ring?

The answer implied by experiments reported in 2000 is a resounding 'yes', but the interpretation of these experiments is subtle and will require a short diversion.

In these experiments, we have to deal with the fact that interference is not revealed through the appearance of observable interference 'fringes'. Consider instead the macroscopic quantum states created in a superconducting ring in which electrons flow anticlockwise (which we denote as $|\circlearrowleft\rangle$) and clockwise ($|\circlearrowright\rangle$) around the ring. These states sit in separate, distinct, potential energy 'wells'. To cross from one well to another (i.e. to go from the $|\circlearrowleft\rangle$ state to the $|\circlearrowright\rangle$ state), we would have to raise the energy of the ring (by warming it up), change the applied field, and then cool it back down to its superconducting transition temperature. Applying the notion of macroscopic realism means that we would never expect to detect anything other than one or other of these states.

[11] Clark, John (1994). *Scientific American*, August, 46.

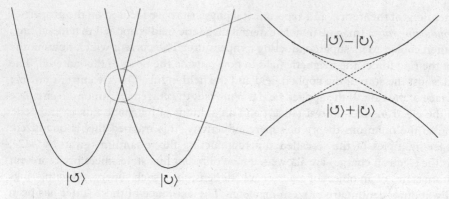

Fig. 12.4 Macroscopic quantum states $|\circlearrowright\rangle$ and $|\circlearrowleft\rangle$ occupy distinct potential energy 'wells' that intersect schematically as shown on the left. The intersection of the potential energy curves is shown in more detail on the right as the dashed lines. With the introduction of a Josephson junction, the two distinct macroscopic states mix in the region of the intersection, and the crossing is 'avoided' by the formation of superposition states $|\circlearrowright\rangle + |\circlearrowleft\rangle$ and $|\circlearrowright\rangle - |\circlearrowleft\rangle$. The size of the splitting between the energies of the superposition states depends on the tunnel splitting, which is a characteristic of the Josephson junction.

Now, suppose that we insert a Josephson junction. The effect of the junction is to allow the $|\circlearrowright\rangle$ and $|\circlearrowleft\rangle$ states to combine in a superposition. In fact, where the energies of the two states are equal or near-equal, the states mix together to form a slightly lower-energy $|\circlearrowright\rangle + |\circlearrowleft\rangle$ combination, and a slightly higher-energy $|\circlearrowright\rangle - |\circlearrowleft\rangle$ combination. Spectroscopists refer to this kind of effect as an 'avoided crossing'—in essence, the potential energy curves of the two states avoid each other at the point where they intersect (see Fig. 12.4). Instead of crossing, two new states are formed in this energy region with characteristics of mixtures of the original states. The amount of splitting between these mixed states is given by the so-called tunnel splitting, which is a characteristic of the Josephson junction. This kind of splitting between quantum states is fairly common in atomic and molecular spectroscopy, but it is perhaps novel to consider its effects in quantum states involving large numbers of electrons moving in a metal ring of macroscopic dimensions.

The experimental demonstration of the possibility of superpositions of distinct macroscopic states then comes down to this: What happens in the energy region where the two macroscopic states $|\circlearrowright\rangle$ and $|\circlearrowleft\rangle$ are of equal or near-equal energy? In practice, the experiments require the application of a bias, ε, to the junction. As ε is varied, microwave radiation is used to probe the size of the splitting between the states. A splitting of the order of $|\varepsilon|$ implies that $|\circlearrowright\rangle$ and $|\circlearrowleft\rangle$ are distinct macroscopic states, and macroscopic realism prevails. A splitting of the order of the square root of $(\varepsilon^2 + \Delta^2)$, where Δ is the tunnel splitting, implies that a superposition is created.

The above description is a considerable simplification of both the problem and the nature of the experiments performed to date. The results of two different kinds of experiments were reported in 2000 by two different groups of researchers. In the results reported by Jonathan Friedman, Vijay Patel, W. Chen, S. K. Tolpygo, and J. E. Lukens at the State University of New York, Stony Brook, the SQUID ring measured 140 μm by 140 μm, and

Leggett estimates that for these experiments, both the extensive difference and the disconnectivity were of a similar order of magnitude, around 10^{10}. Both researchers reported observation of superpositions of distinct macroscopic states.

Leggett sees no reason to suppose that these experiments cannot be done with larger SQUID rings of the order of 1 cm in size.[12] He compares the current situation with early experiments to test Bell's inequality and proposes an exactly analogous inequality that would provide a test of quantum theory versus the whole class of macroscopically realistic theories. Decoherence of some form or another will likely ensure that Schrödinger's cat is either alive or dead. But the possibility of creating superpositions of some kinds of distinct macroscopic states is real and limited only by current experimental technology. And, as the experiments described in Chapters 9 and 10 amply demonstrate, it is not wise to underestimate how quickly the technological barriers can be overcome.

[12] Leggett, A. J. (2002). *Journal of Physics: Condensed Matter*, **14**, R415. See his footnote 36 on page R447, in which he comments on previous claims of observation of superpositions in SQUID rings of this size.

13

I think, therefore . . .

The orthodox Copenhagen interpretation of quantum theory is silent on the question of the collapse of the wave function. The field is therefore wide open. If we choose to reject the strict Copenhagen interpretation, we are, given our present level of understanding, free to choose exactly how we wish to fill the void. Any suggestion, no matter how strange, is in principle legitimate, provided that it does not produce a theory inconsistent with the predictions of quantum theory known to have been upheld so far by experiment. We are not constrained to a specific solution. Our choice is therefore a matter of personal taste.

Now we can try to be objective about how we change quantum theory to make the collapse explicit, and the GRW theory and Penrose's proposal are good examples of that approach. But we should remember that there is no *a priori* reason why we should distinguish between the observed quantum object and the measuring apparatus based on the size of the object under study or the number of particles involved or the curvature of space–time or any other inherent physical property, other than the fact that we seem to possess a theory of the microscopic world that sits very uncomfortably in our macroscopic world of experience. However, macroscopic measuring devices are indisputably made of microscopic quantum entities, and should therefore obey the rules of quantum theory—unless we add something to the theory specifically to change those rules. If the consequences were not so bizarre, we would, perhaps, have no real difficulty in accepting that quantum theory should be no less applicable to large objects than it is to atoms and molecules. In fact, this was something that John von Neumann was perfectly willing to accept.

Von Neumann's theory of measurement (revisited)

As stated repeatedly in this book, John von Neumann's *Mathematical foundations of quantum mechanics* was an extraordinarily influential work. It is important to recall that the language most commonly used to describe and discuss the measurement process in terms of a collapse or projection of the wave function essentially originates with this classic work.

It was von Neumann who so clearly distinguished (in the mathematical sense) between the continuous time-symmetric quantum mechanical equations of motion and the discontinuous, time-asymmetric measurement process. Although much of his contribution to the development of the theory was made broadly within the boundaries of the Copenhagen view, he stepped beyond those boundaries in his interpretation of quantum measurement.

Von Neumann clearly distinguished between two fundamentally different types of process. The first, which he referred to as process 1, which we have called the collapse of the wave function, is the discontinuous, irreversible transformation of a pure quantum state into a mixture involving the projection of some initial state into one of a set of possible eigenstates. The second, process 2, is the continuous, deterministic, reversible evolution of a quantum state governed by the Schrödinger equation. He then discussed the measurement process in terms of three fundamental components, labelled I, II, and III. Component I is the quantum system under observation, II is the physical measuring device, and III is the actual observer who observes and records the measurement outcome. He demonstrated that if a quantum system is present in some eigenstate of a measuring device, the product of this eigenstate and the state vector of the measuring device should evolve in time in a manner quite consistent with both the quantum-mechanical equations of motion and the expected measurement probabilities. In other words, there is no mathematical reason to suppose that quantum theory does not account for the behaviour of macroscopic measuring devices—there is therefore no mathematical reason to expect to find the collapse of the wave function in the composite system I plus II. Process 2 applies equally to macroscopic measuring devices as it does to quantum systems.

So how does the collapse of the wave function arise? Von Neumann's book was published in German in Berlin in 1932, 3 years before the publication of the paper in which Schrödinger introduced the world to his cat, but it is clear that von Neumann was aware of the implications for an infinite regress that were later made so apparent by Schrödinger's 1935 publication. Von Neumann's resolution of the problem was as simple as it is alarming. If quantum theory exemplified by process 2 applies equally well to macroscopic measuring devices, there is no good reason to suppose that it ceases to apply when considering the function of the human sense organs, their connections to the brain, and the brain itself. In this situation, we can consider the component I to be the quantum system plus measuring device, II is then the system of human organs involved in perception, including the brain, and III is the consciousness of the observer.[1] Process 2 applies equally to the composite system formed from I and II, and yet, by the time the measurement has registered in the consciousness of the observer, process 1 has clearly occurred, and only one measurement outcome is subsequently recorded. There is no evidence to suggest that process 2 applies to human consciousness, as it lies outside the physical, 'observed' parts of the world. According to von Neumann, it is here that the boundary should be drawn: the wave function collapses when it interacts with a *consciousness*.

This is where von Neumann goes beyond the Copenhagen interpretation. It is difficult to fault the logic behind this conclusion, at least in mathematical terms. Quantum particles are known to obey the laws of quantum theory: they are described routinely in terms of superpositions of the measurement eigenstates of devices designed to detect them. Those

---

[1] See Von Neumann, John (1955). *Mathematical foundations of quantum mechanics*. Princeton University Press, Princeton, NJ, p. 421. In this English translation, Von Neumann refers to the 'abstract ego' of the observer.

devices are themselves composed of quantum entities and should, in principle, behave similarly. This leads us to the presumption that linear superpositions of macroscopically different states of measuring devices (different pointer positions, for example) are possible. But the observer never consciously registers such superpositions.

Von Neumann argued that photons scattered from the pointer and its scale enter the eye of the observer and interact with their retina. This is still a quantum process. The signal which passes (or does not pass) down the observer's optic nerve is in principle still represented in terms of a linear superposition. Only when the signal enters the brain and from there into the conscious mind of the observer does the wave function encounter a 'system' which we can suppose is not subject to the time-symmetrical laws of quantum theory, and the wave function collapses. We still have a basic dualism in nature, but this is now a dualism of *matter* and conscious *mind*.

Wigner's friend

But whose mind? In the early 1960s, the physicist Eugene Wigner addressed this problem using an argument based on a measurement made through the agency of a second observer. This argument has become known as the paradox of Wigner's friend.

Wigner reasoned as follows. Suppose a measuring device is constructed which produces a flash of light every time a quantum particle is detected to be in a particular eigenstate, which we will denote as $|\psi_+\rangle$. The corresponding state of the measuring device (the one giving a flash of light) is denoted $|\varphi_+\rangle$. The particle can be detected in one other eigenstate, denoted $|\psi_-\rangle$, for which the corresponding state of the measuring device (no flash of light) is $|\varphi_-\rangle$. Initially, the quantum particle is present in a superposition state comprising an equal mixture of $|\psi_+\rangle$ and $|\psi_-\rangle$. The combination (particle in state $|\psi_+\rangle$, light flashes) is given by the product $|\psi_+\rangle|\varphi_+\rangle$. Similarly, the combination (particle in state $|\psi_-\rangle$, no flash) is given by the product $|\psi_-\rangle|\varphi_-\rangle$. If we now treat the combined system—particle plus measuring device—as a single quantum system, we must express the state vector of this combined system as a superposition of the two possibilities: $|\psi_+\rangle|\varphi_+\rangle$ and $|\psi_-\rangle|\varphi_-\rangle$ (see the discussion of entangled states and Schrödinger's cat in Appendix 15).

Wigner can discover the outcome of the next quantum measurement by waiting to see if the light flashes. However, he chooses not to do so. Instead, he steps out of the laboratory and asks his friend to observe the result. A few moments later, Wigner returns and asks his friend if he saw the light flash.

How should Wigner analyse the situation before his friend speaks? If he now considers his friend to be part of a larger measuring 'device', with states $|\varphi'_+\rangle$ and $|\varphi'_-\rangle$, then the total system of particle plus measuring device plus friend is represented by a superposition state comprised of an equal mixture of $|\psi_+\rangle|\varphi'_+\rangle$ and $|\psi_-\rangle|\varphi'_-\rangle$. Wigner can therefore anticipate that there will be a 50 per cent chance that his friend will answer 'Yes' and a 50 per cent chance that he will answer 'No'. If his friend answers 'Yes', then as far as Wigner himself is concerned, the wave function collapses at that moment, and the probability that the alternative result was obtained is reduced to zero. Wigner thus infers that the particle was detected in the eigenstate $|\psi_+\rangle$ and that the light flashed.

But now Wigner probes his friend a little further. He asks 'What did you feel about the flash before I asked you?', to which his friend replies: 'I told you already, I did [did not] see a flash.' Wigner concludes (not unreasonably) that his friend must have already made up his mind about the measurement before he was asked about it. Wigner wrote that the superposition involving $|\varphi'_+\rangle$ and $|\varphi'_-\rangle$ '... appears absurd because it implies that my friend was in a state of suspended animation before he answered my question.'[2] And yet we know that if we replace Wigner's friend with a simple physical system such as a single atom, capable of absorbing light from the flash, then the mathematically correct description is in terms of the superposition and not either of the collapsed states. Wigner wrote that 'It follows that the being with a consciousness must have a different role in quantum mechanics than the inanimate measuring device: the atom considered above.' Of course, there is nothing in principle to prevent Wigner from assuming that his friend was indeed in a state of suspended animation before answering the question. 'However, to deny the existence of the consciousness of a friend to this extent is surely an unnatural attitude.' That way also lies *solipsism*—the view that all the information delivered to your conscious mind by your senses is a figment of your imagination, that is, nothing exists but your consciousness.

Wigner was therefore led to argue that the wave function collapses when it interacts with the *first* conscious mind it encounters. Are cats conscious beings? If they are, then Schrödinger's cat might again be spared the discomfort of being both alive and dead: its fate is already decided (by its own consciousness) before a human observer lifts the lid of the box.

Conscious observers would therefore appear to violate the physical laws that govern the behaviour of inanimate objects. Wigner calls on a second argument in support of this view. Nowhere in the physical world is it possible physically to act on an object without some kind of reaction. Should consciousness be any different? Although small, the action of a conscious mind in collapsing the wave function produces an immediate reaction—knowledge of the state of a system is irreversibly (and indelibly) generated in the mind of the observer. This reaction may lead to other physical effects, such as the writing of the result in a laboratory notebook or the publication of a research paper. In this hypothesis, the influence of matter over mind is balanced by an influence of mind over matter.

Although the possible role of consciousness in quantum measurement tends to be identified with von Neumann, Wigner was a significant influence on von Neumann in drafting his now famous chapter on measurement in the *Mathematical foundations of quantum mechanics*.

The ghost in the machine

This cannot be the end of the story, however. Once again, we see that a proposed solution to the quantum measurement problem is actually no solution at all—it merely shifts the focus from one thorny problem to another. In fact, the approach adopted by von Neumann and Wigner forces us to confront one of science and philosophy's oldest problems: What is consciousness? Just how does the consciousness (mind) of an observer relate to the corporeal structure (body) with which it appears to be associated?

Although our bodies are outwardly different in appearance, it is our consciousness that allows us to perceive ourselves as individuals and to relate that sense of self to the world outside. Consciousness defines who we are. It is the storehouse for our memories, thoughts, feelings, and emotions, and governs our personality and behaviour.

[2] Wigner, Eugene in Good, I. J. (ed.) (1961). *The scientist speculates: an anthology of partly-baked ideas.* Heinemann, London.

The seventeenth-century philosopher René Descartes chose consciousness as the starting point for what he hoped would become a whole new philosophical tradition. In his *Discourse on method*, published in 1637, he spelled out the criteria he had set for himself in establishing a rigorous approach based on the apparently incontrovertible logic of geometry and mathematics. He would accept nothing that could be doubted[3]:

as I wanted to concentrate solely on the search for truth, I thought I ought to...reject as being absolutely false everything in which I could suppose the slightest reason for doubt...

In this way, he could build his new philosophical tradition with confidence in the absolute truth of its statements. This meant rejecting information about the world received through his senses, since our senses are easily deceived and therefore not to be trusted.

Descartes argued that as he thinks independently of his senses, the very fact that he thinks is something about which he can be certain. He further concluded that there is an essential contradiction in holding to the belief that something that thinks does not also exist, and so his existence was also something about which he could be certain. *Cogito, ergo sum*, he concluded—I think, therefore I am.

While Descartes could be confident in the truth of his existence as a conscious entity, he could not be confident about the appearances of things revealed to his mind by his senses. He therefore went on to reason that the thinking 'substance' (consciousness or mind) is quite distinct from the unthinking 'machinery' of the body. The machine is just another form of extended matter (it has extension in three-dimensional space) and may—or may not—exist, whereas the mind has no extension and must exist. Descartes had to face up to the difficult problem of deciding how something with no extension could influence and direct the machinery—how a thought could be translated into movement of the body. His solution was to identify the pineal gland, a small pear-shaped organ that lies deep in centre of the brain, as the 'seat' of consciousness through which the mind gently nudges the body into action. This was not an entirely arbitrary selection. The brain is symmetric, with two distinct cerebral hemispheres joined by a bundle of fibres called the *corpus callosum*. The pineal gland at least has the characteristic of uniqueness in a brain that otherwise looks like a duplication of left and right hemispheres.

This mind–body dualism (Cartesian dualism) in Descartes's philosophy is entirely consistent with the medieval Christian belief in the soul or spirit, which was prevalent at the time he published his work. The body is thus merely a shell, or host, or mechanical device used for giving outward expression and extension to the unextended thinking substance. My mind defines who I am, whereas my body is just something I use (perhaps temporarily). Descartes believed that, although mind and body are joined together, connected through the pineal gland, they are quite capable of separate, independent existence.

In his seminal book *The concept of mind*, first published in 1949, the philosopher Gilbert Ryle wrote disparagingly of this dualist conception of mind and body, referring to it as the 'ghost in the machine'. He wrote[4]:

As thus represented, minds are not merely ghosts harnessed to machines, they are themselves just spectral machines. Though the human body is an engine, it is not quite an ordinary engine, since some of its workings are governed by another engine inside it—the interior governor-engine being one of a very special sort. It is invisible, inaudible and it has no size or weight. It cannot be taken

---

[3] Descartes, René (1968). *Discourse on method and the meditations*. Penguin, London, p. 53.
[4] Ryle, Gilbert (1976). *The concept of mind*. Penguin, London, p. 21.

to bits and the laws it obeys are not those known to ordinary engineers. Nothing is known of how it governs the bodily engine.

Descartes's reasoning has been heavily criticized. He had wanted to establish a new philosophical tradition by adhering to some fairly rigorous criteria regarding what he could and could not accept to be beyond doubt. And yet his most famous statement—'I think, therefore I am'—was arrived at by a process which seems to involve assumptions that, by his own criteria, appear to be unjustified. The statement is also a linguistic nightmare and, as the logical positivists later demonstrated to their obvious satisfaction, consequently quite without meaning.

Multiple drafts

The problem with the ghost in the machine is that it gives us absolutely nothing to work with. It sidesteps the issue of explaining consciousness by descending into mysticism. How exactly does the ghost—in Descartes's language the unextended thinking 'substance'—interact with the physical structure of the body? Almost by definition, this cannot be an interaction describable by any physics that we currently understand. Even in the Hollywood movie *Ghost*, Patrick Swayze's character has to come to terms with the fact that, as a ghost, he can pass unseen through walls and doors (we'll leave aside for the moment just exactly how he walks on floors) but cannot interact physically with his environment and therefore cannot communicate with his widow, played by Demi Moore. He eventually learns the trick from another ghost: by channeling his emotions (initially anger) and focusing his mental (?) energy on physical objects, he learns to be able to interact with them. A flight of fancy in an entertaining film, perhaps, but nonetheless illustrative of the difficulties of Cartesian dualism.

In his book *Consciousness explained*, the philosopher Daniel Dennett wrote[5]:

This fundamentally antiscientific stance of dualism is, to my mind, its most disqualifying feature, and is the reason why...I adopt the apparently dogmatic rule that dualism is to be avoided *at all costs*. It is not that I think I can give a knock-down proof that dualism, in all its forms, is false or incoherent, but that, given the way dualism wallows in mystery, *accepting dualism is giving up*.

He illustrates this point by reference to a cartoon taken from the magazine *American scientist*. The cartoon shows two scientists talking about a mathematical derivation chalked up on a blackboard. The second step of the derivation reads: 'Then a miracle occurs.' One scientist is saying to the other: 'I think you should be more explicit here in step two.'

The real problem is not so much the antiscientific stance of dualism, it is that the underlying concept that there exists some kind of central 'control room' of the mind is so seductive. After all, we are all unique individuals, with a unique set of experiences, memories, thoughts, feelings, and emotions that are somehow locked up inside our minds and inaccessible to those outside except when we express them physically. We accept the physicality of the events that take place in the outside world. We accept that our perceptions—sight, touch, smell, hearing—are the results of physical processes. We accept that these perceptions are processed by a brain whose physical function—in terms of chemistry, physiology, and electricity—we will one day be able to grasp in full. But then

---

[5] Dennett, Daniel (1991). *Consciousness explained*. Penguin, London, p. 37. The emphasis is Dennett's.

we start to reach for something, someone, to sit above all this physicality to observe it and to pull the levers that shape our conscious responses. Something or someone inside the brain that makes us *us*. Even scientific theories of consciousness fall prey to the seductive charms of what Dennett calls the 'Cartesian Theatre', a place where 'it all comes together', a place where consciousness 'happens'.

What is the alternative? There are several. Dennett's proposed solution is called the *multiple drafts* model of consciousness. This model recognizes that our sensory apparatus is bombarded with inputs, much of which do not appear to contribute to our direct conscious experience. Dennett argues that the brain is constantly processing these inputs in parallel, multitrack sequences. Each of these sequences is interpreted and contributes to a 'draft': a representation or part representation of a conscious experience. At any one time, the brain is generating multiple drafts, based on multiple interpretations. These vie with each other in a kind of Darwinian-style evolution that does not, however, lead to some final, edited draft (for that takes us back to the Cartesian Theatre, the place where 'it all comes together'). Instead:

These [multitrack processes] yield, over the course of time, something *rather like* a narrative stream or sequence, which can be thought of as subject to continual editing by many processes distributed around in the brain, and contributing indefinitely into the future. Contents arise, get revised, contribute to the interpretation of other contents or to the modulation of behaviour (verbal and otherwise), and in the process leave their traces in memory, which then eventually decay or get incorporated into or overwritten by later contents, wholly or in part. This skein of contents is only rather like a narrative because of its multiplicity; at any point in time there are multiple drafts of narrative fragments at various stages of editing in various places in the brain.

No draft is ever finished. This metaphor for consciousness is a publisher's worst nightmare.

Among many illustrations of the multiple drafts model in action, Dennett makes extensive reference to a series of simple cognitive experiments. If two small spots a small distance apart are illuminated rapidly in succession, an observer will report the experience of seeing a single spot move back and forth, for precisely the same reason that a sequence of still photographs run rapidly in front of our eyes creates the illusion of movement, as in a 'movie'. If we now make these two spots of different colour, the observer reports the experience of seeing a single spot move back and forth, *changing colour as it does so*. Furthermore, the spot is seen to change colour in mid-trajectory, despite the fact that, in reality, there are two spots and no movement.

This seemingly innocuous optical illusion raises some tricky questions for our understanding of consciousness. If we deny the possibility of precognition, then the observer cannot be conscious of the colour change until the second spot has been illuminated (how else does the observer know what colour the spot should turn into?). Does this mean that the observer is first conscious of the spot (on the left, say), then becomes conscious of the spot of different colour on the right, and his consciousness then retrospectively 'fills the gap', resulting in some kind of revisionist experience of movement and colour change? Or does it rather mean that there is really a delay between perception of the spots and development of a conscious experience—the inputs are first recorded and sent to some kind of 'editing suite', which takes the first frame (spot on the left) and the last frame (spot of different colour on the right), puts in the missing frames, and sends the whole thing back as a single conscious experience?

Of course, both of the above attempts at explanation imply that there is some form of 'final' draft, no matter how this is arrived at. The multiple-drafts model denies this. The difference between these explanations is one of semantics. Dennett writes that 'there are no functional differences that could motivate declaring all prior stages and revisions to be unconscious or preconscious adjustments, and all subsequent emendations to the content (as revealed by recollection) to be post-experiential memory contamination.'

Dennett makes no bones about the fact that the multiple-drafts model seeks to replace one set of metaphors with another but argues that metaphors are tools with which we build understanding. With a better set of tools, we are perhaps in a position to work at the problem a little more productively. Is there, then, a sense in which the metaphor of multiple drafts can be traced back to aspects of physical processes occurring in the brain?

The physical basis of consciousness

It is now appropriate to be a little more careful in distinguishing what we mean when we talk of *consciousness*, on the one hand, and *mind* on the other. Consciousness is immediate: it is the short-term receiving, processing, and responding to sensory perceptions and the creation of the words, images, and patterns that constitute thought. Consciousness is something that ceases every night as we drift into deep sleep, when we succumb to aneasthetic or suffer some blow to the head. Mind is (arguably) the consequence of the long-term operation of consciousness, the result of laying down long-term memory, the synthesis of experience that defines us as individuals and gives us an enduring sense of self. We may lose consciousness, but we do not so readily lose our minds or sense of identity.

As so defined, consciousness remains a significant problem for neuroscience, and mind is an even bigger puzzle. But there has been some considerable progress in understanding the workings of the physical and chemical mechanics of the brain that takes us a few steps towards an understanding of the physical basis of consciousness. Much of this progress comes from studying patients with a variety of brain disorders. In *blindsight*, for example, certain stoke victims can no longer see objects in front of them but nevertheless seem strangely conscious of their existence and can even point them out with a probability of success well in excess of random guesswork, even though they believe they are just guessing. The opposite affliction, called *prosopagnosia*, is the conscious awareness of objects (especially people) but without the associations provided by long- or short-term memory and therefore without recognition, as memorably described in the case of Dr P. in Oliver Sacks' *The man who mistook his wife for a hat*.

Yet more insight comes from studying patients suffering severe epilepsy involving abnormal electrical activity on one side of the brain, for which an effective remedy is to cut the corpus callosum, the bundle of fibres connecting the brain's two hemispheres. The result is a 'split brain'. In truth, the two halves are still connected by central regions of the brain, and the patients therefore retain a strong sense of a single self and, in fact, suffer no loss of faculties that is otherwise characteristic of brain damage. However, the left and right hemispheres have very different functions, the left hemisphere tending to dominate in verbal, analytical (reasoning), and sequential processing tasks, and the right hemisphere tending to specialize in spatial and creative, synthetic tasks. The right hemisphere would appear to be the place where we enjoy music. In a very real sense the split-brain patients are conscious in two very distinct ways, distinguished by the specialization of the left and right hemispheres.

Visual input from the left side of the visual field—the right side of the retina—is processed by the right visual cortex, located towards the back of the right hemisphere and *vice versa* for the right side of the visual field. Tactile input from the left hand is processed in the right hemisphere and *vice versa* for the right hand. In a right-handed patient, speech is handled in the left hemisphere. A word flashed into a split-brain patient's right visual field, and hence to the left hemisphere, is immediately recognized: the patient can say the word and write it down. However, a word flashed into the patient's left visual field, and hence to the right hemisphere, is recognized, but the patient can now no longer say it or write it down.

This kind of result tells us that the right hemisphere cannot communicate in speech or writing, but we might not yet be convinced that this is evidence for a separate consciousness belonging to the two hemispheres. Consider, then, the extraordinary results from another split-brain patient who had previously suffered damage to his left hemisphere. As had by now become clear in brain research, the brain has a truly marvelous ability to recover from damage to specialized areas (for speech, vision, motor control, etc.) by recruiting other areas to take on the missing functions. In this case, the patient had over time developed significant speech function in his right hemisphere and could write with his left hand. When his brain was split, it was therefore possible for him to communicate from *both* sides of his now separated brain. In a personal history, Michael Gazzaniga (who, with Roger Sperry, pioneered split-brain research, with Sperry winning a Nobel prize in 1981) wrote[6]:

Instead of wondering whether or not [the patient's] right hemisphere was sufficiently powerful to be dubbed conscious we were now in a position to ask [the patient's] right side about its view on matters of friendship, love, hate, and aspirations . . . When [the patient] first demonstrated this skill, my student, Joseph LeDoux, and I . . . just stared at each other for what seemed an eternity. A half-brain had told us about its own feelings and opinions, and the other half-brain, the talkative left, temporarily put aside its dominant ways and watched its silent partner express its views.

Needless to say, the feelings and opinions of the patient's right hemisphere did not always coincide with those of the left.

Through research such as this, and painstaking work on the neuronal architecture and the underlying chemistry and physics of the brain, we are approaching understanding of many of the details of brain function and activity. We are, however, still tantalizingly short of a fully developed physical theory of consciousness. To illustrate what we do know, we will return to Wigner's friend (whom we have neglected terribly if he really is in a state of suspended animation) and draw on an outline of a scientific theory of consciousness popularized recently by neuroscientist Susan Greenfield.

Having been left in the laboratory to observe the result of the quantum measurement, Wigner's friend is initially in a state of high alertness, focused on the apparatus and waiting for the moment when the light will flash, or not, as the case may be. We could say that he is in a state of focused 'arousal'. This has some very clear neurochemical and neurophysical consequences. His focus on the apparatus serves as an 'epicentre' for a series of what Greenfield calls *transient neuronal assemblies*. Put simply, the focus results in the release of fountains of chemical neurotransmitters which modulate his brain cells—neurons—to

[6] This quotation by Gazzaniga is taken from Thompson, Richard F. (1993). *The brain: a neuroscience primer* (2nd edn.). W. H. Freeman, New York, p. 417.

become more readily excitable. The greater the focus, the more neurons can be 'recruited' to the process, increasing their sensitivity by marshalling the neurons into a temporary working group. The larger the group, the deeper the level of consciousness for a brief moment in time.

Although he is focused on the apparatus, he is also aware (he is conscious) of many other things besides. These other things serve as epicentres for other neuronal assemblies, some recruiting the same neurons, some recruiting neurons in different parts of his brain. As time goes by, his attention starts to wander. Perhaps he is replaying an argument with his girlfriend from the night before. Competition among the assemblies grows more intense, the size of the assemblies falls, and he becomes somewhat less conscious of the apparatus. In all this time, fragments of consciousness—let us call them drafts—flicker in and out and undergo constant revision. There are multiple drafts involving different groups of neurons, yet Wigner's friend has only one consciousness at any moment in time. That one consciousness is an *emergent property* of the neuronal assemblies.

Suddenly, he breaks from his reverie and renews his focus on the apparatus. After all, this is a singularly important measurement that Wigner has asked him to make and he is only going to get one shot at it. The neuronal assembly stimulated by this focused arousal strengthens as it recruits more neurons, and his consciousness of the apparatus deepens. We now suppose that the quantum system under study forms its superposition state and is measured by the apparatus. The apparatus enters a superposition state of particle-in-state-$|\psi_+\rangle$/light flashes and particle-in-state-$|\psi_-\rangle$/no flash. This superposition is cascaded to his retina, optic nerve, and brain. We could speculate that this creates two predominant epicentres, two neuronal assemblies—one corresponding to the composite quantum state involving the light flashing, the other corresponding to the composite state involving no flash. Speculating further, we might say there are two drafts, both equally probable, corresponding to the different measurement outcomes. These drafts—fragments of consciousness—are potentially describable in terms of physical and chemical events unfolding in the brain and would therefore be included within von Neumann's composite system I plus II. We are therefore left to assume that only when a single conscious state *emerges* from the competing neuronal assemblies does the wave function collapse. Wigner's friend is then conscious of only one result—the light flashed (or did not flash)—and this is what he writes down in his laboratory notebook.[7]

We are left with this escape clause for von Neumann's theory of measurement because the emergence of consciousness from the physicality of brain chemistry and neuronal assemblies is not currently describable in physical terms. Greenfield writes[8]:

the one part of the formal description that we have been unable to justify in physical terms is that consciousness is an *emergent property*. This shortcoming is scarcely surprising inasmuch as if we could explain consciousness as the mere aggregation of its constituent parts—arousal and [transient neuronal assemblies]—it would not be an emergent property after all.

---

[7] If the collapse is the result of the emergence of consciousness from competing neuronal assemblies (which we might be tempted to ascribe to the mental state of the observer), then it does not immediately follow that two observers recording a single measurement simultaneously would necessarily record the same result. We would have to assume further that the role of consciousness (any consciousness) is simply to make the result real in some mechanistic way, rather than to influence the nature of the result itself. More on this below.

[8] Greenfield, Susan A. (1995). *Journey to the centres of the mind*. W. H. Freeman, New York, p. 162. The emphasis is Greenfield's.

It is tempting to think of this explanation as not much further advanced from the statement: 'Then a miracle occurs'. But we are surrounded every day by the emergent properties of bulk materials that we would not wish (and, in any case, are not in a position) to begin to describe in terms of the physical properties of fundamental building blocks of matter. Consciousness is an emergent property of the brain, which is an intrinsic part of the body. Describing consciousness in terms of the functions of chemical neurotransmitters and neurons is likely to be no easier than describing the structure of a snowflake in terms of the basic functions of protons, neutrons, and electrons. Or, to stretch the analogy to breaking point, describing consciousness is likely to be no easier than describing the enigmatic smile in da Vinci's *Mona Lisa* in terms of the physical and chemical properties of the oils with which it was painted.

As it turns out, this was a profoundly important measurement, and as a consequence, Wigner's friend is rewarded with a glittering scientific career at the leading edge of quantum physics. He replays the moment that he observed the result over and over to himself and to others, synaptic connections between certain neurons are strengthened as a result (owing to a mechanism first identified by neuroscientist Donald Hebb), and the events enter his long-term memory. The girlfriend is forgotten.

AI

We may lack a formal scientific explanation for consciousness, but we do now have a rather detailed understanding of brain function and activity at a fairly fundamental level. This understanding has led some scientists to speculate that one day we will be able to build an artificial brain that not only demonstrates what we would regard as intelligence but is, to all intents and purposes, conscious. The primary objectives of research on artificial intelligence (AI) are based on a more pragmatic need to develop *robotic systems* to carry out repetitive (but nonetheless intelligent) tasks to replace humans (who are often more costly, get bored, and are less reliable) and to develop so-called *expert systems*, which are essentially methods of packaging expert knowledge in a piece of software. On a somewhat less pragmatic (but potentially more profound) level, research on AI is also thought to be a route to developing a better understanding of human intelligence, consciousness, and mind.

The community of scientists and philosophers that share interests in these subjects appears quite sharply polarized between those on the one hand who believe that the brain is simply hardware on which consciousness or mind is the operating system and the software, and those on the other hand who reject the notion that the brain and consciousness or mind can be disentangled this easily. If the brain is a biological computer, it follows that any system (it does not have to be constructed out of the same materials) that can be operated in an exactly analogous way will replicate what the brain does, including the generation of intelligence and consciousness if these are emergent properties of information-processing activities in the brain. But how could we tell if such an artificial brain were really exhibiting behaviour consistent with our interpretation of human intelligence? One answer was provided in 1950 by mathematician Alan Turing, perhaps best known for his wartime activities as part of the team of 'code-breakers' at Bletchley Park in England. The machine and a human volunteer are put behind a screen, and both are asked a series of questions. If the interrogator cannot distinguish the machine from the human being based on their responses, then the machine is exhibiting intelligence.

But would such a machine really be conscious? Given that communication is the only way to ascertain the depth of understanding and the range of emotional responses that we might identify with consciousness or mind, there is always the possibility that we can be fooled by a cleverly constructed algorithm. In other words, the machine could be programmed to give responses that create the impression of human-like consciousness without having any real consciousness itself. The American philosopher John Searle has emphasized this shortcoming in a concept that has become known as Searle's 'Chinese room'. Suppose the questions require a series of yes/no responses and are asked in Chinese. Instead of using the machine, Searle himself (who does not understand Chinese) sits behind the screen and takes each input character, analyses the sequence according to some prescribed algorithm, and responds with the Chinese for 'yes' or 'no' in a way that does not contradict any previous questions. However, Searle is simply *processing* the inputs, he is not *understanding* the questions in the way a native Chinese speaker would understand them. Searle argues that even a machine programmed with an algorithm sufficiently sophisticated to fool the interrogator and pass the Turing test is far from understanding anything, let alone displaying characteristics of human intelligence or consciousness.

It is possible to imagine how a machine could be developed not only to mimic but to exceed greatly the brain's capacity for the rational, logical, analytical processing of input information. It is not quite so easy to see how such a machine could develop the full range of emotional content—feelings, desires, loves, hates—characteristic of any individual human being. It is obviously not enough to reduce this to a problem for rational processing—to *model* the emotional content by simply programming the right responses—as this defeats the object (and in any case begs the question as to what the 'right' responses are). It is also clear that both the rational and emotional content are needed to define fully what it means to have conscious experiences and a sense of self. To appreciate this point, it is useful to recall the terrible story of Phineas Gage.

Gage was working as a foreman for the Rutland and Burlington Railroad, blasting rock to clear the way prior to laying down new track. One afternoon in the summer of 1848, he suffered a terrible accident. Having drilled a hole in a rocky outcrop and having filled it with gunpowder, fuse, and sand, he was gently 'tamping' the sand with a specially fashioned iron rod (something he had done many times before) when the power exploded, driving the rod through his left cheek, up through the base of his skull, and out the top of his head. Surprisingly, this did not kill him. Even more surprisingly, he actually remained conscious, talking coherently and even making jokes as his colleagues rushed him to a doctor in the nearby small town of Cavendish. Sophisticated analysis conducted by computer simulation some 120 years later showed that the rod had missed regions of the brain that specialize in language and motor function, but had damaged areas of his prefrontal cortices. Gage recovered from his injuries within 2 months but emerged a 'different person'. The damage to his brain...[9]

compromised his ability to plan for the future, to conduct himself according to the social rules he previously had learned, and to decide on the course of action that ultimately would be most advantageous to his survival.

[9] Damasio, Antonio R. (1994). *Descartes's error*. Macmillan, London, p. 33.

Gage had become temperamental, abusive, and shiftless, in marked contrast to his behaviour and personality before the accident. He lost his job at the railroad, drifted into farm labour, and spent time as a circus exhibit. He died 12 years after his accident.

Following studies of similarly afflicted patients, the neurologist Antonio Damasio suggests that Gage had not suffered any injuries that had reduced his powers of reason—his brain could still function as an analytical machine. But he was now 'unbalanced'. Stripped of the emotional checks and balances that had previously guided his behaviour, Gage could no longer make the 'right' choices. We are left to conclude that conscious decisions are *not* the result of simply running an analytical simulation on the hardware of the brain but are also strongly shaped by emotions. Generating artificial emotional content in the 'mind' of a machine—and not just simulating it with a clever algorithm—is well beyond the reach of present-day AI research. And yet all the evidence suggests that consciousness is not possible without it.

Consciousness and objective reduction

Roger Penrose's two books, *The emperor's new mind* and *Shadows of the mind*, are major polemical works ranged against the notion of what Penrose calls *strong AI*, the view that thinking is computation and that consciousness is the result of information processing in the brain. His arguments are complex but basically come down to what he believes is the inherent non-computability of consciousness. Just as Gödel's theorem denies that we can logically prove all propositions that can be formulated from some consistent set of axioms from within this axiomatic structure (see Chapter 6), so Penrose argues that mathematical understanding is not possible within the basic logical 'rules' of mathematics. Mathematical understanding requires other elements such as insight and intuition that, because they are not part of the formal logical framework of mathematics, are non-computable. Penrose then generalizes this to consciousness as a whole.

In seeking then to provide a physical rationale for consciousness, Penrose joins the pieces of two modern puzzles together and suggests that we identify conscious events with an objective reduction of the wave function. Together with anaesthetist Stuart Hameroff, Penrose has identified *microtubules*—tiny tubes of protein that lie at the centre of virtually every cell in the body, including neurons—as the place where quantum superpositions might undergo spontaneous, objective reduction. Objective because, here in the brain at least, there is no observer. To make the connection, it is necessary to suppose that a coherent superposition of microtubules in specific protein conformations can be created, spanning large numbers of neurons, forming what I guess we might call a transient neuronal assembly. At this point, through the effects of space–time curvature or whatever other mechanism we can develop, the coherent superposition collapses, and a conscious event occurs—Wigner's friend is conscious of the light flash (or conscious of the fact that there was no flash). This seems to fit, Penrose argues, because the non-computability of consciousness is traced back to the basic non-computability of the reduction process. This is going much further than von Neumann and Wigner, who pointed to consciousness as the 'outside' agency not describable in terms of physical laws, which is responsible for the collapse of the wave function. Penrose is turning this on its head and suggesting that the collapse is itself responsible for consciousness.

Penrose's arguments against strong AI have provoked much debate and, if anything, have probably helped to polarize the community further into distinct 'for' and 'against' camps. His speculations on a quantum-mechanical foundation for consciousness have provoked similar debate but appear to have garnered little support. There is the counter-argument that there is too much environmental 'noise' inside a warm brain for a quantum superposition to be maintained for the kinds of timescales indicative of the registering of conscious events; that environmental decoherence would wash out such a neuronal superposition very rapidly, even if it was assumed to form in the first place. There is also the observation that microtubules are present in a variety of cells, not just in neurons. To be sure, Penrose does not advertise these views as anything more than speculations. The first step towards assessing the plausibility of such a quantum basis for consciousness will likely come from experiments designed to identify whether or not objective reduction is itself a feature of quantum physics.

Free will and determinism

You may have been tempted from time to time in your reading of this book to cast your mind back to the good old days of Newtonian physics, where everything seemed to be set on much firmer ground. Classical physics was based on the idea of a grand scheme: a mechanical clockwork universe where every effect could be traced back to a cause. Set the clockwork universe in motion under some precisely known initial conditions, and it should be possible to predict its future development in unlimited detail.

However, apart from the reservations we now have about our ability to know the initial conditions with sufficient precision, there are two fairly profound philosophical problems associated with the idea of a completely deterministic universe. The first is that if every effect must have a cause, there must have been a *first* cause that brought the universe into existence. The second is that, if every effect is determined by the behaviour of material entities conforming to physical laws, what happens to the notion of *free will*? We will defer discussion of the first problem and turn our attention here to the second problem.

The Newtonian vision of the world is essentially reductionist: the behaviour of a complicated object is understood in terms of the properties and behaviour of its elementary constituent parts. If we apply this vision to the brain and consciousness, we are ultimately led to the modern view that both should be understood in terms of the complex (but deterministic) physical and chemical processes occurring in the machinery, and we slip into the territory of strong AI.

One consequence of this completely deterministic picture is that our individual personalities, behaviour, thoughts, actions, emotions, and so on are effects that we should in principle be able to trace back to a set of one or more material causes. For example, my choice of words in this sentence is not a matter for my individual freedom of will; it is a necessary consequence of the many physical and chemical processes occurring in my brain. That I should decide to boldly split an infinitive in this sentence was, in principle, dictated by my genetic make-up and physical environment, integrated up to the moment that my 'state of mind' led me to 'make' my decision.

We should differentiate here between actions that are essentially instinctive (and which are therefore *reactions*) and actions based on an apparent freedom of choice. I would accept that my reaction to pain is entirely predictable, whereas my senses of value, justice, truth, and beauty seem to be matters for me to determine as an individual. Ask an individual exhibiting some pattern of conditioned behaviour, and they will tell you (somewhat indignantly) that, at least as far as accepted standards of behaviour and the law are concerned, they have their own mind and can exercise their own free will. Are they suffering a delusion?

Before the advent of quantum theory, the answer given by the majority of philosophers would have been 'Yes'. As we have seen, Einstein himself was a realist and a determinist, and consequently rejected the idea of free will. In choosing at some apparently unpredictable moment to light his pipe, Einstein saw this not as an expression of his freedom to will a certain action to take place, but as an effect which has some physical cause. One possible explanation is that the chemical balance of his brain is upset by a low concentration of nicotine in his bloodstream, a chemical on which his body had come to depend. A complex series of chemical changes takes place, which is translated by his mind as a desire to smoke his pipe. These chemical changes therefore cause his mind to will the act of lighting his pipe, and that act of will is translated by the brain into bodily movements designed to achieve the end result. If this is the correct view, we are left with nothing but physics and chemistry.

In fact, is it not true that we tend to analyse the behaviour patterns of everyone (with the usual exception of ourselves) in terms of their personalities and the circumstances that lead to their acts? Our attitude towards an individual may be sometimes irreversibly shaped by a 'first impression', in which we analyse the physiognomy, speech, body language, and attitudes of a person and come to some conclusion as to what 'kind' of person we are dealing with. How often do we say: 'Of course, that's just what you would expect him to do in those circumstances'? If we analyse our own past decisions carefully, would we not expect to find that the outcomes of those decisions were entirely predictable, based on what we know about ourselves and our circumstances at the time? Is anyone truly *un*predictable?

Classical physics paints a picture of the universe in which we are nothing but fairly irrelevant cogs in the grand machinery of the cosmos. However, quantum physics may paint a rather different picture, possibly allowing us to restore some semblance of self-esteem. Out go causality and determinism, to be replaced by the indeterminism embodied in the uncertainty relations. Now the future development of a system becomes impossible to predict except in terms of probabilities. Furthermore, if we accept von Neumann's and Wigner's arguments about the role of consciousness in quantum physics, then our conscious selves become the most important things in the universe. Quite simply, without conscious observers, there would be no physical reality. Instead of tiny cogs forced to grind on endlessly in a reality not of our design and whose purpose we cannot fathom, we become the *creators* of the universe. *We* are the masters.

However, we should not get too carried away. Despite this apparent changed role, it does not necessarily follow that we have much freedom of choice in quantum physics. If and when the wave function collapses, it does so unpredictably in a manner which would seem to be beyond our control. Although our minds may be essential to the realization of a particular reality, we cannot know or decide in advance what the result of a quantum measurement will be. We cannot choose what kind of reality we would like to perceive beyond choosing the measurement eigenstates. In this interpretation of quantum measurement, our only influence over matter is to make it real. Unless we are prepared to accept the possibility of a variety of paranormal phenomena, it would seem that we cannot bend matter to our will.

Of course, the notion that a conscious mind is necessary to sustain reality is not new to philosophers, although it is perhaps a novel experience to find it advocated as an explanation of one of the most important and successful of modern scientific theories.

The mind of God?

Einstein's comment that 'God does not play dice' is one of the best known of his many remarks on quantum theory and its interpretation. Niels Bohr's response is somewhat less well known: 'But still, it cannot be for us to tell God, how he is to run the world.'

Is it possible that after centuries of philosophical speculation and scientific research on the nature of the physical world, we have, in quantum theory, finally run up against nature's grand architect? Is it possible that the fundamental problems of interpretation posed by quantum theory in its present form arise from our inability to fathom the mind of God? Are we missing the 'ultimate' hidden variable? Could it be that behind every apparently indeterministic quantum measurement, we can discern God's guiding hand?

Away from the cut and thrust of their scientific research papers, Einstein, Bohr, and their contemporaries spoke and wrote freely about God and his designs. To a limited extent, this habit continues with modern-day scientists. For example, in *A brief history of time*, Stephen Hawking writes in a relaxed way about a possible role for God in the creation of the universe, and in *The emperor's new mind*, Roger Penrose writes of 'God-given' mathematical truth. Among scientists, speculating about God in the open literature appears to be the preserve of those who have already established their international reputations. We would surely raise our eyebrows on discovering that the research programme of a young, struggling academic scientist in the twenty-first century is organized around their desire to know how God created the world. We would at least anticipate that such a scientist may have difficulties securing the necessary funding to carry out their research.

But discovering more about how God created the world was all the motivation Einstein needed for his work. Admittedly, Einstein's was not the traditional medieval God of Judaism or Christianity, but an impersonal God *identical* with Nature: *Deus sive Natura*—God or Nature—as described by the seventeenth-century philosopher Baruch Spinoza.

And herein lies the difficulty. In modern times, it is almost impossible to resist the temptation to equate belief in God with an adherence to a religious philosophy or orthodoxy. Scientists are certainly taught not to allow their scientific judgement to be clouded by their personal beliefs. Religious belief entails blind acceptance of so many dogmatic 'truths' that it negates any attempt at detached, rational, scientific analysis. In saying this, I do not wish to downplay the extremely important sociological role that religion plays in providing comfort and identity in an often harsh and brutal world. But once we accept God without religion, we can ask ourselves the all-important questions with something approaching intellectual rigour. The fact that we have lost the habit or the need to invoke the existence of God should not prevent us from examining this possibility as a serious alternative to the interpretations of quantum theory discussed previously. It is, after all, no less metaphysical or bizarre than some of the other possibilities we have considered so far.

We might start by asking ourselves: Does God exist? There is a timelessness about this question. It has teased the intellects of philosophers for centuries and weaves its way through the entire history of philosophical thought. Even in periods where it may have been generally accepted to be a non-question, it has lurked in the shadows, biding its time.

For many centuries, philosophical speculation regarding the existence of God was so closely allied to theology as to be essentially indistinguishable from it. In the thirteenth century, Thomas Aquinas helped to restore Aristotelian philosophy and science, an ancient learning that had been buried and all but forgotten during the 'Dark Ages'. But Aquinas was a scholar of the Roman Catholic Church, and he took great pains to ensure that pagan and other heretical elements were carefully weeded out of Aristotle's philosophy. The Church elevated Aristotelianism to the exalted status of a religious dogma and so pronounced on all matters not only of religious faith, but also of science. To contradict the accepted wisdom of the Church was to court disaster, as Galileo discovered on the 22 June 1633, when, at the end of his trial for heresy, he was forced to abjure the Copernican doctrine and was placed under house arrest.

Against this background, a seventeenth-century philosopher wishing to establish a new philosophical tradition had to tread warily. René Descartes had just completed a major work, which he had called *De mundo*, when in November 1633, he received news of Galileo's trial and condemnation. Descartes was dismayed: the Copernican system formed the basis of his work, and it became clear that if he published it, it would not have the effect he had hoped. Instead, he chose to 'leak' bits of it out, hoping always to stay on the right side of the Church authorities. It is perhaps not surprising that when he published his *Meditations*, in which he offered three different proofs for the existence of God, he decided to dedicate this work to the Dean and Doctors of the Sacred Faculty of Theology of Paris.

Descartes's aim was not to subvert the teachings of the Church, but to demonstrate that the orthodox conclusions regarding the soul of man and the existence of God could be reached using the power of reason. His intention was to bring something approaching mathematical rigour to bear on these philosophical questions. Having said that, it is apparent that Descartes's arguments fall somewhat short of the ideal which he had set for himself in his *Discourse on method*, published 4 years earlier. Nevertheless, his approach marked a distinct break with the past.

Descartes advanced three proofs for the existence of God. Two are to be found in his 'Third meditation', but the one from which he seemed to derive most pleasure is found in his 'Fifth meditation'. This is the so-called ontological proof or ontological argument. Remember that Descartes had already established (with certainty) that he is a thinking being and that, therefore, he exists. As a thinking being, he recognizes that he is imperfect in many ways, but he can conceive of the *idea* of a supremely perfect being, possessing all possible perfections. Now it goes without saying that a being that is imperfect in any way is not a supremely perfect being. Descartes assumed that existence is a perfection, in the sense that a being that does not exist is imperfect. Therefore, he reasoned, it is self-contradictory to conceive of God as a supremely perfect being that does not exist and so lacks a perfection. Such a notion is as absurd as trying to conceive of a triangle that has only two angles. Thus, God must be conceived as a being who exists. Hence, God must exist.

The methods used by Descartes were picked up by other philosophers of his time, although many did not always feel it necessary to indulge in the kind of systematic doubting that Descartes had thought to be important. Thus, the German philosopher Gottfried Wilhelm Leibniz was happy to accept as self-evident much of what Descartes had taken great pains to prove, and adapted and extended many other elements in Descartes's line of reasoning. For example, in developing his own philosophical position, Leibniz was happy to accept the existence of the world also to be self-evident, although its nature might not be.

Like his predecessor, Leibniz also presented three proofs for the existence of God. Two of these are similar to two of Descartes's proofs. The third, which is usually known as the cosmological proof or the cosmological argument, was published in 1697 in Leibniz's essay *On the ultimate origination of things*. Leibniz's argument is based on the so-called principle of *sufficient reason*, which he interpreted to mean that if something exists, there must be a good reason. Thus, the existence of the world and of the eternal truths of mathematics and logic must have a reason. Something must have caused these things to come into existence. He claimed that there is within the world itself no sufficient reason for its own existence. As time elapses, the state of the world evolves according to certain physical laws of change. It could be argued, then, that the cause of the existence of the world at any one moment is to be found in the existence of the world just a moment before. Leibniz rejected this argument[10]: 'however far you go back to earlier states, you will never find in those states a full reason why there should be any world rather than none, and why it should be such as it is.'

The world cannot just *happen* to exist, and whatever (or whoever) caused it to exist must also exist, since the principle of sufficient reason demands that something cannot come from nothing: *ex nihilo, nihilo fit*. Furthermore, the ultimate, or first, cause of the world must exist outside the world. Of course, this first cause is God. God is the only sufficient reason for the existence of the world. The world exists; therefore, it is necessary for God also to exist.

The cosmological proof has a long history. Plato used something akin to it in his discussion of God as creator in the *Timaeus*. It also has an entirely modern applicability. We now have good reason to believe that the world (which, in its modern context, we take to mean the universe) was formed about 12 billion years ago in the big bang space–time singularity. The subsequent expansion of space–time has produced the universe as we know it today, complete with galaxies, stars, planets, and living creatures. Modern theories of physics and chemistry allow us to deduce the reasons for the existence of all these things (possibly including life) based on the earlier states of the universe. In other words, once the universe was off to a good start, the rest followed from fundamental physical and chemical laws. Scientists are generally disinclined to suggest that we need to call on God to explain the evolution of the post big bang universe. But the universe had a *beginning*; which implies that it must have had a first cause. Do we need to call on God to explain the big bang? Stephen Hawking writes[11]: 'An expanding universe does not preclude a creator, but it does place limits on when he might have carried out his job!'

To be sure, there are a number of theories that suggest that the big bang might not have been the beginning of the universe but only the beginning of the present phase of the universe. These theories invoke endless cycles, each consisting of a big bang, expansion, contraction, and collapse of the universe in a 'big crunch', followed by another bang and expansion. It has even been suggested that the laws of physics might be redefined at the

[10] Leibniz, Gottfried Wilhelm (1973). *Philosophical writings*. J. M. Dent, London.
[11] Hawking, Stephen W. (1988). *A brief history of time*. Bantam Press, London.

beginning of every cycle. However, this does not solve the problem. In fact, we come right back to Leibniz's argument about previous states of the world not providing sufficient reason for the existence of the current state of the world.

Although Baruch Spinoza was a contemporary of Leibniz, his views concerning God could not have been more different. The work of Spinoza represents a radical departure from the pseudo-religious conceptions of God advanced by both Descartes and Leibniz. A Dutch Jew of Portugese descent living in a largely Christian society, Spinoza was ostracized by both the Jewish and Christian communities as an atheist and a heretic. This isolation suited his purposes well, since he wished to work quietly and independently, free of more 'earthly' distractions. It is not that Spinoza did not believe there to be a God, but his reasoning led him to the conclusion that God is *identical* with nature rather than its external creator.

Spinoza's argument is actually based on his ideas regarding the nature of *substance*. He distinguished between substances that could exist independently of other things and those that could not. The former substances provide in themselves sufficient reason for their existence—they are their own causes (*causa sui*)—and no two substances can possess the same essential attributes. He then defined God to be a substance with infinite attributes. Since different substances cannot possess the same set of attributes, it follows logically that if a substance with infinite attributes exists, this must be the *only* substance that can exist: 'Whatever is, is in God.'

Spinoza's seventeenth-century conception of God is quite consistent with twenty-first-century thinking. His is not the omniscient, omnipresent God of Judao-Christian tradition, who is frequently imagined to be an all-powerful being with many human-like attributes (such as mind and will). Rather, Spinoza's God is the embodiment of everything in nature. The argument is that when we look at the stars, or on the fragile earth and its inhabitants, we are seeing the physical manifestations of the attributes of God. God is not outside nature—he did not shape the fundamental physical laws by which the universe is governed—he *is* nature. Neither is he a free agent in the sense that he can exercise a freedom of will outside fundamental physical laws. He is free in the sense that he does not rely on an external substance or being for his existence (he is *causa sui*). He is a deterministic God in that his actions are determined by his nature.

This is the kind of God with which most western scientists would feel reasonably comfortable, if they had to accept that a God exists at all. The fact that modern physics has been so enormously successful in defining the character of physical law does not reduce the power of the argument in favour of the existence of Spinoza's God. Indeed, in his book *God and the new physics*, Paul Davies suggests that science 'offers a surer path to God than religion'.[12] Although scientists tend not to refer in their papers to God as such, with the advent of modern cosmology and quantum theory, some have argued that the need to invoke a 'substance with infinite attributes' is more compelling than ever.

As mentioned earlier, Einstein's frequent references to God were references to Spinoza's God. In his studies, he was therefore concerned to discover more about 'God or Nature'. This does not mean to say that Einstein did not believe that there must be some kind of divine plan or order to the universe. This much is obvious from his adherence to strict causality and determinism, and his later opposition to the Copenhagen interpretation. He

[12] Davies, Paul (1984). *God and the new physics*. Penguin, London.

expected to find *reason* in nature, not the apparent trusting to luck suggested by quantum indeterminism.

The triumphs of seventeenth century science clearly demonstrated that the Aristotelian dogma espoused by the Church was completely untenable. As the grip of the Church relaxed, and public opinion became generally more liberal, so it became possible for a final parting of the ways between philosophy and theology. This transition was achieved by two giants of eighteenth-century philosophy—David Hume and Immanuel Kant.

Hume demolished both the ontological and cosmological proofs for the existence of God in his *Dialogues concerning natural religion*. That this work was still controversial is evidenced by the fact that Hume preferred to arrange for its publication after his death (it was published in 1779). Some sense of Hume's situation can be gleaned from a quotation which appeared on the title page of his substantial work *A treatise of human nature*, published in 1739: 'Seldom are men blessed with times in which they may think what they like, and say what they think'.[13]

Most of Hume's arguments, which are made through the agency of a dialogue between three fictional characters, hinge around the contention that there is an inherent limit to what can be rationally claimed through metaphysical speculation and pure reason. He presents the case that the earlier conclusions regarding the existence of God made by Descartes and Leibniz and others simply do not stand up to close scrutiny. They fail because too many assumptions are made without justification. Why should existence be regarded as a perfection, as Descartes assumed? Why is it necessary for the world to have a cause, whereas God does not (indeed, cannot). Why not simply conclude that the world itself needs no cause, eliminating the need for God? Surely, the wretched state of mankind is itself sufficient evidence that the benevolent God of Christian tradition *cannot* exist?

Although Kant, coming a few years after Hume, did not entirely accept Hume's outright rejection of metaphysics, the die was effectively cast. Kant's *Critique of pure reason*, published in 1781, picked up more or less where Hume left off. In this work, Kant concluded that all metaphysical speculation about God, the soul, and the natures of things cannot provide a path to knowledge. True knowledge can be gained only through experience, and since we appear to have no direct experience of God as a supreme being, we are not justified in claiming that he exists. However, unlike Hume, it was not Kant's intention to develop a purely empiricist philosophy, in which all things that we cannot know through experience are rejected. We must *think* of certain things as existing in themselves, even though we cannot know their precise natures from the ways in which they appear to us. Otherwise, we would find ourselves concluding that an object can have an appearance without existence, which Kant argued to be obviously absurd.

Thus, Kant did not reject metaphysics *per se* but redefined it and placed clear limits on the kind of knowledge to be gained through speculative reasoning. There is still room for religion in Kant's philosophy, and he argues that there are compelling practical reasons why *faith*, as distinct from knowledge, is important: 'I must, therefore, abolish *knowledge* to make room for *belief*',[14] meaning that belief in God and the soul of man is not founded on knowledge of these things gained through speculative reason, but requires an act of faith. This does not have to be religious faith in the usual sense; it can be a very practical

[13] Hume, David (1969). *A treatise of human nature*. Penguin, London.
[14] Kant, Immanuel (1934). *Critique of pure reason* (translated by Meiklejohn, J. M. D.). J. M. Dent, London.

faith which is necessary to make the connection between things as they appear and the things in themselves of which we can have no direct experience.

Like Hume, Kant also demolished the ontological and cosmological proofs for the existence of God, because these arguments necessarily transcend experience. Thus, any attempt to prove the existence of God requires assumptions that go beyond our conscious experience and cannot therefore be justified. Belief in the existence of God is not something that can be justified by pure reason but may be justified through faith. This does not make God unnecessary, but it does limit what we can know of him.

The fundamental shift in the direction of philosophical thought which was initiated by Hume and Kant was continued and reinforced by philosophers in the nineteenth century. The divorce of philosophy from religion became permanent. Hume's outright rejection of metaphysical speculation as meaningless was eventually to provide one of the inspirations for the Vienna Circle. Indeed, logical positivism represents the ultimate development of the kind of empiricism advocated by Hume. As discussed in Chapter 6, the positivist philosophy is based on what we can say meaningfully about what we experience. With the positivists of the twentieth century, philosophy essentially became an *analytical* science. Wittgenstein once remarked that the sole remaining task for philosophy is the analysis of language.

Despite the positivists' efforts to eradicate metaphysics from philosophy, the old metaphysical questions escaped virtually unscathed. I find it rather fascinating to observe that although the possibility of the existence of God and the relationship between mind and body no longer form part of the staple diet of the modern philosopher of science, they appear to have become increasingly relevant to discussions on modern quantum physics. Three centuries of gloriously successful physics have brought us right back to the kind of speculation that it took three centuries of philosophy to reject as meaningless. It may be that the return to metaphysics is really a grasping at straws—an attempt to provide a more 'acceptable' world view until such time as the further subtleties of nature can be revealed in laboratory experiments and this agonizing over interpretation thereby relieved. But we have no guarantee that these subtleties will be any less bizarre than quantum physics as it stands at present.

And what of God? Does quantum theory provide any support for the idea that God is behind it all? This is, of course, a question that cannot be answered here, and I am sure that readers are not expecting me to try. Like all of the other possible interpretations of quantum theory discussed in this chapter, the God hypothesis has many things to commend it, but we really have no means (at present) by which to reach a logical, rational preference for any one interpretation over the others. If some readers draw comfort from the idea that either Spinoza's God or God in the more traditional religious sense (Western or Eastern) presides over the apparent uncertainty of the quantum world, then that is a matter for their own personal faith.

14

Many worlds, one universe

The concept of the collapse of the wave function was introduced into quantum theory by John von Neumann in the early 1930s and has since become an integral part of the theory's orthodox interpretation. But if we were to ask ourselves what evidence we have that this collapse is a real physical phenomenon, we would have to admit that we have none. The notion of a collapse is necessary to explain how a quantum system initially present in a linear superposition state before the process of measurement is converted into a quantum system present in one, and only one, of the measurement eigenstates after the process has occurred. It was introduced into the theory because it is our experience that pointers point in only one direction at a time.

The Copenhagen solution to the measurement problem is to say that there is no solution. Pointers point because they are part of a macroscopic measuring device which conforms to the laws of classical physics. The collapse is therefore the only way in which the 'real' world of classical objects can be related to the 'unreal' world of quantum particles. It is simply a useful invention, an algorithm, that allows us to predict the outcomes of measurements. If we wish to make the collapse a real physical change occurring in a real physical property of a quantum system, we must *add* something to the theory, if only the suggestion that consciousness is somehow involved.

The simplest solution to the problem of quantum measurement is to say that there is no problem. Over the last 70 years, quantum theory has proved its worth time and time again in the laboratory: Why change it or add extra bits to it? Although it is overtly a theory of the microscopic world, we know that macroscopic objects are composed of protons, neutrons, electrons, atoms, and molecules, so why not accept that quantum theory applies equally well to pointers, cats, and human observers? Finally, if we have no evidence for the collapse of the wave function, why introduce it?

At first sight, such a suggestion seems counter-productive. If it is indeed our experience that pointers point in only one direction, and cats are either alive or dead, then giving up the collapse would seem to be taking us in the wrong direction. Surely, we would be left with an infinite regress, with an endless complexity of superposition states of measuring devices, cats, and human observers? Well, not necessarily. By foregoing the collapse, we are acknowledging that *all* possible measurement results are realized. But perhaps they are not necessarily realized in the world that we happen to be observing. This was the proposal made over 40 years ago by Hugh Everett III in his Princeton University Ph.D. thesis. In this interpretation, the act of measurement splits the world[1] into a number of branches, with a different result being recorded in each. This is the so-called 'many-worlds' interpretation of quantum theory.

Relative states

Everett discussed his original idea, which he called the 'relative state' formulation of quantum mechanics, with John Wheeler while at Princeton. Wheeler encouraged Everett to submit his work as a Ph.D. thesis, which he duly did in March 1957. Everett went on to become a member of the Weapons System Evaluation Group at the Pentagon and a multi-millionaire. A shortened version of this thesis was published in July 1957 in the journal *Reviews of Modern Physics* and was 'assessed' by Wheeler in a short paper published in the same issue. Everett set out his interpretation in a much more detailed article which was eventually published in 1973, together with copies of some other relevant papers, in the book *The many-worlds interpretation of quantum mechanics*, edited by Bryce S. DeWitt and Neill Graham. Everett's original work was largely ignored by the physics community until DeWitt and Graham began to look at it more closely and to popularize it some 10 years later.

Everett insisted that the pure Schrödinger wave mechanics is *all that is needed to make a complete theory*. Thus, the state vector obeys the deterministic, time-symmetric equations of motion at all times in all circumstances. Initially, no interpretation is given for the state vector. Rather, its meaning emerges from the formalism itself. Without the collapse, the measurement process occupies no special place in the theory. Instead, the results of the interaction between a quantum system and an external observer are obtained from the properties of the larger composite system formed from them, analogous to von Neumann's systems I and II discussed in the last chapter.

In complete contrast to the special role given to the observer in von Neumann's and Wigner's theory of measurement, in Everett's interpretation the observer is nothing more than an elaborate measuring device. In terms of the effect on the physics of a quantum system, a conscious observer is no different from an inanimate, automatic recording device which is capable of storing an experimental result in its memory.

Everett went on to show that his relative states formulation of quantum mechanics is entirely consistent with the way quantum theory is used in its orthodox interpretation to derive probabilities. Instead of talking about projection amplitudes and probabilities, it is necessary to talk about conditional probabilities: the probability that a particular result will be obtained in a measurement given certain conditions. The name is different, but the procedure is the same.

All this is reasonably straightforward and relatively non-controversial. However, the logical extension of Everett's formulation of quantum theory leads inevitably to the conclusion that, once entangled, the relative states can *never* be disentangled.

---

[1] I'm using 'world' here deliberately in preference to 'universe', for reasons that will hopefully become apparent.

The branching world

In Everett's formulation of quantum theory, there is no doubt as to the reality of the quantum system. Indeed, his theory is quite deterministic in the way that Schrödinger had originally hoped that his wave mechanics was deterministic. Given a certain set of initial conditions, the state vector of the quantum system develops according to the quantum laws of (essentially wave) motion. The state vector describes the real properties of a real system and its interaction with a real measuring device: all the speculation about determinism, causality, quantum jumps, and the collapse of the wave function is unnecessary. However, the restoration of reality in Everett's formalism comes with a fairly large trade-off. If there is no collapse, each term in the superposition of the total state vector is real—*all* experimental results are realized.

Each term in the superposition corresponds to a state of the composite system and is an eigenstate of the observation. Each describes the correlation of the states of the quantum system and measuring device (or observer). Everett argued that this correlation indicates that the observer perceives only one result, corresponding to a specific eigenstate of the observation. In his July 1957 paper, he wrote[2]:

Thus with each succeeding observation (or interaction), the observer state 'branches' into a number of different states. Each branch represents a different outcome of the measurement and the *corresponding* eigenstate for the [composite] state. All branches exist simultaneously in the superposition after any given sequence of observations.

Thus, in the case where an observation is made of the linear polarization state of a photon known to be initially in a state of circular polarization, the act of measurement causes the world to split into two separate worlds. In one of these worlds, an observer measures and records that the photon was detected in a state of vertical polarization. In the other, the same observer measures and records that the photon was detected in a state of horizontal polarization. The observer now exists in two distinct states in the two worlds. Looking back at the paradox of Schrödinger's cat, we can see that the difficulty is now resolved. The cat is not simultaneously alive and dead in one and the same world; it is alive in one branch of the world and dead in the other.

With repeated measurements, the world, together with the observer, continues to split in the manner shown schematically in Fig. 14.1. In each branch, the observer records a different sequence of results. In early many-worlds theories, the observer is assumed to be unaware of the branching, and the results appear entirely consistent with the notion that the wave function of the circularly polarized photon collapsed into one or other of the two measurement eigenstates.

Why does the observer not retain some sensation that the world splits into two branches at the moment of measurement? The answer given by early proponents of the Everett theory is that the laws of quantum mechanics simply do not allow the observer to make this kind of observation. DeWitt argued that if the splitting were to be observable, it should be possible in principle to set up a second measuring device to obtain a result from the memory of the first device which differs from that obtained by its own direct observation. Wigner's friend could respond with an answer which differs from the one that Wigner could check for himself. This not only never happens (except where a genuine human

---

[2] Everett III, Hugh (1957). *Reviews of modern physics*, **29**, 454.

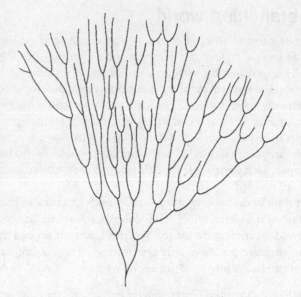

Fig. 14.1 Representation of a branching world. A repeated measurement for which there are two possible outcomes continually splits the world. The path followed from the beginning of the 'tree' to the end of one of its branches corresponds to a particular sequence of results observed in one of the split worlds.

error occurs) but also is not allowed by the mathematics. The branching of the world is unobservable.

In a footnote added to the proofs of his July 1957 paper, Everett accepted that the idea of a branching world appears to contradict our everyday experience. However, he defended his position by noting that when Copernicus first suggested that the earth revolves around the sun (and not the other way around), this view was criticized on the grounds that the motion of the earth could not be felt by any of its inhabitants. In this case, our inability to sense the earth's motion was explained by Newtonian physics. Likewise, our inability to sense a splitting of the world into different branches is explained by quantum physics.

'Schizophrenia' with a vengeance

If the act of measurement has no special place in the many-worlds interpretation, there is no reason to define measurement to be distinct from any process involving a quantum transition between states. Now, there have been a great many quantum transitions since the big bang origin of the universe, some 12 billion years ago. Each transition will have involved the development of a superposition of state vectors with each term in the super-position corresponding to different final states of the transition. Each transition will have therefore split the world into as many branches as there were terms in the superposition. DeWitt has estimated that by now there must be more than 10^{100} branches.

Presumably, some of these branches will be almost indistinguishable from the one that I (and presumably you) currently inhabit. Some will differ only in the way in which the polarized photons scattered from the surface of the monitor of the computer on which I am composing these words interact with the light-sensitive cells in my eyes. Many of these branches will contain almost identical copies of this book, being read by almost identical copies of you. Many of these worlds bear witness to the different outcomes of decisions you have made, compared with this world filled only with things you could have done (or should have done). No wonder DeWitt called the many-worlds interpretation 'schizophrenia with a vengeance'.

That there may exist 'out there' a huge number of different worlds is a rather eerie prospect. We can speculate that many of these worlds will contain the same arrangement of galaxies that we can see in our 'own' world. Some will contain a small, rather insignificant G2-type star identical to our own sun with a beautiful, but fragile-looking blue–green planet third from the centre in its planetary system. But in some of these branches, the kinds of quantum transitions involving cosmic rays and giving rise to chance mutations in living creatures will have turned out differently from those which occurred in 'our' earth's past history. Perhaps in some of these branches, mankind has not evolved, and life on earth is dominated by a different species. An individual quantum transition may appear an unimportant event, but perhaps it can have ultimately profound consequences.

Parallel worlds and 'schizophrenic' neutrons

When Everett presented his theory, he wrote of the observer state 'branching' into different states and drew an analogy with the branches of a tree. However, more recent variants of Everett's original interpretation have been proposed, in which the world with which we are familiar is but one of a very large number (possibly an infinite number) of *parallel* worlds. Thus, instead of the world splitting into separate branches as a result of a quantum transition, the different terms of the superposition are partitioned between a number of already-existing parallel worlds.

Perhaps the major difference between this interpretation and the Everett original is that it allows for the possibility that the parallel worlds may interact and *merge*. Indeed, it has been argued that we obtain indirect evidence for such a merging every time we perform an interference experiment. For example, the physicist Lev Vaidman has discussed the many-worlds interpretation in the context of an interference experiment involving (purely hypothetical) *sentient* neutrons. A classic interferometer experiment is illustrated in Fig. 14.2. In this experiment, a beam of neutrons impinges on a beamsplitter (BS_1). In a classical, particulate model, the neutrons either pass through this beamsplitter or are reflected. One possible trajectory is therefore from BS_1 to the mirror M_1, which reflects the neutron towards a second beamsplitter, BS_2. A second possible trajectory is from BS_1 to the mirror M_2, which also reflects the neutron towards BS_2. In this classical picture, there would be an equal chance of reflection and transmission of the neutrons at BS_2 and

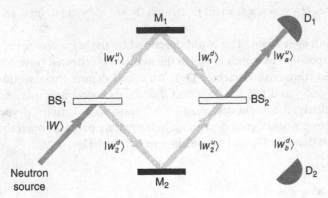

Fig. 14.2 A neutron interferometer. A neutron beam is split by beamsplitter BS_1 and recombined following reflection at mirrors M_1 and M_2 at another beamsplitting surface, BS_2. As a result of interference between the split beams at BS_2, the neutrons are subsequently detected only by detector D_1 and not by detector D_2.

therefore an equal probability of detection by detectors D_1 and D_2. Of course, in practice, neutrons are not classical particles. The conventional quantum theoretical explanation for what happens is that the neutron state vector encounters BS_1, splits and develops along both paths in the interferometer, and recombines at BS_2. With no phase difference between the paths, the state vector interferes with itself at BS_2 with the result that detection occurs *only* at D_1 (constructive interference), with *no* detection at D_2 (destructive interference).

If an individual neutron were a sentient being, Vaidman suggests, then the conventional explanation gives it some unpleasant experiences: 'it experiences being in two places and moving in two different directions simultaneously. Inside the interferometer the neutron must therefore have schizophrenic experiences.'[3] He admits that the term 'schizophrenic' does not properly describe the neutron's experiences but cannot find a better alternative.

Rather than confound the problem, as implied by the quote from DeWitt given above, Vaidman believes that the problem of the neutron's (or, more generally, the observer's) apparent schizophrenia is *resolved* in the many-worlds interpretation. On encountering BS_1, the neutron is essentially partitioned between two 'neutron worlds'. In one world, the neutron passes through BS_1 and is reflected from M_1. In a second world, the neutron is reflected from BS_1 and from M_2. To follow what happens next, it is useful to cast the problem in the language of the state vectors of the *worlds* involved.

Let us suppose that the state vector of the world prior to the interaction of a neutron with BS_1 is given by $|W\rangle$. After this interaction, the world has split into two, denoted as world 1 and world 2. In the first of these worlds, described by the state vector $|w_1^u\rangle$, the neutron has passed through BS_1 into the upper arm of the interferometer and is following an 'upwards' trajectory. On reflection at M_1, it is transformed to the state $|w_1^d\rangle$ and follows a 'downwards' trajectory. In the second world, described by the state vector $|w_2^d\rangle$, the neutron enters the lower arm of the interferometer and on reflection at M_2 is transformed to the state $|w_2^u\rangle$. Therefore, by the time the neutron arrives at BS_2, the state $|W\rangle$ has been transformed into a linear superposition of neutron worlds described by $|w_1^d\rangle$ and $|w_2^u\rangle$. Clearly, the device we have called a beamsplitter is really acting here as a 'world-splitter'.

Instead of describing these two possible routes to BS_2 in terms of quantum probabilities, Vaidman suggests that it is more appropriate in the many-worlds interpretation to think of the 'measure of existence' of the different worlds that the neutron enters. These can be calculated in the usual way from the modulus-squares of the coefficients of the different worlds that appear in the superposition. In these terms, both $|w_1^d\rangle$ and $|w_2^u\rangle$ have equal existence. This does *not* mean that the neutron has an equal probability of being in *either* $|w_1^d\rangle$ or $|w_2^u\rangle$. Rather, the neutron is in *both* worlds simultaneously, and both worlds have an equal measure of existence. However, a sentient neutron is aware of being in one, and only one, world.

At BS_2, the two worlds both split again. The world described by the state vector $|w_1^d\rangle$ transforms into a linear superposition of worlds in which the neutron is reflected (towards detector D_1) and transmitted (through to detector D_2). We could denote these worlds in terms of the state vectors $|w_{1a}^u\rangle$ and $|w_{1b}^d\rangle$, with u and d again denoting upwards and downwards trajectories. Similarly, the world described by $|w_2^u\rangle$ also splits into $|w_{2a}^u\rangle$ and $|w_{2b}^d\rangle$. We now have four neutron worlds, each describing four possible paths through the interferometer, two leading to detector D_1 and two leading to detector D_2.

---

[3] Vaidman, Lev (1996). Quant-ph/9609006 v1.

Now here is the rub. The neutron, whilst experiencing no identity crisis as it travels along two paths in two different worlds, also has no 'memory' of the path it has just taken. So, the states describing the trajectory from BS_2 to D_1, for example, cannot be distinguished on the basis of the path taken prior to BS_2. The states $|w_{1a}^u\rangle$ and $|w_{2a}^u\rangle$ therefore describe the same world (which we will denote using the state vector $|w_a^u\rangle$). Similarly, $|w_{1b}^d\rangle$ and $|w_{2b}^d\rangle$ describe the same world, which we denote $|w_b^d\rangle$. Finally, although the neutron carries no memory of the path it has taken, the nature of the superposition does depend on the direction in which it approached BS_2, according to a simple phase convention (see Appendix 27). These two superpositions combine to reduce the measure of existence of $|w_b^d\rangle$ to zero (through a cancellation of terms involving $|w_b^d\rangle$—see Appendix 27). Thus, the effect of the neutron's passage through the entire apparatus—from BS_1 to the detectors—is to transform $|W\rangle$ into $|w_a^u\rangle$. The neutron worlds exhibit interference effects, and an observer therefore perceives the neutrons to be detected only by D_1.

Clearly, this explanation holds up only if we invoke a loss of memory on the part of the neutron as it passes through the apparatus in the different neutron worlds. Indeed, we know well enough that experiments performed in the laboratory that are specifically designed to gain 'which way' information will ultimately destroy the possibility of interference. In terms of the above example, tagging the neutron somehow in order to distinguish between the worlds $|w_{1a}^u\rangle$ and $|w_{2a}^u\rangle$ and, similarly, $|w_{1b}^d\rangle$ and $|w_{2b}^d\rangle$ would mean that there would be no cancellation of terms. All four worlds would have equal measures of existence. In two of these worlds, the neutron is detected by D_1, and in the other two it is detected by D_2. With repeated measurements, the observer in any one of these worlds would conclude that the neutrons are detected by D_1 and D_2 with equal probability, and interference is lost.

This loss of memory had led theoretician David Deutsch to propose another experiment, which we will modify slightly to remain consistent with the neutron interferometer example used above. Imagine that we set up the experiment in such a way that we somehow tag the neutron so that its trajectory can be followed through either the upper or lower arm of the interferometer. We persuade Wigner's friend to note down that he definitely perceives the neutron to pass through either the upper or lower arm of the interferometer but he does not tell us which. The experiment is performed, and the result enters the memory of Wigner's friend. He writes in his notebook that he definitely saw the neutron pass through one or other arm. Now, according to the von Neumann–Wigner interpretation of the quantum measurement process, the wave function of the neutron collapsed when it encountered the observer's consciousness. The observed result is therefore the *only* result, and the other has 'disappeared' in the sense that its probability has been reduced to zero by the act of measurement. However, in the many-worlds interpretation, *both* results are obtained in two different neutron worlds, and in principle, an interference effect can still be obtained if we can somehow merge these worlds together. This would be equivalent to somehow merging the quantum memory states of Wigner's friend in the two different worlds to produce an interference. Now interference is associated with a loss of memory. Thus, having noted that he saw the neutron pass through one or other of the arms of the interferometer, Wigner's friend 'forgets' which one it was.

The observer must feel very odd under these circumstances. He remembers that he saw the neutron pass through one arm but cannot remember which one it was. This is in complete contrast to the situation obtained if the wave function collapses since, in this case, the observer will remember which arm the neutron passed through. Here, then, is one proposal for a laboratory test of the many-worlds interpretation of quantum theory (a more brutal proposal is considered below). Unfortunately, the brain does not appear to function at the level of individual quantum events. If it did, we might be able to 'feel' every quantum transition occurring inside our brains—not a very appealing prospect, perhaps. However, Deutsch has suggested that it may be possible one day to construct an artificial brain capable of functioning at the quantum level. Instead of performing this experiment with a human observer, we would ask this artificial brain to perform the experiment for us, and simply ask it what it felt.

The non-existence of non-locality

You may have been intrigued by the fact that, in the discussion of the neutron interferometer experiment above, the neutron in each of the different worlds was described in terms of its 'trajectory' through the apparatus. This is not necessarily a return to a purely particulate description of quantum entities and events, but rather a return to Schrödinger's original concept of a localized wave-packet state (where we again set aside our reservations about rapid dispersion of such a wave-packet as it moves through space). The key point is that the neutron is *localized* as it makes its way from BS_1 to either of the two detectors, with interference arising as a result of an interaction between worlds in which the neutron takes different paths. What does this mean for quantum non-locality and Bell's theorem?

Let us return for a moment to the discussion of quantum versus local hidden variable correlations in Chapter 8 and Appendix 18. We now need to reinterpret this kind of experimental test in terms of many worlds. Recall the scenario. A quantum transition results in the emission of two photons constrained by the physics to be emitted in opposite states of circular polarization. The two photons, A and B, propagate across the laboratory. Photon A encounters a polarization analyser oriented vertically with respect to some arbitrary laboratory axis, and photon B encounters a polarization analyser with its vertical axis orientated at an angle $(b - a)$ to that of the first. The two analysers are space-like separated, meaning that it is impossible to send information between them within the timescale of detection of both photons without exceeding the speed of light and hence violating one of the key principles of special relativity.

There are four possible measurement outcomes, corresponding to situations in which (1) both A and B are detected in vertical polarization states, (2) A is detected in a vertical and B a horizontal polarization state, (3) A horizontal and B vertical, and (4) both A and B horizontal. It does not matter which photon is detected first (indeed, if they are space-like separated, then the concept of simultaneity and the relative ordering of events loses its meaning). According to the many-worlds interpretation, when either one of the two photons encounters its polarization analyser, the world splits into two. When the second photon encounters its analyser, the two worlds are split further to give a total of four. The first of these worlds, which we could denote by the state vector $|w_{++}\rangle$, corresponds to a world in which both A and B are detected in vertical polarization states. The other three worlds, denoted $|w_{+-}\rangle, |w_{-+}\rangle$, and $|w_{--}\rangle$, correspond to worlds in which the alternative measurement outcomes are realized. The measurement outcomes in these separate worlds are quite deterministic. They are prescribed the moment the photons are emitted and do not change as the photons (described individually by localized wave-packet states) propagate towards their respective detectors. There is no collapse—no 'spooky action at a distance'— when the photons encounter the polarization analysers, and there is therefore no mystery.

Suppose you are the observer. You initiate the experiment, and when the photon encounters the polarization analysers, the world splits into four. You are not conscious of the split. You are conscious of the fact that the apparatus has recorded a '+−' result. In the other worlds your other selves are conscious of '++', '−+' and '−−' results, as appropriate. Now you repeat the measurement, the world splits again, and you note that the result is now '−−'. You continue to repeat the measurement to build up a picture of the relative frequency with which you obtain each of the four different possible measurement outcomes. (In practice, you do not follow individual measurement outcomes in quite this way—you leave this task to a box of electronics called a coincidence counter and observe only the coincidence *rates* when you have accumulated a large number of individual results.)

With each splitting, you as an observer build up a 'history' determined by the sequence of seemingly random measurement results—'+−', '−−', '−+', '−−', '++', '−+', '+−', and so on. With each measurement, the number of alternative histories quickly grows. Having collected a large, statistically significant, number of results you are now in a position to calculate the relative frequencies with which you obtained each result and hence the probabilities of obtaining each result in any future experiment. What determines these relative frequencies? They are determined by the measure of existence of each of the worlds you have sampled as you built up your history. They are given by the modulus-squares of the coefficients of the state vectors of each world as they appear in the linear superposition and are subject to the kind of phase convention given in Table 2, Appendix 12. You conclude that the relative frequencies depend on the angle between the vertical axes of polarizers, $(b - a)$. You repeat the experiment for a number of different angles and publish a paper claiming a violation of Bell's inequality.

According to the many-worlds interpretation, you have misinterpreted the meaning of the relative frequencies (and hence probabilities) of the different measurement outcomes. You have taken them as a measure of the (quantum) probability of the two-photon wave function collapsing (in one world only) into one of the four possible measurement eigenstates and are left to conclude that this quantum world is inherently non-local. Instead, the proponents of many-worlds would argue, you should have concluded that the relative frequencies are measures of the existence of different, but entirely local, worlds. The hidden assumption in the derivation of Bell's inequality is that—once the wave function has collapsed—the photons are present in one, and only one, quantum state in one world. In the many-worlds interpretation, there is no collapse, and the photons are present in all the different possible quantum states in different worlds.

Those who favour the many-worlds interpretation argue that the experimental observation of a violation of Bell's inequality (and, by analogy, the results of experiments on three-photon GHZ states) is not evidence for quantum non-locality. It is evidence for many worlds.

Quantum suicide: dead again?

In his book *The quark and the jaguar*, Murray Gell-Mann comments that[4]:

One distinguished physicist, well versed in quantum mechanics, inferred from certain commentaries on Everett's interpretation that anyone who accepts it should want to play Russian roulette for high stakes, because in some of the 'equally real' worlds the player would survive and be rich.

Princeton physicist Max Tegmark has developed this observation to its logical conclusion. He has proposed an experiment to test the many-worlds interpretation, which is not for the faint of heart. Indeed, experimentalists of weak disposition should look away now.

Imagine that we are able to connect two photomultipliers to a machine gun. The apparatus is set up in such a way that when the first photomultiplier (PM_1) detects a photon, a bullet is loaded into the chamber of the machine gun, and the gun fires. When the second photomultiplier (PM_2) detects a photon, no bullet is loaded into the chamber, and the gun instead just gives an audible 'click'. We now place a linear polarization analyser in front of the photomultipliers, set up so that circularly polarized photons that impinge on it are either transmitted as vertically polarized photons and are detected by PM_1 or reflected as horizontally polarized photons and detected by PM_2. We stand well back, turn on the source of circularly polarized photons, and satisfy ourselves that the apparatus fires bullets and gives audible clicks with equal frequency in an apparently random sequence. Now for the grisly bit.

You now stand with your head in front of the machine gun. (I am afraid I'm not so convinced by arguments for the many-worlds interpretation that I am prepared to risk my life in this way, and *somebody* has got to do this experiment.) Of course, as a believer in many worlds, you know that all you will hear is a long series of audible clicks. You are aware that there are worlds in which your brains have been liberally distributed around the laboratory, but you are not bothered by this because there are other worlds where you are spared. By definition, if you are not dead, then your history is one in which you have heard only a long series of audible clicks. You can check that the apparatus is still working properly by moving your head to one side, at which point you will start to hear gunfire again. If, however, the collapse is a physically real phenomenon, triggered perhaps by spontaneous random localizations (GRW theory) or by microtubules in your brain (Penrose, Hameroff), you might be lucky with the first few measurements, but, make no mistake, you will soon be killed. Your continued existence (indeed, you appear to be miraculously invulnerable to an apparatus that really should kill you) would appear to be convincing evidence that the collapse does not occur.

Apart from the obvious risk to your life, the problem with this experiment becomes apparent as soon as you try to publish a paper describing your findings to a sceptical physics community. There may be worlds in which all you recorded was a long series of audible clicks. There are, however, many other worlds where I was left with a very unpleasant mess and a lot of explaining to do. The possibility of entering one of these worlds when you repeat the experiment does not disappear, and you will find that you

---

[4] Gell-Mann, Murray (1994). *The quark and the jaguar.* Little Brown, London, p. 138.

have a hard time convincing your peers that you are anything less than quite mad. Tegmark wrote:

Perhaps the greatest irony of quantum mechanics is that . . . if once you feel ready to die, you repeatedly attempt quantum suicide: you *will* experimentally convince yourself that the [many-worlds interpretation] is correct, but you can never convince anyone else!

Time travel

It might seem unusual to go looking for paradoxes at the leading edge of theoretical physics in Hollywood 'blockbuster' movies, but the notion and consequences of time travel have always featured prominently in the literary and visual arts, from well before H. G. Wells' *The time machine* and increasingly ever since. The plotlines of these books and movies tend to draw on a number of paradoxes that, it has been argued, are evidence for the physical impossibility of time travel. The first of these is the so-called *grandfather paradox*, which is only one of any number of *inconsistency paradoxes* that we could think up. If a time-traveller goes back in time to the past and murders his own maternal grandfather before his mother is born, this will clearly prevent his mother from being born, preventing his own birth in turn. This is inconsistent. A second, more subtle *knowledge paradox* is illustrated by stories in which some futuristic technology is sent back from the future and used either directly or indirectly for the creation and establishment of that technology. As a result, the knowledge required for the creation of the design of the technology is never actually gained: nobody has solved the kinds of problems usually required to generate new knowledge. This is the technological equivalent of the ultimate 'free lunch'.

It is hard to see how such paradoxes in themselves might *prevent* time travel, although we might have to accept that we must modify our notion of free will on the part of the time-traveller if the paradoxes are to be avoided. The time-traveller would appear to be able to exercise a certain freedom in terms of an independence of thought or action but might be prevented from being able to realize certain outcomes from those actions. This is also the stuff of movie scripts. In the recent remake of *The time machine*, the central character, Alexander Hartdegen (played by Guy Pearce), is driven to build his machine in order to go back in time to prevent his fiancée from being killed. He builds the machine, goes back in time, and saves her from death at the hands of a robber, only for her to be killed in a road accident. He subsequently realizes that he cannot save her, since her death was the reason he built the time machine in the first place. Without her death, there is no machine.

In an enjoyable example of art provoking new science, it was the arrival of the manuscript of Carl Sagan's novel *Contact* on theoretical physicist Kip Thorne's desk that prompted Thorne to think more deeply about how travel over vast interstellar distances might be accomplished. Sagan had originally proposed sending his heroine, Eleanor Arroway (played by Jodie Foster in the movie version of Sagan's novel) through a black hole, but Thorne believed that travel through a *wormhole* in space–time might be more 'practical'. As his tinkering with wormhole physics developed into a formal research program, a comment made by a colleague at an astrophysics symposium made him realize that wormholes could be used not only to travel vast distances but also in principle to travel through time.

Modern physical theories do not appear to deny the possibility of time travel outright, and in recent years, a number of proposals have been advanced in which space–time wormholes could be created and navigated to exploit aspects of Einstein's relativity and move forward or backward in time. Of course, these are all highly speculative proposals, but they are consistent within the framework of current theory, and they serve to make the time–travel paradoxes as much a matter for serious scientific debate as they are plot twists for Hollywood scriptwriters.

In order to strip the inconsistency paradoxes of their more egotistical elements, Thorne picked up on a challenge from physicist Joe Polchinski that keeps the paradox firmly within the realms of physics. In this paradox, a billiard ball rolls across a billiard table and drops into the mouth of a wormhole. It emerges from the other side of the wormhole a short distance away and a short time *earlier* and continues its trajectory across the table. This trajectory brings it into collision with its later self, knocking its later self off-course so that it no longer enters the wormhole. This is, of course, inconsistent: if the ball did not enter the wormhole then it could not have emerged at an earlier time to knock its later self off-course.

Thorne and his colleagues Fernando Echeverria and Gunnar Klinkhammer were able to demonstrate that it is possible to devise fully self-consistent trajectories that satisfy all the rules of (in this case) classical physics. These differ from the paradoxical trajectory proposed by Polchinski, but they nevertheless demonstrate that time travel can be consistent with *some* trajectories. In fact, Thorne realized that for every classical trajectory, it is possible to devise an infinite number of alternatives involving time travel through a wormhole. These were subsequently rationalized as potential quantum-mechanical trajectories, each with a quantum probability of occurring.

There might be another way to avoid the inconsistency paradoxes, however. If we are prepared fully to embrace the many-worlds interpretation of quantum theory, it becomes possible to contemplate unfettered time travel, with no constraints on free will, individual determinism, or our ability to realize specific outcomes. Oxford theoretician David Deutsch is one of the staunchest defenders of the many-worlds interpretation expressed in the form of parallel universes (or what Deutsch calls *the multiverse*) and uses this approach extensively in his work on quantum computing. In such a multiverse, there is nothing to prevent a time-traveller returning to the past of a *different parallel universe* and murdering his grandfather. Cause and effect would dictate that he is 'erased' from the future of that particular parallel universe, but there are plenty of others in which his future is assured, in much the same way that you were seemingly invulnerable in the quantum suicide experiment.

Time travelling in the multiverse enables any number of bizarre possibilities, such as meeting a younger version of yourself (or, indeed, gathering together for a 'reunion' of many versions of yourself). Despite this, Deutsch argues, there is nothing implicit in either the idea or the mathematical description of time travel through the multiverse that would render it physically impossible in terms of the inconsistency paradoxes. Problems arising from knowledge paradoxes cannot be ruled out until the concept of knowledge is properly accounted for in the theoretical structure. Deutsch therefore concludes that, as far as current theoretical principles are concerned, time travel is not impossible.

So why, we might ask, are we not constantly being visited by travellers from the future? The reason could be that engineering a wormhole in a way that would allow safe passage of humans (whether through one universe or a multiverse) is a technology that lies far in the future, and although travel to the past would in principle be possible using such a machine, it could only be used to travel back to a time no earlier than the machine itself was built.

We should note that not everyone shares the same level of enthusiasm for time travel as Deutsch and other theoreticians, such as Princeton astrophysicist Richard Gott. Stephen Hawking has proposed a 'chronology protection conjecture', in effect a kind of temporal censorship in which nature prevents travel to the past, leaving the universe 'safe for historians'. In this case, it is suggested that quantum vacuum fluctuations—a constant creation and annihilation of virtual particles out of nothing (the vacuum) within timescales allowed by the uncertainty principle—may circulate through the wormhole and destroy it before it can become a time machine. Kip Thorne thinks that Hawking might be right on this one. Having opened up the physics of time travel for serious scientific debate with the publication of a research paper on the subject in 1988, Thorne writes in his 1994 book *Black holes and time warps* that he is not so confident that he is prepared to bet against Hawking's conjecture.

Many minds

Many proponents of the many-worlds interpretation do not believe that the world splits into physically distinct worlds every time we create a quantum superposition. This then begs the question that if the worlds are not *physically* distinct, in what way are we meant to distinguish them?[5]

If the brain is just a complicated machine, as the proponents of strong AI would have us believe, then presumably it acts just like another measuring device, as von Neumann reasoned. In fact, the dark-adapted eye is a very good example of a detection device capable of operating at the quantum level. It can respond to the absorption of a single photon by the retina. We do not 'see' single photons because the brain has a mechanism for filtering out such weak signals as peripheral 'noise' (but we can see as few as 10 photons if they arrive together). In the von Neumann–Wigner theory of quantum measurement, the wave function of the photons and what we might consider as the 'wave function of the brain' presumably combine in a superposition state which is then somehow collapsed by the conscious mind 'contained' in the brain. It follows that, prior to the collapse, the physical brain of a conscious observer exists in a superposition of states.

There is an alternative possibility. What if, in analogy with the many-worlds interpretation, the 'stream of consciousness' of the observer is split by the measurement process? In his book *Mind, brain and the quantum*, philosopher Michael Lockwood puts forward the proposal that the *consciousness* of the observer is split on entering a superposition state. Each of the different measurement possibilities is therefore realized, registered in different versions of the observer's conscious mind. Instead of the observer splitting the world every time they make a quantum measurement, the world splits the observer whenever a superposition state is created. Presumably, each version will be statistically weighted according to the modulus-squares of the coefficients in the superposition in the usual way. But the observer is aware of, and remembers, only one result. Observation of a particular succession of measurements, for example '+−', '−−', '−+', '−−', '++', '−+', '+−', ..., creates a particular *biography* that becomes increasingly differentiated from other possible biographies over time.

---

[5] I appreciate that this is a question of semantics: if, as conscious entities, we are prevented from ever experiencing these different worlds, then we clearly have no means to distinguish them. This does not prevent us from asking the question in what way they should be distinguished *theoretically*.

This approach, which is sometimes referred to as the *many-minds* interpretation, has at least the advantage that it is more economical than many worlds. Instead of creating a multiplicity of different physical worlds in which to realize all the different measurement possibilities, we create instead a multiplicity of alternative biographies. In each of these biographies, the observer has, in principle, a kind of quantum memory of the measurement process in which different possibilities are recalled in different parallel states of consciousness. Over time, we might expect these parallel selves to develop into distinctly different individuals as a multitude of quantum events washes over the observer's senses. Within one brain may be not one, but many ghosts, all unaware of each other.

The quantum theory of the universe

It has been said that the many-worlds interpretation of quantum theory is cheap on assumptions but expensive with worlds. In order to avoid the collapse postulate, for which we have no direct physical evidence and no clue as to where it might occur in the chain linking a quantum event with the recording of a measurement result, we appear to be obliged to reach for a mechanism involving a multiplicity of different worlds, or minds. We can avoid the collapse and still account for interference as the result of interactions between the worlds, but clearly as we can gain no direct evidence for the existence of these worlds, we appear to be really no further forward. This leaves us to scramble for options that are all really rather metaphysical. Although John Wheeler was an early champion of Everett's approach, he later rejected the theory 'because there's too much metaphysical baggage being carried along with it'. That such a bizarre interpretation can result from the simplest of solutions to the quantum measurement problem demonstrates how profound the problem is.

On the surface, we might conclude that the many-worlds interpretation does not seem to have much going for it. And yet, in an informal poll of physicists carried out at a quantum mechanics workshop in August 1997, the many-worlds interpretation was second only to Copenhagen in terms of popularity.[6] Why?

Part of the answer lies in developments in theoretical cosmology over the last decade or so. Cosmology is, in essence, the application of physical theories to study the origin and evolution of the universe. Whether we approach cosmology from a classical or quantum perspective, explaining the process and dynamics of evolution usually requires us to specify the system's initial and final conditions. When applied to the universe as a whole, we can imagine that we know something (though perhaps not everything) of the final condition, but we know nothing of the initial condition. This problem is solved in most approaches by taking what we know of the final condition and of the physics of dynamical evolution and working backwards. It was this kind of approach that led Robert Dicke, Jim Peebles, David Wilkinson, and Peter Roll to predict that the radiation left over from the big bang origin of the universe should have been cooled by subsequent expansion to a few degrees above absolute zero.[7] The near simultaneous discovery in 1965 of the 3 K cosmic microwave background radiation, by Arno Penzias and Robert Wilson, was the first real evidence for the big bang and a significant milestone in modern cosmology.

---

[6] Tegmark, Max (1997). Quant-ph/9709032 v1.

[7] Dicke, Peebles, Wilkinson, and Roll were not aware of earlier published work by George Gamow, Ralph Alpher, and Robert Herman in which the latter predicted a temperature of 5 K for this background radiation.

Applying quantum theory leads to a description of the *quantum state of the universe* and establishes a *quantum cosmology*. Your initial reaction might be to express surprise that it is even thought sensible to seek to apply quantum theory to the entire universe, given that we struggle to interpret the theory when we start to consider applying it to even the smallest of macroscopic objects. But, of course, the universe was not always as it is now. In the early stages of its evolution following the big bang, it was hot and dense, and had dimensions we would happily associate with quantum entities. Furthermore, if the universe as a whole can be considered to be in a 'pure' quantum state (rather than a mixture), then every quantum entity contained in the universe is entangled with every other quantum entity.

Even thinking about the definition of a quantum wave function of the universe leads immediately to all sorts of problems. All of the wave functions or state vectors that we write down in conventional quantum theory relate to some set of variables that refer to some form of *external* framework. For example, we could choose to write the wave function of an atom in terms of its centre of mass in relation to some inertial frame of reference, or the wave function of a photon in terms of its polarization orientation in relation to some arbitrary laboratory vertical axis. Most importantly, the notion of time itself starts to lose its meaning without reference to an *external clock*. Obviously, when considering the universe, there is no inertial frame of reference, no laboratory axis, and no external clock.

No quantum theory of the universe is possible without taking due consideration of mass and hence gravity. Quantizing Einstein's general relativity (an approach that Einstein always thought of as 'childish') leads straight into some profound conflicts that we will consider further below. In some of the earliest attempts, progress was found to be possible only by undoing the structure of the four-dimensional space–time that Einstein had originally established in his general theory of relativity. In 1967, Bryce DeWitt published an equation (which has since become known as the Wheeler–DeWitt equation) which is, in effect, the *stationary* Schrödinger equation of the universe for zero total energy. This is an equation in a space of three dimensions and in which time has disappeared, with different possible configurations of the wave function of the universe thought to represent different instants in time. DeWitt did not attach much physical significance to his approach, which also goes by the name of *canonical quantum gravity*, and later referred to the Wheeler–DeWitt equation as 'that damned equation'. But it served to highlight the tremendous problems associated with the fusion of quantum theory and general relativity, and the implied loss of the concept of time.

Actually, whilst there is much made of the need to treat time on an equal footing with space in general relativity, in fact space and time are still quite distinct. Intervals in space–time can be imaginary, in that they may contain i, the square root of -1.[8] The appearance of i is indicative of a time interval and can be taken as evidence that space and time do not become indistinguishable dimensions in four-dimensional space–time. Multiplying time intervals by i therefore has the effect of turning them into spatial intervals indistinguishable in a mathematical sense from 'real' spatial intervals and turns four-dimensional space–time into a four-dimensional space. This allows quantum approaches, such as Feynman's sum over histories, to be applied in a consistent manner. We then need to consider *all* the

---

[8] Space–time intervals are given by $\sqrt{[d^2 - (ct)^2]}$, where d is the difference in spatial coordinates (simplified here to one dimension), t is the time difference, and c is the speed of light. The space–time interval between your current position and this same position in five minutes' time is therefore roughly $90i$ million kilometres.

different four-dimensional *spatial* geometries (called four-geometries) that are consistent with the initial and final *space–time* states of the universe. A technique called 'analytic continuation' is used to turn one of the four spatial dimensions (designated as 'imaginary time') into a time dimension. The calculations can be simplified by neglecting geometries that are very different from those that most obviously 'join' initial and final states and that are therefore likely to be characterized by low probabilities, in much the same way that we can neglect the contributions of light paths far from the straight-line direction of a light ray (see Chapter 3).

But a knowledge of the initial state is still required. Stephen Hawking and James Hartle proposed a way out of this impasse in 1983. In their 'no boundary proposal', they consider only four-geometries that map onto the space–time geometry of the final state. The result is the wave function and corresponding probability of a universe with these final state characteristics *created from nothing*, where nothing here means the *absence* of space and time.

The geometries that give particularly large contributions in the sums are called *instantons*. They can be used to describe the spontaneous creation of a small universe from nothing. Different types of instantons are possible and provide the initial conditions for a variety of 'realistic' universes. Whether these provide appropriate descriptions of the initial conditions of the visible universe remains to be seen. Different instantons do predict different (but small) temperature variations in the cosmic microwave background radiation, arising from vacuum fluctuations during the so-called inflationary period of expansion following the big bang, and it is hoped that these variations may be detected by the next generation of cosmic background explorer satellites.

Clearly, assuming that outside the universe there is no external inertial frame of reference, no laboratory axis and no external clock implies that there are no external observers or measuring devices either. A quantum description of the universe becomes untenable if we have to invoke some sort of measuring device, sitting outside the universe, whose function is to collapse the wave function.[9] Many quantum cosmologists work to the assumption that there is only *one universe*, and everything that is, is in the universe. There is therefore no 'outside' in which a measuring device can be located. Tegmark has summarized the opposing 'inside' and 'outside' views as follows:

The outside view (the mathematical structure) is physically real, and the inside view and all the human language we use to describe it is merely a useful approximation for describing our subjective perceptions . . . [alternatively] . . . The subjectively perceived inside view is physically real, and the outside view and all its mathematical language is merely a useful approximation.

Quantum cosmologists clearly prefer the outside view; the mathematical structure obtained by applying the deterministic Schrödinger equation to the wave function of the universe is taken to be the reality. Under this restriction, there can be no collapse and no addition to the unitary time evolution described by the Schrödinger equation, and we are left with no choice but to opt for some form of many-worlds interpretation.

[9] The more theologically minded who may have no problem with an omnipresent, omniscient 'measuring device' that collapses the wave function of the universe are referred to the discussion on the existence of God included in the last chapter.

Consistent histories

There are many variants on this particular theme, however. If you are unnerved by the idea of many physically distinct worlds containing many versions of your conscious self, or of one world with many variants of your own consciousness, then be reassured that there are other approaches that do not invoke this kind of apparent schizophrenia. One of these approaches has been described by Murray Gell-Mann[10]:

We believe Everett's work to be useful and important, but we believe that there is much more to be done. In some cases too, his choice of vocabulary and that of subsequent commentators on his work have created confusion. For example, his interpretation is often described in terms of "many worlds", whereas we believe that "many alternative histories of the universe" is what is really meant. Furthermore, the many worlds are described as being "all equally real", whereas we believe it is less confusing to speak of "many histories, all treated alike by the theory except for their different probabilities."

The approach that Gell-Mann refers to is based on some original ideas by physicist Robert Griffiths, who elaborated the notion of 'history' in the context of quantum theory. The term history is used here much as we use it in common language, as a summary of a series of connected events unfolding in time. In this respect, Griffiths' histories are similar to Feynman's histories as described in Chapter 3, except that Feynman's approach was focused on the specific dynamics associated with the motion of quantum entities from one place to another, whereas Griffiths' approach is more concerned with the logical framework of quantum theory. Alternative Feynman histories refer to different paths a quantum entity might take from here to there, whereas Griffiths' histories refer to different descriptions of different kinds of events unfolding in time that, individually, serve as internally consistent ways of looking at the process.

The best way to illustrate the consistent histories interpretation of quantum theory is by reference to a simple example. Suppose a photon (from some conventional light source or laser) is incident on a beamsplitter, as shown in Fig. 14.3(a). The state vector of the photon is given by some arbitrary initial state, $|\Psi_0\rangle$. After interacting with the beamsplitter, conventional quantum theory would describe the state vector as a superposition of two states, one in which the photon is transmitted, which we denote $|\varphi_t\rangle$, and one in which the photon is reflected, $|\varphi_r\rangle$. This superposition evolves in time and, again according to conventional wisdom, collapses on encountering either of the two detectors, denoted D_1 and D_2, at which point the photon is projected into either of the two measurement eigenstates—$|\varphi_t^+\rangle|D_1^+\rangle$ (photon transmitted and detected by D_1) or $|\varphi_r^+\rangle|D_2^+\rangle$ (photon reflected and detected by D_2). In this description, the superscript + denotes the final state of the photon/measuring device following detection.

The consistent histories interpretation takes a very different view. Under this interpretation, all possible alternative histories must be considered. These histories describe the evolution of the system as a 'storyline'—a time sequence of events that form a 'family'. For the sake of simplicity, we will consider here events at three distinct times: an initial time, t_0, before the photon encounters the beamsplitter, an intermediate time, t_1, after interaction with the beamsplitter and a time t_2, after detection of the photon. The events within each family of histories are assigned probabilities, and each is internally consistent

---

[10] Gell-Mann, Murray (1994). *The quark and the jaguar*. Little Brown, London, p. 138.

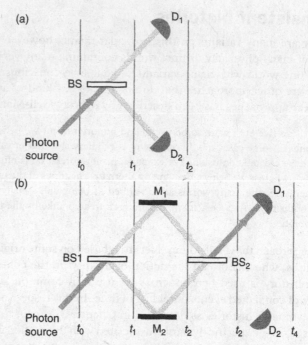

Fig. 14.3 (a) Measurement involving interaction of a photon with a beamsplitter, BS, and its subsequent detection by detectors D_1/D_2. Families of consistent histories are constructed based on quantum events that have occurred at three moments in time, t_0, t_1, and t_2. (b) Extension of (a) which is now an interferometer experiment capable of producing interference effects. Families of consistent histories can be constructed which extend to events that have occurred at times t_3 and t_4, but the histories at t_0 and t_1 will be the same as in (a).

Table 14.1 Families of histories with non-zero probabilities

Time	Family F_1	Family F_2	Family F_3
t_0	System is in initial state $|\Psi_0\rangle$	System is in initial state $|\Psi_0\rangle$	System is in initial state $|\Psi_0\rangle$
t_1	The system is in *either* state $|\varphi_t\rangle$ *or* state $|\varphi_r\rangle$	The system is in a superposition state consisting of an equal mixture of $|\varphi_t\rangle$ *and* $|\varphi_r\rangle$	The system is in a superposition state consisting of an equal mixture of $|\varphi_t\rangle$ *and* $|\varphi_r\rangle$
t_2	The system is in *either* state $|\varphi_t^+\rangle|D_1^+\rangle$ *or* $|\varphi_r^+\rangle|D_2^+\rangle$, and the photon is therefore detected by *either* D_1 *or* D_2	The system is in *either* state $|\varphi_t^+\rangle|D_1^+\rangle$ *or* $|\varphi_r^+\rangle|D_2^+\rangle$, and the photon is therefore detected by *either* D_1 *or* D_2	The system enters a macroscopic superposition state consisting of an equal mixture of $|\varphi_t^+\rangle|D_1^+\rangle$ *and* $|\varphi_r^+\rangle|D_2^+\rangle$

in that the family keeps faith with the sequence of events and the logic of probability calculus. Mutually exclusive outcomes are captured within a family, and different families of histories are incompatible in that it is not possible to take events or logical reasoning from one family and apply it to another. For the example considered, there are three families of histories that have non-zero probabilities (there are many others with zero probability that we will not consider). These are given in Table 14.1.

We recognize the first family of histories, F_1, as a description in which 'which way' information is available and in which no superposition is created—the photon either passes through the beamsplitter or is reflected by it. In this family, the probability of the system evolving from $|\Psi_0\rangle$ to either $|\varphi_t\rangle$ or $|\varphi_r\rangle$ as time passes from t_0 to t_1 is clearly $\frac{1}{2}$. As time passes from t_1 to t_2, the system in $|\varphi_t\rangle$ evolves to $|\varphi_t^+\rangle|D_1^+\rangle$ with unit probability. Similarly, the system in $|\varphi_r\rangle$ evolves to $|\varphi_r^+\rangle|D_2^+\rangle$ with unit probability. The evolution of the system from $|\varphi_t\rangle$ to $|\varphi_r^+\rangle|D_2^+\rangle$ has zero probability (and *vice versa*).

The second family, F_2, is perhaps most familiar from all our discussion so far. In this case, the system evolves from $|\Psi_0\rangle$ into a superposition state consisting of an equal mixture of $|\varphi_t\rangle$ and $|\varphi_r\rangle$ with unit probability. It then evolves into either the state $|\varphi_t^+\rangle|D_1^+\rangle$ or the state $|\varphi_r^+\rangle|D_2^+\rangle$ with a probability of $\frac{1}{2}$. The third, and final, family is also now reasonably familiar and captures the essence of the Schrödinger cat paradox. In this case, the system evolves from $|\Psi_0\rangle$ into a superposition state with unit probability and thence into a macroscopic superposition at t_2, also with unit probability.

We might be tempted at this stage to ask: What is the *correct* family? However, the consistent histories interpretation denies that there is some rule or law of nature that determines what the right history is. Instead, *all* alternative histories are equally valid (they are considered to be equally 'real' in the context of the theory) and, in this respect (and *only* in this respect), is the interpretation similar to many worlds in its approach. There is only one correct family—one single *framework*, as Griffiths describes it—but this depends on the *questions* we put to the system in the form of measurements we might make.

We can see this most clearly by expanding the arrangement in Fig. 14.3(a) to include a couple of mirrors and a second beamsplitter, giving us Fig. 14.3(b), an interferometer experiment exactly analogous to that which we considered in our discussion of schizophrenic neutrons. Obviously, the families of histories that we need to consider for this expanded arrangement are different from those considered so far and extend further in time to t_3 and t_4. However, the first event, from t_0 to t_1, is the same and described by the histories included in families F_1 and F_2, above. If we now *ask* which path an individual photon follows through the apparatus depicted in Fig. 14.3(b) and with what probability, and set up the experiment to find out this kind of 'which way' information, then a family of histories similar to the early events in F_1 is the appropriate family to consider. If, instead, we ask questions pertaining to subsequent interference effects, then a family of histories similar to F_2, with an evolution into a superposition state at t_1, is appropriate, and it is *meaningless* to ask in this situation which path the photon followed as this question relates to an *incompatible* family. Griffiths writes[11]:

Whether the events of interest occur at intermediate or final times, checking the predictions given by different incompatible frameworks always involves alternative experimental arrangements which are either mutually exclusive... or of a sort in which one set of measurements makes it impossible to discuss the system using some alternative consistent family. In any case, just as there is no single 'correct' choice of consistent family for describing the system, there is no single arrangement of apparatus which can be used to verify the predictions obtained using different families.

If we recognize some of the possible alternative histories as 'particle histories' (with 'which way' trajectories) and some as 'wave histories' (resulting in interference effects), then the consistent-histories interpretation is a restatement of Bohr's principle of complementarity in the language of probability, perhaps lending some weight to Griffiths' claim that consistent histories is 'Copenhagen done right'.

[11] Griffiths, Robert B. (1998). *Physical Review A*, 57, 1604.

The interpretation has undergone many refinements since Griffiths first introduced it in 1984. In 1991, Murray Gell-Mann and James Hartle extended the consistency conditions and introduced the concepts and principles of decoherence, and the consistent histories interpretation is now often referred to in the literature as a 'decoherent histories' interpretation. There is now something of an international group, consisting of Griffiths, Roland Omnès, Gell-Mann, Hartle, Eric Joos, Dieter Zeh, and Wojciech Zurek, that broadly sponsors the consistent-histories approach (though each member of the group appears to have a slightly different understanding and interpretation of the details).

This all seems very reasonable. All of the disconcerting metaphysical baggage associated with many-worlds has been left behind, yet here is an approach that avoids the need for any kind of notional measuring device to 'realize' the different possible outcomes of quantum events. The consistent histories interpretation puts probability and indeterminism firmly back on centre stage. The problems of objectification and the preferred basis discussed in the context of decoherence in Chapter 12 remain, yet these problems collectively represent a relatively small price to pay for the opportunity to progress theoretical developments in quantum cosmology and, as we will see below, quantum gravity. For a time, a sense of calmness and order prevailed, and much of the 'nonsense' associated with the more extreme interpretations of quantum theory evaporated.

But it does not do to underestimate the way quantum theory consistently undermines our conception of physical reality. The community of theorists were jolted back to a sense of unreality by theorists Fay Dowker and Adrian Kent, at a conference on quantum gravity at Durham in England, in the summer of 1995. In his book *Three roads to quantum gravity*, Lee Smolin described what happened[12]:

When Fay Dowker began her presentation on the consistent histories formulation, that approach was generally regarded as the best hope for resolving the problems of quantum cosmology... In a masterful presentation, she built up the theory, elucidating along the way some of its most puzzling aspects. The theory seemed in better shape than ever. Then she proceeded to demonstrate two theorems that showed that the interpretation did not say what we thought it did. While the 'classical' world we observe... may be one of the consistent worlds described by a solution to the theory, Dowker and Kent's results showed that there had to be an infinite number of other worlds too. Moreover, there are an infinite number of consistent worlds that have been classical up to this point but will not be anything like our world in five minutes' time. Even more disturbing, there were worlds that were classical now that were arbitrarily mixed up superpositions of classical at any point in the past. Dowker concluded that, if the consistent-histories interpretation is correct, we have no right to deduce from the existence of fossils now that dinosaurs roamed the planet a hundred million years ago.

Because there is no 'correct' family of histories that emerges as a result of the exercise of some law of nature, the theory regards all possible families to be equally real, and the choice of framework depends on the questions we ask. This leaves us with a significant context dependence, in which our ability to make sense of the theory depends on our ability to ask the 'right' questions. Rather like the vast computer Deep Thought, built to answer the ultimate question of 'life, the universe and everything' in Douglas Adams's *Hitch-hiker's guide to the galaxy*, we are furnished with the *answer*[13] but can only make sense of this if we can be more specific about the *question*.

[12] Smolin, Lee (2000). *Three roads to quantum gravity*. Perseus Books, New York, p. 43.
[13] 42.

Quantum gravity

The problems that the quantum cosmologists are experiencing can be summarized in Fig. 14.4, which is a simplified version of a diagram used by Penrose in his book *The large, the small and the human mind*. In Fig. 14.4, the principal theories of the physical world are related by a cube structure. The top face of the cube represents 'classical' theories, the bottom face represents quantum theories, the back face represents non-relativistic theories, the front face represents relativistic theories, and the left and right faces represent non-gravitational and gravitational theories, respectively. Starting at the top left with classical mechanics (which Penrose referred to as 'Galilean physics', a term I have retained here), then accounting for the effects of gravity (represented by a non-zero gravitational constant, G) takes us along the edge of the cube to Newtonian physics. Accounting for relativistic effects through a finite speed of light c (and hence a non-zero value of $1/c$) brings us forward to Einstein's special relativity. Accounting for quantum effects through a non-zero value of $\hbar$ takes us downwards to non-relativistic quantum mechanics.

We can, in principle, play the same game from any corner of the cube, although drawing straight arrows as I have done here should *not* be taken to imply a straightforward link between theories that result just from the introduction of a physical constant. (In

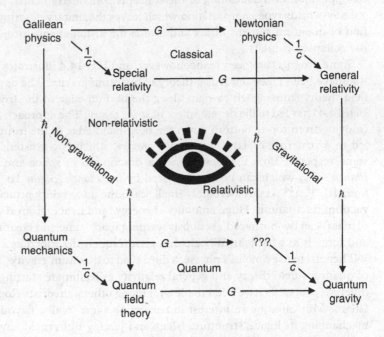

Fig. 14.4 Theories of physics and the relationships between them expressed in terms of a simple cube. Incompatibilities arise between classical and quantum theories because the status of the observer is different, and the interpretation of quantum theory is largely anti-realist in nature. The ??? represents a quantized but non-relativistic version of Newtonian physics. Adapted with permission from Penrose, Roger (1997). *The large, the small and the human mind*. Cambridge University Press, Cambridge, p. 91.

Penrose's original diagram, he drew increasingly twisting and convoluted arrows leading to theories towards the bottom right of the cube—I have omitted these for simplicity.) The primary purpose of this figure is to focus on the status of the *observer* in each of these theories (represented in Fig. 14.4 as an ever-watchful eye). In Galilean and Newtonian physics, space and time are absolute, and the observer has no special or privileged role. In Einstein's special and general theories of relativity, the nature of space and time is radically changed, but the status of the observer remains broadly the same (despite the important role of 'observation' in relativity). In non-relativistic quantum mechanics, the Newtonian concepts of space and time are retained, but the status of the observer is now potentially dramatically changed. Shoehorning special relativity and quantum mechanics together to create quantum field theory caused the structure literally to blow up in the faces of the theorists. The considerable difficulties encountered in forging a full account of gravitational effects in a quantum theory—creating a theory of quantum gravity—may be resolved only when the conflicting demands of the status of the observer are resolved.

A parallel argument is that the most common interpretation of quantum theory is fundamentally anti-realist in nature, but both the special and general theories of relativity are realist. It might be imagined that these philosophical issues of interpretation are irrelevant to the creation of a self-consistent theory, but the problems that arise when we attempt to fuse quantum field theory and general relativity stem in part from a fundamental opposition of the meaning of the concepts contained in each, hence the attraction of a many-worlds type interpretation which leaves the unitary description of the time evolution of quantum systems intact and avoids the instantaneous projection associated with the collapse postulate.

Some progress has been made, however, and Fig. 14.4 illustrates the various directions that have been taken towards a theory of quantum gravity. The approach from quantum field theory (moving left to right along the bottom edge of the front face of the cube in Fig. 14.4) has led to the development of *string theory*. The approach from general relativity (moving from top to bottom down the right-hand edge of the front face of the cube) has led to a structure called *loop quantum gravity*, which is consistent with string theory in some respects. Both theories provide a description of space and time at the so-called *Planck scale*, which can be characterized by the *Planck length*, 10^{-33} cm and the *Planck time*, 10^{-43} s.[14] At this incredibly small scale, the uncertainty principle permits very large vacuum fluctuations. Huge amounts of energy (and hence, from $E = mc^2$, huge amounts of mass) can be 'borrowed', seriously warping space–time and creating all manner of tubes and tunnels in what John Wheeler has picturesquely described as a 'space–time foam'.

There is, to use Smolin's phrase, a third road to quantum gravity. This involves rejection of quantum-field theory and general relativity as legitimate starting points. A number of different proposals have been made by, among others, theorists Roger Penrose and Chris Isham. Motivated by an interest in recovering some 'realist flavour' in quantum theory by changing its logical structure, Isham and Jeremy Butterfield have applied a version of category theory called *topos theory*, originally developed by mathematician Bill Lawvere and others in the early 1960s. The result is a more realist version of quantum theory (thereby reducing problems of incompatibility with general relativity) which retains the

---

[14] The Planck length and Planck time can be calculated from the fundamental constants $G, \hbar$, and c. The Planck length is given by $\sqrt{(G\hbar/c^3)}$, and the Planck time is given by $\sqrt{(G\hbar/c^5)}$ (the Planck length divided by c).

context dependence of the Copenhagen interpretation whilst attributing values to physical quantities that Copenhagen would deny. All this is reconciled by adopting a logic that propositions regarding the values of physical quantities can be something other than 'true' or 'false'.[15]

How far are we away from a robust theory? Smolin believes that the present problems *will* be overcome and that we shall have a basic framework for the quantum theory of gravity by around 2010, or 2015 at the latest, and that quantum gravity will be taught in high school by the end of the twenty-first century.

[15] See, for example, Isham, C. J. and Butterfield, J. (2000). *Foundations of Physics*, **30**, 1707. This parallels a similar attempt to develop a 'quantum logic' by Garrett Birkhoff and John von Neumann. In a paper published in 1936, they showed that quantum theory demanded a non-classical logic and sought to introduce an alternative which obeyed a non-distributive law of identities among set combinations. The logic applied in Isham and Butterfield's application of topos theory differs from the Birkhoff–von Neumann logic in that it is distributive.

Closing remarks

Quantum theory emerged from a desperate measure designed to shoehorn inherently non-classical phenomena into 'impossible' classical structures. Those structures were eventually swept away in the early 1930s, to be replaced with a quantum formalism that was strong on mathematical rigour but light on physical intuition, and an interpretation that gradually became strongly anti-realist as a result of Einstein's constant probing and incisive criticisms. For some considerable time during the twentieth century, it was not possible for a voice to be raised against the dogma of the Copenhagen school and still be taken as a serious contribution to the development of quantum physics. According to the Copenhagen interpretation, any hope of restoring some sense of an underlying physical reality had to be abandoned as futile and meaningless metaphysics. The physics of the microworld had reached the end of the road. There was no deeper knowledge to be gained. The world of physics was driven to a schism between realist and anti-realist philosophies, with the anti-realists very much in the ascendancy.

But the stranglehold of the 'Kopenhagener Geist' was gradually relaxed, and the natural inquisitiveness of a few overcame the cautiousness of the many. Anyone who holds to a belief that real progress can be made within a climate of unquestioning acceptance of dogmatic 'truths' should understand the lessons of quantum physics of the 1950s and 1960s. First David Bohm, then John Bell began asking awkward questions, refused to accept the standard answers, and refused to go away. Without their efforts, there would have been no ideas for new experimental tests of quantum theory. With their help, what had been rather semantic arguments about philosophical principles became disturbing arguments about experimental facts. At stake was quantum non-locality and its implications for our grasp on reality.

Remarkably, although the principle of non-locality has been upheld by experiments of ever-increasing sophistication, these results have *not* led to the dismissal of alternatives to the Copenhagen interpretation. Far from it. The experiments have served to bring home the inherent weirdness of the quantum world to many more scientists and non-scientists. And almost every time anyone reaches for the Copenhagen interpretation as an explanation or as a way out, it is usually found wanting. In his book *The character of physical law*, first published in 1965, Richard Feynman wrote[1]:

I think I can safely say that nobody understands quantum mechanics... Do not keep saying to yourself, if you can possibly avoid it, 'But how can it be like that?' because you will get 'down the drain', into a blind alley from which nobody has yet escaped. Nobody knows how it can be like that.

The new generation of physicists and cosmologists struggle with quantum concepts no less than their predecessors and regularly find themselves going down innumerable blind alleys, or even 'down the drain'. But, despite everything, they appear more determined than ever to understand just how it can be like that. The schism in physics remains, as evidenced by the realism of theorists such as Roger Penrose versus the anti-realism of others such as Stephen Hawking, and the Copenhagen interpretation remains the entrenched, default view of many physicists. But the debate is now much more open, and there is greater tolerance of theoretical speculations that are more metaphysical in nature. The proponents of a more metaphysical outlook are not (paraphrasing Ayer) branded as criminals, nor even as patients, but as advocates of legitimate theoretical constructions that deserve to be taken seriously and put to the test wherever and whenever this is possible.

We have reached the end of our guided tour of the meaning of quantum theory. I hope you have enjoyed it. I have tried to be a reasonably impartial guide in the sense that I have tried to resist the temptation to argue from a particular position, in favour of a particular interpretation. In fact, I hope that I have argued for all the different positions described in this book with something approaching equal force. This has been necessary to capture the lively nature of a debate that has been going on for over 70 years. It has been necessary, moreover, to get across the important message that quantum theory has *more than one interpretation*.

It is usual at the end of a tour such as this for the guide to be asked their opinion. I have read a number of books written by mathematicians and physicists in which all the experimental evidence against the notion of local reality has been carefully weighed, but which then close with some kind of final plea for an independent reality. I hope I have done enough in this book to demonstrate that, no matter where we start from, we always return to the central philosophical arguments of the anti-realist versus the realist. The conflict between these philosophical positions formed the basis of the Bohr–Einstein debate and continues today in lively exchanges between theoreticians such as Hawking and Penrose. No matter what the state of experimental science, the conflict between the anti-realists' conception of an empirical reality and the realists' conception of an independent reality can never be resolved. The experimental results described in Chapters 9 and 10 cannot shake the realists' deeply held belief in an independent reality, although they certainly make this reality more obscure. Thus, any final plea for an independent reality is really an appeal to *faith*, in the sense that the realist must ultimately accept the logic of the anti-realists' argument but will still not be persuaded. In Einstein's words, this is a 'trust in the rational character of reality and in its being accessible, to some extent, to human reason'.

I myself am not deeply troubled by the prospect of a reality which is not independent of the observer or the measuring device. However, I do not share the uncompromising views characteristic of the anti-realist. I am convinced that the desire to relate their theories to

---

[1] Feynman, Richard (1965). *The character of physical law*. MIT Press, Cambridge, MA, p. 129.

elements of an independent reality is part of the psychological make-up of many scientists. They feel it is necessary to try continually to go beyond the symbols in a mathematical formalism and attach a deeper meaning to them. Without this continual attempt to penetrate to an underlying reality, science would be a sterile, passive, and rather unemotional activity. This it certainly is not. Like all acts of faith, the search for an independent reality involves striving for a goal that can never be reached. This does not mean that the effort is any less worthwhile. On the contrary, when free of the straitjacket of dogma, it is through this process of striving for the unachievable that real progress in science is made.

With regard to quantum theory, my personal view is that we still do not yet know enough about the physical world to make a sound judgement about its meaning. The anti-realist says that the theory is all there is, but the realist says: Look again, we do not yet have the whole story. As to where we might look, it is fairly obvious that the profound incompatibilities between general relativity and quantum-field theory cannot be tolerated and must be resolved in any workable structure for quantum cosmology and quantum gravity. My recommendation is to watch *time* closely: we do not yet seem to have a good explanation of it. Indeed, there are some who argue that time itself is really an illusion. This is not to say that a better understanding of time will automatically solve all the conceptual problems of quantum theory. Time, I suppose, will tell.

I am reasonably certain of one thing. The unquestioning acceptance of the Copenhagen interpretation has served to hold back progress on the development of alternative approaches. Blind acceptance of the orthodox position cannot produce the challenges needed to push the theory eventually to its breaking point. And break it will, probably in a way nobody can predict, to produce a theory nobody can imagine. The arguments about reality will undoubtedly persist, but at least we will have a better theory.

I have tried to argue that quantum theory is a difficult subject for modern students of physical science because its interpretation is so firmly rooted in philosophy. If, in arguing this case, I have only made the subject seem even more confusing, then I apologize. However, my most important message is a relatively simple one: quantum theory is rife with conceptual problems and contradictions, and its most common interpretation is anti-realist in nature. If you find the theory difficult to understand, this is the theory's fault—not yours.

Appendices

Appendices

Appendix 1

Maxwell's equations and the speed of light

The first of Maxwell's equations comes from Coulomb's law and Gauss's law for the electric field. It says simply that the strength (or intensity or flux) of the electric field passing through a particular enclosed region of space depends on the net electrical charge contained in that space. This is intuitively obvious—twice as much charge produces twice the field strength. Mathematically, this is expressed in terms of the *divergence* of the field, written:

$$\operatorname{div} \mathbf{E} = \frac{\partial E_x}{\partial x} + \frac{\partial E_y}{\partial y} + \frac{\partial E_z}{\partial z} = \frac{\rho}{\varepsilon},$$

where $\mathbf{E}$ is the electric field vector, E_x, E_y, E_z are the vector components of $\mathbf{E}$ in the x, y, z directions, ρ is the electric charge density, and ε is a constant characteristic of the medium through which the field is passing, called the electric permittivity.

The electric field generates an associated magnetic field, but as there are no independent sources of magnetic field strength (we would say today that there are no magnetic monopoles—at least so far as we know), there is nothing that can cause the magnetic field strength to diverge in an analogous way, so:

$$\operatorname{div} \mathbf{B} = 0,$$

where $\mathbf{B}$ is the magnetic field vector.

The remaining two equations refer to a property of the fields that, at first sight, might appear to be a rather odd thing to be interested in. Their origin relates to the approach that Maxwell took when he first derived the equations, which was to visualize the fields as resulting from forces generated by whirlpool-like vortices in the ether. Maxwell was therefore interested in the rotation (or circulation, or 'vorticity') of the fields which today we represent in terms of the *curl* of the field vectors.[1] Maxwell was also well aware of

---

[1] In fact, those familiar with vector calculus will recognize that the divergence of a vector field is the scalar or 'dot' product of the gradient operator (∇) with the field, that is, $\operatorname{div} \mathbf{E} = \nabla \cdot \mathbf{E}$, and the curl is the vector

the results of experiments in which electric current is passed along a straight wire. In this case, a magnetic field is generated with lines of force rotating clockwise around the wire, an effect predicted by the 'right-hand rule'.

Like divergence, curl has a complex mathematical form, but it is relatively easy to understand the physical effects that these are describing. A uniform field moving in a single direction has zero divergence and zero curl. A field moving radially outwards from a single point source has a divergence related to the strength of the source, but zero curl. A uniformly circulating field has zero divergence and a curl related to the direction of rotation. Mathematically, the curl of a vector field (e.g. the electric field $\mathbf{E}$) along the z-direction is:

$$(\text{curl}\,\mathbf{E})_z = \frac{\partial E_y}{\partial x} - \frac{\partial E_x}{\partial y}$$

$(\text{curl}\,\mathbf{E})_x$ and $(\text{curl}\,\mathbf{E})_y$ can be obtained simply by permuting the coordinates x, y, z. Maxwell obtained:

$$\text{curl}\,\mathbf{E} = -\frac{\partial \mathbf{B}}{\partial t},$$

which describes the generation of a 'vortex' in the electric field resulting from a time-varying magnetic field, and

$$\text{curl}\,\mathbf{B} = \mu\sigma\mathbf{E} + \mu\varepsilon\frac{\partial \mathbf{E}}{\partial t},$$

where μ is the permeability of the medium, and σ is its electrical conductivity. The first term in the equation for $\text{curl}\,\mathbf{B}$ is often written $\mu\mathbf{J}$, where $\mathbf{J}$ is the total density of electric current, and $\mathbf{J} = \sigma\mathbf{E}$ is Ohm's law. This term describes the generation of a vortex in the magnetic field resulting from the passage of an electric current through the medium. The second term represents a further contribution to the vortex resulting from a time-varying electric field. These two expressions cannot be symmetric in their representation of $\mathbf{E}$ and $\mathbf{B}$ because it is not possible to generate currents by the flow of magnetic charge.

If we now consider electric and magnetic fields moving in free space, there is no medium and hence no charge density, $\text{div}\,\mathbf{E} = 0$, $\text{div}\,\mathbf{B} = 0$, $\sigma = 0$, $\varepsilon = \varepsilon_0$, and $\mu = \mu_0$ where ε_0 and μ_0 are, respectively, the permittivity and permeability of free space. Under these conditions, Maxwell's equations reduce to:

$$\text{curl}\,\mathbf{E} = -\frac{\partial \mathbf{B}}{\partial t}$$

and

$$\text{curl}\,\mathbf{B} = \mu_0\varepsilon_0\frac{\partial \mathbf{E}}{\partial t}.$$

These equations can be restructured to give[2]:

$$\nabla^2\mathbf{E} = \mu_0\varepsilon_0\frac{\partial^2 \mathbf{E}}{\partial t^2}$$

or 'cross' product of the gradient operator with the field, that is, $\text{curl}\,\mathbf{E} = \nabla \times \mathbf{E}$. This clearly brings out the symmetry of Maxwell's equations.

[2] It is true. Trust me.

and

$$\nabla^2 \mathbf{B} = \mu_0 \varepsilon_0 \frac{\partial^2 \mathbf{B}}{\partial t^2},$$

where $\nabla^2 = \partial^2/\partial x^2 + \partial^2/\partial y^2 + \partial^2/\partial z^2$ is called the Laplacian operator, after the mathematician Pierre Laplace. Not only do these last equations demonstrate the symmetry in the time dependences of the fields, but also they are rather obviously equations describing *wave motion*. The equation for three-dimensional wave motion, well known in Maxwell's time, is:

$$\nabla^2 \Psi = \frac{1}{v^2} \frac{\partial^2 \Psi}{\partial t^2},$$

where Ψ is a generalized *wave function*, and v is the wave velocity. Maxwell was led to the conclusion that the velocity of what he could now regard as electromagnetic waves travelling in free space is directly related to the permittivity and permeability of free space according to the relation:

$$v = \frac{1}{\sqrt{(\mu_0 \varepsilon_0)}}.$$

He used the most recently available results from the 1856 experiments of Wilhelm Weber and Rudolph Kohlrausch to calculate the value for v and found it[3]:

so nearly that of light, that it seems we have strong reason to conclude that light itself (including radiant heat, and other radiations if any) is an electromagnetic disturbance in the form of waves propagated through the electromagnetic field according to electromagnetic laws.

[3] Quotation reproduced in Hecht, Eugene and Zajac, Alfred (1974). *Optics*. Addison-Wesley, Reading, MA.

Appendix 2

Black-body radiation and the origin of the quantum

Towards the end of the nineteenth century, several theoretical models had been constructed to describe how the spectral density of black-body radiation, written as $\rho(v, T)$, depends on the values of the radiation frequency, v, and temperature, T. These expressions were moderately successful, but provided only parts of the story that Max Planck was eventually to complete. Wien's law from 1896 is an example:

$$\rho(v, T) = \alpha v^3 e^{-\beta v/T},$$

where α and β are constants. This does a reasonably good job of reproducing the experimentally observed variation of $\rho(v, T)$ with v at high radiation frequencies (short wavelengths) but fails in the infrared, as illustrated in Fig. 1.3(a).

In contrast, the Rayleigh–Jeans law was derived more directly using basic thermodynamic arguments in which the available energy was assumed to be evenly distributed over a set of 'oscillators' of largely unspecified nature whose purpose was to ensure that equilibrium was reached (see text). In Rayleigh's derivation, each oscillator was assumed to carry an internal energy equal to kT, where k is Boltzmann's constant. The Rayleigh–Jeans law can be written:

$$\rho(v, T) = \frac{8\pi v^2}{c^3} U = \frac{8\pi v^2}{c^3} kT,$$

where U is the internal energy of an individual oscillator, and the term $8\pi v^2/c^3$ can be derived from Maxwell's equations and scales the internal energy of a single oscillator to that of the entire radiation field. The Rayleigh–Jeans law does much better in the infrared but, of course, predicts that the spectral density should increase with v without limit in what later came to be known as the ultraviolet catastrophe.

In October 1900, Planck was unaware of Rayleigh's efforts published in June of the same year but was guided by the latest experimental results to deduce the formula:

$$\rho(v, T) = \frac{8\pi v^2}{c^3} \frac{hv}{e^{hv/kT} - 1}$$

or, perhaps more familiarly,

$$\rho(v, T) = \frac{8\pi h v^3}{c^3} \frac{e^{-hv/kT}}{1 - e^{-hv/kT}},$$

where h is Planck's constant. We can see from this equation that at high frequencies, the term $1 - e^{-hv/kT}$ tends to 1, and Planck's radiation law becomes functionally equivalent to Wien's law, where Wien's $\alpha = 8\pi h/c^3$ and Wien's $\beta = h/k$. At low frequencies, $e^{hv/kT}$ can be approximated as $1 + hv/kT$, and Planck's radiation law reproduces the Rayleigh–Jeans law.

In itself, this result was perhaps relatively unremarkable. Whilst it provided an excellent fit to the experimental data, it suffered from the simple fact that Planck had not derived it strictly from 'first principles'. In the meantime, the seemingly innocuous term hv had entered the physics literature for the first time.

In searching to provide a derivation of his law that carried more physical meaning, Planck was forced to use Boltzmann's statistical thermodynamic approach. A connection between energy distributions and the radiation law was available through the entropy of the radiation field. From basic principles (and comparison with the Rayleigh–Jeans law, though as mentioned above Planck was unaware of this law at the time), it is clear that the internal energy of an individual oscillator required by Planck's result is not kT, but rather:

$$U = \frac{hv}{e^{hv/kT} - 1}.$$

This equation can be rearranged to give an expression for $1/T$ as a function of U, from which it is possible to use the thermodynamic equality $\partial S/\partial U = 1/T$ to yield:

$$S = k \log \left[\frac{(1 + (U/hv))^{1+U/hv}}{(U/hv)^{U/hv}} \right] + C,$$

where C is a constant of integration. Aside from the fact that it did not make much sense at the time to talk of the entropy of a single oscillator, this result for S begs comparison with Boltzmann's famous equation $S = k \log W$, where W is the probability of the most probable energy distribution. Planck recognized the functional form of the expression in square brackets as the result of the combinatorial relation for the number, P, of *indistinguishable* energy elements ε, distributed over N oscillators. To see this clearly, we simply need to obtain the total entropy for N oscillators (which is simply N times the result for one oscillator, above) and recognize that the total energy is given by N times the individual average oscillator energy U and also by the total number of energy elements P times the size of each element, ε. In other words: $NU = P\varepsilon$, or $U = P\varepsilon/N$.

Putting all this together gives:

$$S_N = k \log \left[\frac{(N + (P\varepsilon/hv))^{N+P\varepsilon/hv}}{N^N (P\varepsilon/hv)^{P\varepsilon/hv}} \right] + C,$$

where S_N is the total entropy of N oscillators. We are now very close. All that Planck needed to do was identify the term in square brackets with the most probable distribution of energy over N oscillators, W_N, and assume that the *energy elements* ε *are fixed in size and equal to* $h\nu$. By taking $\varepsilon/h\nu = 1$, the term in square brackets simplifies to:

$$W_N = \frac{(N+P)^{N+P}}{N^N P^P}.$$

As Planck realized, this is a modified version of the combinatorial expression for the distribution of P indistinguishable energy elements over N oscillators, which is given by:

$$W_N = \frac{(N-1+P)!}{(N-1)!P!} \cong \frac{(N+P)!}{N!P!} \cong \frac{(N+P)^{N+P}}{N^N P^P},$$

where we have used Stirling's formula, $N! = (N/e)^N$, to approximate the factorials of the large numbers N and P.

Appendix 3

Atomic theory and the emergence of quantum numbers

Balmer had measured one series of emission lines from atomic hydrogen and had discovered that the pattern of frequencies of the lines conformed to the relation:

$$\nu_n = R \left(\frac{1}{2^2} - \frac{1}{n^2} \right), \quad n = 3, 4, 5 \dots$$

where ν_n is the frequency of the emitted radiation, and R became known as the Rydberg constant. It was the involvement of integer numbers that gave Bohr a clue to the explanation of Balmer's formula. Within a few weeks of learning (or being reminded) of its existence, he had finished a paper explaining that the integers were, in fact, *quantum numbers*.

In the first of three papers setting out his new theory, Bohr adopted a model for the hydrogen atom based on an electron forced to move in a stable elliptical orbit around a singly positively charged nucleus. To simplify matters, it is better to jump straight to the model of circular orbits that Bohr later adopted, following the arguments first presented by J. W. Nicholson. In classical terms, the problem is the familiar one of a mechanical system involving motion under a central force. In this case, the central force is the electrostatic attraction in free space between the positively charged nucleus and the negatively charged electron, as given by Coulomb's law, $-e^2/4\pi\varepsilon_0 r^2$, where e is the magnitude of the electron charge, ε_0 is the permittivity of free space, and r is the fixed radius of the orbit. This is balanced by the centrifugal force of the moving electron, $m_e v^2/r$, where m_e is the mass of the electron, and v is its orbital velocity (and we have conveniently forgot all about the fact that, classically, this is impossible). In other words:

$$\frac{m_e v^2}{r} - \frac{e^2}{4\pi\varepsilon_0 r^2} = 0,$$

which can be rearranged to give:

$$v = \frac{e}{\sqrt{(4\pi\varepsilon_0 m_e r)}}.$$

At this stage, Bohr imposed the condition that defined the stable electron orbits. He chose to impose this condition not on the total orbital energy (which we will come to later), but on the orbital *angular momentum*. The angular momentum, **L**, of the electron in a circular orbit around the nucleus is given by the product of the electron position vector **r** and momentum vector **p** according to $\mathbf{L} = \mathbf{r} \times \mathbf{p}$. For a circular orbit, the **L** vector points in a direction perpendicular to the plane of the orbital motion, and the vector product can be simplified to $L = rp = m_e vr$, where we have used $p = m_e v$.

Bohr imposed the condition:

$$L_n = m_e v_n r_n = n\frac{h}{2\pi},$$

where n is an integer number, and we have introduced the subscripts to denote the angular momentum, velocity and radius of the specific orbit characterized by n. Using the result for v above, we can rearrange the above expression as follows:

$$r_n = n^2 \frac{\varepsilon_0 h^2}{\pi m_e e^2}$$

showing how the radius of the nth orbit depends on the value of n.

We are now ready to consider the total energy. This is given simply by the sum of the kinetic and potential energies of the electron in the nth orbit according to:

$$E_n = \frac{1}{2}m_e v_n^2 - \frac{e^2}{4\pi \varepsilon_0 r_n} = -\frac{e^2}{8\pi \varepsilon_0 r_n},$$

where we have again substituted for v. E_n is the energy associated with the stable orbit characterized by the integer number n and is negative since it represents a state of lower energy compared with the completely separated stationary electron and nucleus defined as the arbitrary energy zero. This last expression allows us to calculate the energy of the orbit from its radius, so we can use the above equation for r_n to give:

$$E_n = -\frac{m_e e^4}{8\varepsilon_0^2 h^2}\frac{1}{n^2}.$$

Now we see that the energy of the electron orbit increases (becomes less negative) as the integer n increases, and the relation between energy and the inverse square of n is clearly taking us closer to Balmer's formula.

The only thing that remained for Bohr was to impose a final condition. The equation for E_n defines a set of *quantized* energy levels for the hydrogen atom. Bohr assumed that the lines seen in the hydrogen atom emission spectrum correspond to *transitions* between these quantized energy levels. So, in falling from an energy level characterized by a high quantum number (say E_{n_2}) to a level characterized by a lower quantum number (say E_{n_1}), the energy difference would appear as electromagnetic radiation with a frequency proportional to the difference in energy between E_{n_2} and E_{n_1}. Obviously, the constant of proportionality would be Planck's constant:

$$E_{n_2} - E_{n_1} = h\nu.$$

Using the expression for E_n allows us to calculate the frequency of emitted radiation as:

$$\nu = \frac{m_e e^4}{8\varepsilon_0^2 h^3} \left(\frac{1}{n_1^2} - \frac{1}{n_2^2} \right).$$

Balmer's formula is therefore just a special case of a more general expression and describes the series of emission lines sharing a common lower energy level, $n_1 = 2$, and higher energy levels characterized by the quantum numbers $n_2 = 3, 4, 5$, etc. The Rydberg constant is simply a collection of fundamental physical constants. Bohr used the prevailing values of the physical constants to predict a value for the Rydberg constant which was within experimental error of the measured value. Bohr noted that setting $n_1 = 3$ gives the Paschen series, and $n_1 = 1$ and $n_1 = 4$ and 5 predicted further emission series that had not at that time been observed experimentally.

Appendix 4

Special relativity and de Broglie's hypothesis

Einstein's special theory of relativity places constraints on objects in motion that are absent in Newton's mechanics but which are necessary to ensure an equivalence of the laws of physics for all observers in all inertial frames of reference. These constraints are felt most acutely as objects are accelerated to speeds close to that of light, which in Einstein's theory represents an ultimate speed that cannot be exceeded. In a sense, many of the equations of special relativity are all about ensuring that the speed of light retains this status. Thus, a particle moving with velocity v has an inertial mass m given by the equation:

$$m = \frac{m_0}{\sqrt{(1 - v^2/c^2)}}.$$

As v approaches c, the inertial mass m (a measurement of the particle's resistance to further acceleration) tends to infinity, and it becomes increasingly difficult to accelerate the particle further, clearly demonstrating the role of c as an ultimate speed.

The other crucial aspect of special relativity (and the aspect for which Einstein is most popularly known) is, of course, the equivalence of mass and energy as expressed in the equation $E = mc^2$. The relativistic energy of a freely moving particle is therefore obtained by combining this expression with the equation for relativistic mass given above:

$$E = \frac{m_0 c^2}{\sqrt{(1 - v^2/c^2)}}.$$

We can make a connection between this expression and the familiar equations of classical mechanics by rewriting the term $(1 - v^2/c^2)^{-1/2}$ in the form of a power series expansion and neglecting all terms in the series higher than v^2/c^2:

$$\frac{1}{\sqrt{(1 - v^2/c^2)}} \cong 1 + \frac{1}{2}\frac{v^2}{c^2}.$$

This will be applicable for speeds v much slower than the speed of light. In these circumstances, the energy approximates to:

$$E \cong m_0 c^2 + \tfrac{1}{2} m_0 v^2,$$

which is just the energy equivalent of the object's rest mass plus the non-relativistic kinetic energy. In classical mechanics, we deal only with *changes* in the energy of objects as a result of acceleration, and therefore we would consider only the kinetic energy term. The above considerations demonstrate that classical mechanics can always be recovered from the equations of special relativity in circumstances where $v \ll c$.

Returning to the equation for the relativistic energy, this can be rearranged to give:

$$E^2 \left(1 - \frac{v^2}{c^2}\right) = m_0^2 c^4.$$

We can tidy this up by combining the definition of linear momentum, $p = mv$, with $E = mc^2$ to give $v = pc^2/E$ and inserting this result into the above expression. After rearranging, we get:

$$E^2 = p^2 c^2 + m_0^2 c^4.$$

This is a general expression for the relativistic energy of a freely moving particle.

A photon (with energy ε) moves at the speed of light and is thought to have zero rest mass. Thus, the equation for energy above reduces to:

$$\varepsilon = pc,$$

where we have taken the positive root (more generally, $\varepsilon = |p|c$).

Although, in 1923, de Broglie supposed that a light-quantum possesses a small rest mass, we can obtain his result simply by combining the above expression with Planck's $\varepsilon = h\nu$:

$$\varepsilon = h\nu = pc.$$

Since $\nu = c/\lambda$, where λ is the wavelength of the light-quantum, this last expression can be rearranged to give an expression for λ in terms of p:

$$\lambda = \frac{h}{p}.$$

This is the de Broglie relation. It associates a wave with wavelength λ with an associated linear momentum p and, conversely, a particle with linear momentum p with an associated wavelength.

Appendix 5

Schrödinger's wave equation

Recall that it is not possible to 'derive' Schrödinger's quantum wave equation from classical physics. Nevertheless, it is possible to follow Schrödinger's reasoning from notebooks he kept at the time. His starting point was the well-known equation of classical wave motion, which interrelates the space and time dependences of the waves. We first met this equation in Appendix 1. Although Schrödinger began by considering only the time-independent component of this equation, we can start with the full time-dependent version:

$$\nabla^2 \Psi = \frac{1}{v^2} \frac{\partial^2 \Psi}{\partial t^2}.$$

In this equation, $\nabla^2 = \partial^2/\partial x^2 + \partial^2/\partial y^2 + \partial^2/\partial z^2$ is the Laplacian operator, Ψ is a generalized 'wave function', and v is the wave velocity. We will simplify things a little here by considering a wave constrained to propagate only in one direction (the x-direction), for which the wave equation reduces to:

$$\frac{\partial^2 \Psi}{\partial x^2} = \frac{1}{v^2} \frac{\partial^2 \Psi}{\partial t^2}.$$

The solutions to equations of this kind are readily at hand. One set of solutions is provided by the wave functions:

$$\Psi = A e^{i(kx - \omega t)},$$

where A is the wave amplitude (the height of the wave at its peak), k $(= 2\pi/\lambda)$ is the *wave vector*, and $\omega (= 2\pi v)$ is the *angular frequency*. It is straightforward to show that:

$$\frac{\partial^2 \Psi}{\partial x^2} = -k^2 \Psi = -\frac{4\pi^2}{\lambda^2} \Psi$$

and

$$\frac{\partial \Psi}{\partial t} = -i\omega\Psi = -2\pi i v\Psi.$$

You may have noticed that we have taken the second derivative with respect to x but stopped at the first derivative with respect to t. The reason for this will become apparent below.

So far, this has all been straightforward tinkering with the equation of classical wave motion. The next step is therefore to introduce some quantum character through the, by now familiar, simple relations $\lambda = h/p$ and $E = hv$:

$$\frac{\partial^2 \Psi}{\partial x^2} = -\frac{4\pi^2 p^2}{h^2}\Psi$$

and

$$\frac{\partial \Psi}{\partial t} = -\frac{2\pi i E}{h}\Psi.$$

We can also express the total energy of the system, E, as the sum of the kinetic and potential energies:

$$E = \frac{1}{2}mv^2 + V = \frac{p^2}{2m} + V,$$

where we have equated the momentum, p, to mass times velocity, mv, and V represents a general potential energy function. Thus, p can be expressed in terms of E as follows:

$$p^2 = 2m(E - V)$$

and substituting for p^2 gives:

$$\frac{\partial^2 \Psi}{\partial x^2} = -\frac{2m}{\hbar^2}(E - V)\Psi$$

and

$$\frac{\partial \Psi}{\partial t} = -\frac{iE}{\hbar}\Psi,$$

where $\hbar = h/2\pi$. It now becomes clear why we stopped above at the first derivative in t—both equations are now expressions containing the energy E to the first power, and both can be rearranged to give the result $E\Psi$:

$$-\frac{\hbar^2}{2m}\frac{\partial^2 \Psi}{\partial x^2} + V\Psi = E\Psi$$

and

$$i\hbar\frac{\partial \Psi}{\partial t} = E\Psi$$

Finally, these can obviously be combined to give the quantum equivalent of the classical wave equation:

$$-\frac{\hbar^2}{2m}\frac{\partial^2 \Psi}{\partial x^2} + V\Psi = i\hbar\frac{\partial \Psi}{\partial t}$$

or, in three dimensions,

$$-\frac{\hbar^2}{2m}\nabla^2 \Psi + V\Psi = i\hbar\frac{\partial \Psi}{\partial t}$$

As it stands, this equation has several 'peculiarities'. As Peter Atkins has remarked,[1] the equation is 'unbalanced' in terms of its treatment of spatial coordinates and time (unlike the classical wave equation) and consequently has the appearance of a *diffusion* equation. Rebalancing the wave equation so that space and time are treated on an equal footing (and so that it therefore meets the requirements of special relativity) is the subject of Appendix 6.

It is clear from the above that the general wave functions Ψ vary in both space and time. However, the functions can be broken down in such a way that the spatial variation and time variation can be separated:

$$\Psi = \psi e^{-iEt/\hbar}$$

where the functions ψ depend only on the spatial coordinates x, y, z. Much of the concern of undergraduate quantum mechanics is with physical and chemical situations in which the potential energy V is independent of time, and the functions ψ represent *stationary-state* solutions of Schrödinger's wave equation.

Finally, we should note that a common representation of the time-independent component of the wave equation is in terms of the so-called system *Hamiltonian*:

$$\hat{H}\psi = E\psi$$

where

$$\hat{H} = -\frac{\hbar^2}{2m}\nabla^2 + V$$

and the use of the caret (^), is intended to indicate that $\hat{H}$ has the form of a *mathematical operator*. In fact, quantum mechanics can be obtained from classical mechanics by substituting the appropriate classical quantity (such a linear momentum, $p = mv$) with its quantum-mechanical operator equivalent ($\hat{p} = -i\hbar\nabla$). The introduction of the Hamiltonian operator allows the full Schrödinger equation to be written in a particularly succinct form:

$$i\hbar\dot{\Psi} = \hat{H}\Psi$$

where $\dot{\Psi}$ denotes $\partial\Psi/\partial t$. This equation appears on the first-day postmark of stamps issued in Austria to commemorate the 100th anniversary of Schrödinger's birth.

---

[1] Atkins, P. W. (1983). *Molecular quantum mechanics* (2nd edn.). Oxford University Press, Oxford.

Appendix 6

Dirac's relativistic quantum theory of the electron

As written, the Schrödinger wave equation does not meet the requirements of Einstein's special theory of relativity. In fact, Schrödinger had originally embarked on the search for a fully relativistic wave mechanics but found that the predictions of the resulting equations did not match experimental results sufficiently closely, and he abandoned the approach in favour of the non-relativistic version described in Appendix 5. It was eventually Paul Dirac who realized that the problem of 'balancing' the treatment of space and time in Schrödinger's wave equation was not a question of finding a way of introducing a second-order differential in time but of finding a way to treat the spatial coordinates in terms of first-order and not second-order, differentials.

Starting with the time-dependent Schrödinger equation (see Appendix 5):

$$i\hbar\frac{\partial \Psi}{\partial t} = E\Psi,$$

conventional, non-relativistic wave mechanics is recovered by adopting a classical approach to the determination of E in terms of kinetic and potential energy, that is, $p^2/2m + V$. However, for a freely moving particle (no potential energy), the relativistic energy is given by (see Appendix 4):

$$E^2 = p^2c^2 + m_0^2c^4.$$

If we use this last equation for the energy, then a relativistic wave equation can be written as:

$$i\hbar\frac{\partial \Psi}{\partial t} = [c\sqrt{(p^2 + m_0^2c^2)}]\Psi.$$

We know from experience with Schrödinger's wave mechanics that the expression for energy should transform into an operator equation, by replacing the classical p with its quantum-mechanical operator equivalent, $\hat{p} = -i\hbar\nabla$ (see Appendix 5). But we now have a square-root operator which implies a differential equation of infinite order. Somehow,

Dirac needed to transform the square-root expression into a linear equation. Following Pauli's work on electron spin matrices, he therefore tried to find a solution such that:

$$\sqrt{(\hat{p}_x^2 + \hat{p}_y^2 + \hat{p}_z^2 + m_0^2 c^2)} = \alpha_1 \hat{p}_x + \alpha_2 \hat{p}_y + \alpha_3 \hat{p}_z + \alpha_4 m_0 c,$$

where we have used the relation $\hat{p}^2 = \hat{p}_x^2 + \hat{p}_y^2 + \hat{p}_z^2$ where $\hat{p}_x = -i\hbar \partial/\partial x$, and so on, and the coefficients $\alpha_1 - \alpha_4$ are four-by-four matrices:

$$\alpha_1 = \begin{pmatrix} 0 & 0 & 0 & 1 \\ 0 & 0 & 1 & 0 \\ 0 & 1 & 0 & 0 \\ 1 & 0 & 0 & 0 \end{pmatrix}, \quad \alpha_2 = \begin{pmatrix} 0 & 0 & 0 & -i \\ 0 & 0 & i & 0 \\ 0 & -i & 0 & 0 \\ i & 0 & 0 & 0 \end{pmatrix},$$

$$\alpha_3 = \begin{pmatrix} 0 & 0 & 1 & 0 \\ 0 & 0 & 0 & -1 \\ 1 & 0 & 0 & 0 \\ 0 & -1 & 0 & 0 \end{pmatrix}, \quad \text{and} \quad \alpha_4 = \begin{pmatrix} 1 & 0 & 0 & 0 \\ 0 & 1 & 0 & 0 \\ 0 & 0 & -1 & 0 \\ 0 & 0 & 0 & -1 \end{pmatrix}.$$

When re-inserted into the wave equation, Dirac arrived at the result:

$$\frac{i\hbar}{c} \frac{\partial \Psi}{\partial t} = (\hat{\alpha} \cdot \hat{p} + \alpha_4 m_0 c)\Psi,$$

$$\left(\frac{i\hbar}{c} \frac{\partial}{\partial t} - \hat{\alpha} \cdot \hat{p} - \alpha_4 m_0 c \right) \Psi = 0.$$

In this form, Dirac could be assured that his equation conformed to the requirements of special relativity but, as an equation for a free electron, it was not particularly interesting. He therefore used standard techniques to modify the equation to add an electromagnetic field. The result was an additional term representing the spin magnetic moment of the electron and out dropped the fixed value of the spin angular momentum, $\hbar/2$, exactly as required by experiment. This was all the more remarkable as it was not a result that Dirac himself had been aiming for—he had been primarily interested in getting a proper relativistic wave equation, and to have the spin properties of the electron and its magnetic moment emerge naturally from this proper treatment was a great surprise and an unexpected triumph.

The introduction of the four-by-four matrices, necessary to linearize the square-root operator, meant that the wave functions themselves are represented in Dirac's theory not as simple functions but as single-column matrices containing simple functions as their elements:

$$\Psi = \begin{bmatrix} \psi_1 \\ \psi_2 \\ \psi_3 \\ \psi_4 \end{bmatrix}.$$

Two of these functions represent solutions of positive energy and are interpreted as the spin-up and spin-down configurations of the electron. The other two functions represent negative-energy solutions and are interpreted in terms of the (positive-energy) spin-up and spin-down configurations of the anti-electron, or positron (see text).

Appendix 7

The expectation value

When the state vector $|n\rangle$ is an *eigenstate* of the operator $\hat{A}$, the result of operating on the vector with $\hat{A}$ is the eigenvalue, denoted a_n, multiplied by the state vector:

$$\hat{A}|n\rangle = a_n|n\rangle.$$

Both the state vectors and the operators may be complex quantities. Yet, to make any sense, the eigenvalues must be exclusively real if they are to represent the values of observable quantities that we can, in principle, measure in the laboratory. This is an important restriction. Operators whose eigenvalues are exclusively real are called self-adjoint or hermitian operators and have the general property that:

$$\langle m|\hat{A}|n\rangle = \langle n|\hat{A}|m\rangle^*,$$

where $\langle m|\hat{A}|n\rangle$ is the inner product of the state vector $|m\rangle$ and the result of the operation of $\hat{A}$ on $|n\rangle$, that is, $\langle m|\hat{A}|n\rangle = \langle m|(\hat{A}|n\rangle)$. The quantity $\langle n|\hat{A}|m\rangle^*$ is obtained from $\langle m|\hat{A}|n\rangle$ simply by exchanging the indices and taking the complex conjugate and is called the *complex-conjugate transpose*.

Eigenstates have the further property that they are *orthogonal*: $\langle m|n\rangle = 0$ if $|m\rangle$ and $|n\rangle$ are both eigenstates, and if they are both separately normalized, so that $\langle m|m\rangle = 1$ and $\langle n|n\rangle = 1$, then the eigenstates are said to be orthonormal:

$$\langle m|n\rangle = \delta_{nm},$$

where $\delta_{nm} = 0$ when $m \neq n$ and $\delta_{nm} = 1$ when $m = n$.

For a normalized state vector $|n\rangle$, the expectation value of an observable with operator $\hat{A}$ is given by:

$$\langle A_n\rangle = \langle n|\hat{A}|n\rangle$$

This expression is related to an equivalent expression from probability calculus and emphasizes once again that the value of the observable represents an *average* or *mean value* of the observable, often interpreted in the context of the mean of the relative frequencies of a large number of repeated measurements made on an ensemble of identically prepared particles.

If $|n\rangle$ is an eigenstate of $\hat{A}$, the expectation value in these circumstances is, simply:

$$\langle A_n \rangle = \langle n|\hat{A}|n \rangle = \langle n|(a_n|n\rangle) = a_n\langle n|n \rangle = a_n.$$

This follows because, by definition, the state vector is normalized, and so $\langle n|n \rangle = 1$. Thus, when $|n\rangle$ is an eigenstate of $\hat{A}$, the expectation value is *exactly* a_n. We will see later that the use of the term 'mean value' becomes necessary when we consider state vectors that are not eigenstates.

Appendix 8

Complementary observables and the uncertainty principle

Suppose that the state vector $|n\rangle$ is an eigenstate of an operator $\hat{A}$. We know (see Appendix 7) that $\hat{A}|n\rangle = a_n|n\rangle$, where a_n is an eigenvalue of the operator, and the expectation value $\langle A_n \rangle = a_n$. There is, therefore, nothing inherent in the formulation of $|n\rangle$ that limits the precision, or certainty, of the value of the observable corresponding to $\hat{A}$ which in such circumstances is given *exactly* by the eigenvalue a_n. Suppose we wish to measure the value of a second property, given by the expectation value of another operator, $\hat{B}$, with equal precision, or certainty. It follows that the expectation value for this operator, $\langle B_n \rangle$, must therefore be exactly equal to the corresponding eigenvalue b_n. By definition, this requires that the state vector $|n\rangle$ is a *simultaneous* eigenstate of both $\hat{A}$ and $\hat{B}$.

We conclude that in order to measure simultaneously two different observables of a quantum state with arbitrary precision, the state vector describing the quantum state must be a simultaneous eigenstate of both of the operators corresponding to the observables. What does this imply? Consider the action of the *commutator*, $[\hat{A}, \hat{B}] = \hat{A}\hat{B} - \hat{B}\hat{A}$ on $|n\rangle$:

$$[\hat{A}, \hat{B}]|n\rangle = (\hat{A}\hat{B} - \hat{B}\hat{A})|n\rangle = \hat{A}\hat{B}|n\rangle - \hat{B}\hat{A}|n\rangle$$
$$= \hat{A}b_n|n\rangle - \hat{B}a_n|n\rangle = b_n\hat{A}|n\rangle - a_n\hat{B}|n\rangle$$
$$= b_na_n|n\rangle - a_nb_n|n\rangle = 0,$$

that is, $[\hat{A}, \hat{B}] = 0$, and the operators $\hat{A}$ and $\hat{B}$ *commute*.

Put simply, we can measure the values of two observables of some quantum state to arbitrary precision *only if their corresponding operators commute*. We have already seen that simultaneous determination of the position and linear momentum of a quantum particle to arbitrary precision is not possible (uncertainty principle) because their operators do not commute. Such observables are said to be *complementary*: we can measure one or the other with arbitrarily high precision but not both simultaneously.

The connection between commutators and a generalized uncertainty relation was established by the physicist Howard Robertson in 1929. The origin of this relationship can be traced back to a theorem of vectors in Hilbert space, called the Schwarz inequality[1]:

$$|\langle m|n \rangle| \leq \langle m|m \rangle^{1/2} \langle n|n \rangle^{1/2}.$$

The equivalent inequality in the calculus of classical vectors is:

$$|\mathbf{m} \cdot \mathbf{n}| \leq (\mathbf{m} \cdot \mathbf{m})^{1/2}(\mathbf{n} \cdot \mathbf{n})^{1/2}.$$

If the angle between the classical vectors $\mathbf{m}$ and $\mathbf{n}$ is θ, then for classical vectors, the Schwarz inequality simply reduces to $|\cos\theta| \leq 1$.

If $\hat{A}$ and $\hat{B}$ are operators corresponding to two different observables, then application of the Schwarz inequality results in a relationship between the 'uncertainty' (root-mean-square deviation) in the mean value of a multiplied by the uncertainty in b given by:

$$\Delta a \Delta b \geq \tfrac{1}{2} \left| \langle n|[\hat{A}, \hat{B}]|n \rangle \right|.$$

We know that $[x, \hat{p}_x] = i\hbar$, where $\hbar = h/2\pi$. Therefore, the uncertainty in measurements of the position and linear momentum of a quantum particle are given by:

$$\Delta x \Delta p_x \geq \tfrac{1}{2} |i\hbar \langle n|n \rangle| = \tfrac{1}{2}\hbar,$$

which is the result first obtained by Heisenberg.

Interestingly, we can follow a similar route to the same destination using much the same logic but in the language of wave mechanics. A wave function described in terms of a variation of amplitude over spatial coordinates can be translated into an equivalent wave function in 'momentum space' by taking its *Fourier transform*. The Fourier transform is simply a series of mathematical operations applied to the wave function which has the effect of translating a change in position into a shift in the phase of the wave function in momentum space. Conversely, a change in momentum is translated into a phase shift of the wave function in position space.

We can now define Δx and Δp_x as the *half-widths* of the wave function in position and momentum space, respectively. If Δx is the half-width of the wave function in position space, then the Fourier transform of this yields an analogous function in momentum space with a half-width of *at least* Δp_x where the product $\Delta x \Delta p_x \geq \hbar/2$. Obviously, a wave function which is a delta-function in position space (one with amplitude *only* at a single position and zero amplitude everywhere else, pinpointing the position of the associated particle with certainty) has *zero* half-width. The Fourier transform of such a function has an infinite half-width, and therefore no meaningful figure can be assigned as the mean value of the momentum.

---

[1] See Isham, Chris J. (1995). *Lectures on quantum theory*. Imperial College Press, London, p. 141.

Appendix 9

The expansion theorem and quantum projections

Suppose we can specify an arbitrary state vector $|\psi\rangle$ completely in terms of just two eigenstates, $|m\rangle$ and $|n\rangle$, of the operator $\hat{A}$. In this sense, the eigenstates $|m\rangle$ and $|n\rangle$ represent a suitable basis set for the state vector $|\psi\rangle$. We can use the expansion theorem to model $|\psi\rangle$ as a linear combination or superposition of the basis set according to the sum:

$$|\psi\rangle = \sum_i c_i |i\rangle = c_m |m\rangle + c_n |n\rangle,$$

where the coefficients c_m and c_n are *expansion coefficients* which indicate how much of each basis vector is present in the mixture represented by $|\psi\rangle$. If we assume $|\psi\rangle$ is normalized, so that $\langle\psi|\psi\rangle = 1$, then the expectation value of the operator $\hat{A}$ is given by:

$$\langle A_\psi \rangle = \langle\psi|\hat{A}|\psi\rangle = (c_m^*\langle m| + c_n^*\langle n|)\hat{A}(c_m|m\rangle + c_n|n\rangle)),$$

$$= (c_m^*\langle m| + c_n^*\langle n|)(c_m\hat{A}|m\rangle + c_n\hat{A}|n\rangle)),$$

$$= |c_m|^2\langle m|\hat{A}|m\rangle + c_m^*c_n\langle m|\hat{A}|n\rangle + c_n^*c_m\langle n|\hat{A}|m\rangle + |c_n|^2\langle n|\hat{A}|n\rangle,$$

where c_m^* and c_n^* are the complex conjugates of the coefficients c_m and c_n and $|c_m|^2 = c_m^*c_m$, $|c_n|^2 = c_n^*c_n$. We know from Appendix 7 that if $|m\rangle$ and $|n\rangle$ are normalized eigenstates of $\hat{A}$, then $\langle m|\hat{A}|m\rangle = a_m$ and $\langle n|\hat{A}|n\rangle = a_n$. Furthermore, we know that $\langle m|\hat{A}|n\rangle = a_n\langle m|n\rangle = 0$, since the eigenstates are orthogonal. Similarly, $\langle n|\hat{A}|m\rangle = a_m\langle n|m\rangle = 0$. This leaves us with the result:

$$\langle A_\psi \rangle = |c_m|^2 a_m + |c_n|^2 a_n$$

and we see that the expectation value is quite literally the 'mean value' of the observable—it is a weighted average of the contributions arising from the two eigenstates that together form the state vector $|\psi\rangle$.

The above reasoning applies for any arbitrary state vector describable in terms of any number of basis vectors (including an infinite number): the expectation value of an operator corresponding to a particular observable is the weighted sum of the eigenvalues of the basis vectors, where the weighting factors are the modulus-squares of the expansion coefficients.

The question then arises: what are the expansion coefficients? We can answer this question in a relatively straightforward manner using the above example. If we multiply the expression for $|\psi\rangle$ in terms of $|m\rangle$ and $|n\rangle$ from the left by the bra $\langle m|$, we get:

$$\langle m|\psi\rangle = c_m\langle m|m\rangle + c_n\langle m|n\rangle = c_m,$$

where we are again making use of the fact that the basis states are normalized ($\langle m|m\rangle = 1$) and orthogonal ($\langle m|n\rangle = 0$). Similarly, we can show that $\langle n|\psi\rangle = c_n$ and so:

$$|\psi\rangle = |m\rangle\langle m|\psi\rangle + |n\rangle\langle n|\psi\rangle.$$

The inner products $\langle m|\psi\rangle$ and $\langle n|\psi\rangle$ are sometimes called *projection amplitudes* and the products $|m\rangle\langle m|$ and $|n\rangle\langle n|$ are sometimes referred to as 'butterfly operators' (for obvious reasons) or *projection operators*.

Another important mathematical object encountered frequently in quantum theory is the *density matrix*. For the simple two-state expansion considered here, the density matrix has the simple form:

$$\hat{\rho} = \begin{pmatrix} |c_m|^2 & c_m^* c_n \\ c_n^* c_m & |c_n|^2 \end{pmatrix} = \begin{pmatrix} |\langle m|\psi\rangle|^2 & \langle m|\psi\rangle^* \langle n|\psi\rangle \\ \langle n|\psi\rangle^* \langle m|\psi\rangle & |\langle n|\psi\rangle|^2 \end{pmatrix}.$$

It has the following important properties:

1. It is self-adjoint—the density matrix is equal to its own complex-conjugate transpose, and so its eigenvalues are exclusively real numbers.

2. It is positive, semi-definite—$\langle\psi|\hat{\rho}|\psi\rangle \geq 0$ for all $|\psi\rangle$.

3. Its diagonal elements sum to unity. The sum of the diagonal elements of any matrix is called the *trace*, Tr. Hence $\mathrm{Tr}\hat{\rho} = 1$.

The important connection between the density matrix and projection probabilities is brought out clearly when we rewrite the expectation value $\langle A_\psi\rangle$ in terms of $\hat{\rho}$. Recall from our considerations above that we can express the expectation value as follows:

$$\langle A_\psi\rangle = |c_m|^2\langle m|\hat{A}|m\rangle + c_m^* c_n\langle m|\hat{A}|n\rangle + c_n^* c_m\langle n|\hat{A}|m\rangle + |c_n|^2\langle n|\hat{A}|n\rangle.$$

This sum can be thought of as the trace of the matrix obtained by multiplying $\hat{\rho}$ with the matrix elements of the operator $\hat{A}$, that is,

$$\langle A_\psi\rangle = \mathrm{Tr}\begin{pmatrix} |c_m|^2 & c_m^* c_n \\ c_n^* c_m & |c_n|^2 \end{pmatrix}\begin{pmatrix} A_{mm} & A_{mn} \\ A_{nm} & A_{nn} \end{pmatrix},$$

where the matrix elements $A_{mm} = \langle m|\hat{A}|m\rangle$, and so on. In order words, in general $\langle A\rangle = \mathrm{Tr}(\hat{\rho}\hat{A})$. In the case considered above, where $|m\rangle$ and $|n\rangle$ are eigenstates of $\hat{A}$, we know that the diagonal matrix elements A_{mm} and A_{nn} are equal to the eigenvalues a_m and a_n, respectively, and the off-diagonal matrix elements A_{mn} and A_{nm} are zero. And so:

$$\langle A_\psi \rangle = \text{Tr} \begin{pmatrix} |c_m|^2 & c_m^* c_n \\ c_n^* c_m & |c_n|^2 \end{pmatrix} \begin{pmatrix} a_m & 0 \\ 0 & a_n \end{pmatrix} = |c_m|^2 a_m + |c_n|^2 a_n, \quad \text{as before.}$$

The density matrix therefore contains all the probability information available for a specified quantum system, and, according to the quantum rules, it therefore contains all the information there is.

Appendix 10

State vectors and classical unit vectors

A classical vector $\mathbf{v}$ pointing in some arbitrary direction in Euclidean space can be resolved into two components corresponding to its projection onto two orthogonal coordinates (say x and y). We define this length of $\mathbf{v}$ to be unity in some arbitrary unit system. Each component of $\mathbf{v}$ is also a vector which we represent as a coefficient multiplied by the unit vector corresponding to the coordinate. In other words:

$$\mathbf{v} = v_x \mathbf{i} + v_y \mathbf{j},$$

where v_x and v_y are the coefficients, and $\mathbf{i}$ and $\mathbf{j}$ are the unit vectors along the x and y coordinates, respectively.

The coefficients v_x and v_y can be calculated as the inner or dot products of $\mathbf{v}$ and the unit vectors:

$$v_x = (\mathbf{i} \cdot \mathbf{v}) = |\mathbf{i}||\mathbf{v}| \cos\theta,$$

$$v_y = (\mathbf{j} \cdot \mathbf{v}) = |\mathbf{j}||\mathbf{v}| \cos(90 - \theta) = |\mathbf{j}||\mathbf{v}| \sin\theta,$$

where θ is the angle between the direction of the vector $\mathbf{v}$ and the x coordinate. The modulus $|\mathbf{v}|$ represents the magnitude of $\mathbf{v}$ and is independent of its direction. Obviously, $|\mathbf{v}| = |\mathbf{i}| = |\mathbf{j}| = 1$, since these are defined as unit vectors. It follows that:

$$\mathbf{v} = \mathbf{i}(\mathbf{i} \cdot \mathbf{v}) + \mathbf{j}(\mathbf{j} \cdot \mathbf{v}).$$

This relationship parallels the expansion of an arbitrary quantum state vector into two orthogonal basis vectors (see Appendix 9) and provides an analogy between basis vectors and the unit vectors of classical physics. How deep this analogy runs can be ascertained by considering the following properties.

First, the basis vectors and classical unit vectors provide unique *representations* for quantum state vectors and classical vectors, respectively:

$$\textit{Quantum} \quad |\psi\rangle = |m\rangle\langle m|\psi\rangle + |n\rangle\langle n|\psi\rangle,$$

$$\textit{Classical} \quad \mathbf{v} = \mathbf{i}(\mathbf{i} \cdot \mathbf{v}) + \mathbf{j}(\mathbf{j} \cdot \mathbf{v}).$$

These representations can be generalized to situations involving more than two dimensions as follows:

$$Quantum \quad |\psi\rangle = \sum_s |s\rangle\langle s|\psi\rangle,$$

$$Classical \quad \mathbf{v} = \sum_{s=i,j,k} \mathbf{s}(\mathbf{s} \cdot \mathbf{v}).$$

Second, both state vectors and unit vectors have the property of orthogonality:

$$Quantum \quad \langle m|n\rangle = 0,$$

$$Classical \quad (\mathbf{i} \cdot \mathbf{j}) = \cos 90° = 0.$$

Finally, both representations have the property of *completeness*:

$$Quantum \quad |\langle m|\psi\rangle|^2 + |\langle n|\psi\rangle|^2 = 1,$$

$$Classical \quad (\mathbf{i} \cdot \mathbf{v})^2 + (\mathbf{j} \cdot \mathbf{v})^2 = \cos^2 \theta + \sin^2 \theta = 1,$$

where θ is the angle between the vector $\mathbf{v}$ and the x coordinate.

Appendix 11

Quantum indistinguishability: fermions and bosons

When we form a two-particle state vector, we must acknowledge that there is no telling which particle is in which individual quantum state. Suppose that the two quantum states are $|m\rangle$ and $|n\rangle$. The correct two-particle state vector can then be formed from the linear combination:

$$|\Psi\rangle_{12} = c_{mn}|m\rangle_1|n\rangle_2 + c_{nm}|n\rangle_1|m\rangle_2,$$

where the subscripts denote the individual particles and c_{mn} and c_{nm} are expansion coefficients. The superposition must contain equal proportions of each product state as both possibilities are equally likely. If we assume that the two-particle state $|\Psi\rangle_{12}$ is normalized, then $|c_{mn}|^2 + |c_{nm}|^2 = 1$, and it follows therefore that

$$|c_{mn}| = |c_{nm}| = \tfrac{1}{\sqrt{2}}.$$

There are two ways of achieving this equality. The first possibility is that $c_{mn} = -c_{nm}$, in which case the two-particle state vector has the form:

$$|\Psi\rangle_{12} = \tfrac{1}{\sqrt{2}}(|m\rangle_1|n\rangle_2 - |n\rangle_1|m\rangle_2).$$

Now let us consider the effects of exchanging the particles, so that what was labelled 1 now becomes 2 and vice versa. We find that:

$$|\Psi\rangle_{21} = \tfrac{1}{\sqrt{2}}(|m\rangle_2|n\rangle_1 - |n\rangle_2|m\rangle_1),$$

$$= \tfrac{1}{\sqrt{2}}(|n\rangle_1|m\rangle_2 - |m\rangle_1|n\rangle_2),$$

$$= -|\Psi\rangle_{12}.$$

The state vector $|\Psi\rangle_{12}$ is said to be *antisymmetric* (it changes sign) on the exchange of the two particles. Note that, since only the sign of the state vector changes, its modulus-square is indistinguishable from that obtained when the particles are exchanged. The particles are experimentally indistinguishable—their exchange should not (and does not) make any difference to quantities that we can measure experimentally.

Now consider what happens if we imagine a situation where we have two particles in the same quantum state, that is, we set $|n\rangle = |m\rangle$:

$$|\Psi\rangle_{12} = \tfrac{1}{\sqrt{2}}(|m\rangle_1|m\rangle_2 - |m\rangle_1|m\rangle_2) = 0.$$

We conclude from this that quantum particles whose two-particle state vectors are antisymmetric with respect to exchange are *forbidden* from occupying the same quantum state. This is Pauli's exclusion principle. The exclusion principle was developed for electrons, but we can see immediately from the above reasoning that this general rule will apply to all quantum particles with antisymmetric two-particle state vectors. Such particles turn out to have half-integral spin quantum numbers and are collectively called fermions (see Chapter 3).

The second possibility is $c_{mn} = c_{nm}$, or

$$|\Psi\rangle_{12} = \tfrac{1}{\sqrt{2}}(|m\rangle_1|n\rangle_2 + |n\rangle_1|m\rangle_2).$$

In this case, exchange of the two particles gives:

$$|\Psi\rangle_{21} = \tfrac{1}{\sqrt{2}}(|m\rangle_2|n\rangle_1 + |n\rangle_2|m\rangle_1),$$
$$= \tfrac{1}{\sqrt{2}}(|n\rangle_1|m\rangle_2 + |m\rangle_1|n\rangle_2),$$
$$= |\Psi\rangle_{12}.$$

that is, the vector $|\Psi\rangle_{12}$ is symmetric (it does not change sign) on the exchange of particles. Particles whose two-particle state vectors possess this property turn out to have integer spin quantum numbers and are known as *bosons* (see Chapter 3).

Appendix 12

Projection amplitudes for photon-polarization states

Consideration of Malus's law in the context of photon-polarization states forces us to conclude that for suitably normalized state vectors $|v\rangle$ and $|v'\rangle$:

$$|\langle v'|v\rangle|^2 = \cos^2\varphi,$$

where φ is the angle between the vertical, v and v' axes (see Chapter 5, Fig. 5.2). We can furthermore deduce that:

$$|\langle h'|h\rangle|^2 = \cos^2\varphi$$
$$|\langle v'|h\rangle|^2 = \sin^2\varphi$$

and

$$|\langle h'|v\rangle|^2 = \sin^2\varphi.$$

For left- and right-circularly polarized photons, we can observe experimentally that:

$$|\langle v|L\rangle|^2 = \tfrac{1}{2}$$
$$|\langle h|L\rangle|^2 = \tfrac{1}{2}$$
$$|\langle v|R\rangle|^2 = \tfrac{1}{2}$$

and

$$|\langle h|R\rangle|^2 = \tfrac{1}{2}$$

irrespective of the actual orientation of the vertical v and horizontal h axes. We can summarize this information in a table of projection probabilities (see Table A12.1).

Table A12.1 Projection probabilities, $|\langle f|i\rangle|^2$, for photon polarization states

Final state, $|f\rangle$	Initial state, $|i\rangle$					
	$|v\rangle$	$|h\rangle$	$|v'\rangle$	$|h'\rangle$	$|L\rangle$	$|R\rangle$
$|v\rangle$	1	0	$\cos^2\varphi$	$\sin^2\varphi$	$\frac{1}{2}$	$\frac{1}{2}$
$|h\rangle$	0	1	$\sin^2\varphi$	$\cos^2\varphi$	$\frac{1}{2}$	$\frac{1}{2}$
$|v'\rangle$	$\cos^2\varphi$	$\sin^2\varphi$	1	0	$\frac{1}{2}$	$\frac{1}{2}$
$|h'\rangle$	$\sin^2\varphi$	$\cos^2\varphi$	0	1	$\frac{1}{2}$	$\frac{1}{2}$
$|L\rangle$	$\frac{1}{2}$	$\frac{1}{2}$	$\frac{1}{2}$	$\frac{1}{2}$	1	0
$|R\rangle$	$\frac{1}{2}$	$\frac{1}{2}$	$\frac{1}{2}$	$\frac{1}{2}$	0	1

It should have become obvious by now that these three sets of quantum states, $|v\rangle/|h\rangle$, $|v'\rangle/|h'\rangle$ and $|L\rangle/|R\rangle$ are all suitable representations for photon polarization states, and they can therefore be used as basis states. Although we use a convention to assign photons with $m_s = \pm 1$ to states of circular polarization, these states can, in turn, be expressed as linear superpositions of states of linear polarization. For example, we can use the expansion theorem to write:

$$|L\rangle = |v\rangle\langle v|L\rangle + |h\rangle\langle h|L\rangle$$

and

$$|R\rangle = |v\rangle\langle v|R\rangle + |h\rangle\langle h|R\rangle$$

Similar expressions can be written for $|v\rangle$ and $|h\rangle$ in terms of $|L\rangle$ and $|R\rangle$ as basis states. Obviously, we can go no further until we find expressions or the various projection amplitudes.

We can deduce the projection amplitudes for linear polarization states using the axis convention defined in Fig. 5.2 combined with a little vector algebra. From the analogy between state vectors and classical unit vectors described in Appendix 10, we note from Fig. 5.2 that:

$$|v\rangle = |v'\rangle \cos\varphi + |h'\rangle \sin\varphi,$$

where $|v'\rangle$ and $|h'\rangle$ are the equivalent of unit vectors along the v', h' axes. Multiplying both sides of this expression by $\langle v'|$ gives:

$$\langle v'|v\rangle = \cos\varphi$$

since $\langle v'|v'\rangle = 1$ and $\langle v'|h'\rangle = 0$. Similarly, we can deduce that

$$\langle h'|v\rangle = \sin\varphi, \quad \langle v'|h\rangle = \cos(90 + \varphi) = -\sin\varphi \quad \text{and} \quad \langle h'|h\rangle = \cos\varphi.$$

Obviously, it follows that $\langle v|v'\rangle = \langle v'|v\rangle$, and so on; that is, the projection amplitudes for linear polarization states are symmetric to the exchange of initial and final states. A quick glance at Table A12.1 reveals that these projection amplitudes are consistent with the corresponding projection probabilities, and hence they are consistent with experiment.

We now need to go on to consider projection amplitudes involving states of circular polarization. We said above that the $|v\rangle/|h\rangle$, $|v'\rangle/|h'\rangle$ and $|L\rangle/|R\rangle$ states serve as interchangeable sets of basis states for photon polarization. We can therefore express the state $|v'\rangle$ as a linear combination of $|L\rangle$ and $|R\rangle$:

$$|v'\rangle = |L\rangle\langle L|v'\rangle + |R\rangle\langle R|v'\rangle.$$

Multiplying both sides of this expression by $\langle v|$ gives

$$\langle v|v'\rangle = \langle v|L\rangle\langle L|v'\rangle + \langle v|R\rangle\langle R|v'\rangle = \cos\varphi.$$

From Table A12.1, we know that $|\langle v|L\rangle| = |\langle L|v'\rangle| = |\langle v|R\rangle| = |\langle R|v'\rangle| = 1/\sqrt{2}$. There can be no way of reconciling these projection amplitudes with the expression for $\langle v|v'\rangle$ above without recognizing that some of the projection amplitudes must themselves be complex. At this stage it is useful to recall that we can decompose $\cos\varphi$ according to the expression:

$$\cos\varphi = \tfrac{1}{2}(e^{i\varphi} + e^{-i\varphi}) = \tfrac{1}{2}e^{i\varphi} + \tfrac{1}{2}e^{-i\varphi},$$

and so we (quite arbitrarily) identify the first term on the right-hand side of this equation with the term $\langle v|L\rangle\langle L|v'\rangle$ and the second term with $\langle v|R\rangle\langle R|v'\rangle$, that is,

$$\langle v|L\rangle\langle L|v'\rangle = \tfrac{1}{2}e^{i\varphi}$$

and

$$\langle v|R\rangle\langle R|v'\rangle = \tfrac{1}{2}e^{-i\varphi}.$$

Furthermore, since it is logical to associate the terms in $e^{\pm i\varphi}$ with those projection amplitudes involving the state $|v'\rangle$, we can decompose these expressions to give the individual amplitudes as follows:

$$\langle v|L\rangle = \tfrac{1}{\sqrt{2}} \quad \langle L|v'\rangle = \tfrac{1}{\sqrt{2}}e^{i\varphi}$$

and

$$\langle v|R\rangle = \tfrac{1}{\sqrt{2}} \quad \langle R|v'\rangle = \tfrac{1}{\sqrt{2}}e^{-i\varphi}.$$

Notice that $|\langle L|v'\rangle|^2 = |\langle R|v'\rangle|^2 = \tfrac{1}{2}$, as required. If this seems to be a completely arbitrary procedure (we could just as well have taken $\langle v|L\rangle\langle L|v'\rangle = e^{-i\varphi}/2$), it is because this is exactly what it is. Remember that we have no way of knowing the 'actual' signs of the phase factors because this is information that is not revealed in experiments. However, we can adopt a *phase convention* which, if we stick to it rigorously, will always give results that are both internally consistent and consistent with experiment.

Using the phase convention determined by the choices made in the above, we can use the same general procedure to deduce all the projection amplitudes for all of the basis states. These are collected in Table A12.2. Note from this table that our phase convention leads to $\langle f|i\rangle = \langle i|f\rangle^*$, where i and f are any of v, h, v', h', L, and R.

Table A12.2 Projection amplitudes, $\langle f|i \rangle$, for photon polarization states

Final state, $|f\rangle$	Initial state, $|i\rangle$					
	$|v\rangle$	$|h\rangle$	$|v'\rangle$	$|h'\rangle$	$|L\rangle$	$|R\rangle$
$|v\rangle$	1	0	$\cos\varphi$	$\sin\varphi$	$1/\sqrt{2}$	$1/\sqrt{2}$
$|h\rangle$	0	1	$-\sin\varphi$	$\cos\varphi$	$i/\sqrt{2}$	$-i/\sqrt{2}$
$|v'\rangle$	$\cos\varphi$	$-\sin\varphi$	1	0	$e^{-i\varphi}/\sqrt{2}$	$e^{i\varphi}/\sqrt{2}$
$|h'\rangle$	$\sin\varphi$	$\cos\varphi$	0	1	$ie^{-i\varphi}/\sqrt{2}$	$-ie^{i\varphi}/\sqrt{2}$
$|L\rangle$	$1/\sqrt{2}$	$-i/\sqrt{2}$	$e^{i\varphi}/\sqrt{2}$	$-ie^{i\varphi}/\sqrt{2}$	1	0
$|R\rangle$	$1/\sqrt{2}$	$i/\sqrt{2}$	$e^{-i\varphi}/\sqrt{2}$	$ie^{-i\varphi}/\sqrt{2}$	0	1

Appendix 13

Quantum measurement and expectation values

Calcite crystals are naturally birefringent and can spatially separate light into vertically and horizontally polarized components. We find that one 'channel' preferentially transmits vertically polarized light, and the other channel preferentially transmits horizontally polarized light. Placing detectors at the exit of either or both channels allows us to tell which channel the light is passing through. The crystal plus detectors together constitute a measuring device which we express theoretically as a mathematical operation, with operator $\hat{M}$.

Now by definition, if we pass preselected (or 'prepared'—see text) vertically polarized photons through the crystal, we expect them to emerge exclusively from the vertical channel and be detected. This is indeed what happens, and we conclude that $|v\rangle$ is an eigenstate of the measuring device and that the effect of operating on $|v\rangle$ with the operator $\hat{M}$ is to return the measurement eigenvalue R_v. In other words:

$$\hat{M}|v\rangle = R_v|v\rangle.$$

Similarly, $\hat{M}|h\rangle = R_h|h\rangle$, where R_h is the measurement eigenvalue corresponding to horizontal polarization. What happens when we pass left-circularly polarized photons through this device?

Our first step is to express the state vector in the basis of the measurement eigenstates, as it is these states that will be measured, and we are therefore interested in the probability that the measurement will project the incident state vector into the these various possible eigenstates. In this case, the eigenstates of the measuring device are $|v\rangle$ and $|h\rangle$, and we recall that we can use the expansion theorem and the projection amplitudes given in Appendix 12 to write an expression of the state vector for left-circular polarization in terms of these states as follows:

$$|L\rangle = \tfrac{1}{\sqrt{2}}(|v\rangle + i|h\rangle).$$

The effect of $\hat{M}$ operating on $|L\rangle$ is, therefore:

$$\hat{M}|L\rangle = \tfrac{1}{\sqrt{2}}\left(\hat{M}|v\rangle + i\hat{M}|h\rangle\right) = \tfrac{1}{\sqrt{2}}\left(R_v|v\rangle + iR_h|h\rangle\right).$$

If we assume $|L\rangle$ to be normalized, then the expectation value of the operator $\hat{M}$ for the state $|L\rangle$ is given by:

$$\langle\hat{M}_L\rangle = \langle L|\hat{M}|L\rangle = \tfrac{1}{\sqrt{2}}\left(\langle v| - i\langle h|\right) \times \tfrac{1}{\sqrt{2}}\left(R_v|v\rangle + iR_h|h\rangle\right),$$

$$= \tfrac{1}{2}\left(R_v\langle v|v\rangle + iR_h\langle v|h\rangle - iR_v\langle h|v\rangle + R_h\langle h|h\rangle\right) = \tfrac{1}{2}\left(R_v + R_h\right).^{[1]}$$

The expectation value is the average of the two possible outcomes, and we expect to see the photons emerge from the vertical and horizontal polarization channels with equal probability. This is obviously consistent with the projection probabilities given in Appendix 12.

---

[1] Remember, $-i^2 = 1$.

Appendix 14

Complementary observables of two-particle states

An individual quantum particle has a position and momentum that are subject to the quantum-mechanical commutation relation and the position–momentum uncertainty relation. For two independent particles A and B, we have:

$$[q_A, \hat{p}_A] = i\hbar \quad \text{and} \quad [q_B, \hat{p}_B] = i\hbar,$$

where q_A represents the position of particle A, and the conjugate momentum is given by $\hat{p}_A = -i\hbar \partial/\partial q_A$. Similarly for particle B. Now consider the quantities

$$Q = q_A - q_B \quad \text{and} \quad \hat{P} = \hat{p}_A + \hat{p}_B,$$

which represent the difference between the positions of particles A and B and the sum of their momenta. The commutator $[Q, \hat{P}]$ is given by:

$$
\begin{aligned}
[Q, \hat{P}] &= Q\hat{P} - \hat{P}Q = (q_A - q_B)(\hat{p}_A + \hat{p}_B) - (\hat{p}_A + \hat{p}_B)(q_A - q_B), \\
&= q_A\hat{p}_A + q_A\hat{p}_B - q_B\hat{p}_A - q_B\hat{p}_B - (\hat{p}_A q_A - \hat{p}_A q_B + \hat{p}_B q_A - \hat{p}_B q_B), \\
&= (q_A\hat{p}_A - \hat{p}_A q_A) + (q_A\hat{p}_B - \hat{p}_B q_A) - (q_B\hat{p}_A - \hat{p}_A q_B) - (q_B\hat{p}_B - \hat{p}_B q_A), \\
&= [q_A, \hat{p}_A] + [q_A, \hat{p}_B] - [q_B, \hat{p}_A] - [q_B, \hat{p}_B].
\end{aligned}
$$

In this equation, $[q_A, \hat{p}_B] = [q_B, \hat{p}_A] = 0$, since these operators refer to different quantum particles. Hence, $[Q, \hat{P}] = 0$, the operators Q and $\hat{P}$ commute, and there is therefore no restriction in principle on the precision with which we can measure the difference between the positions of particles A and B and the sum of their momenta.

Appendix 15

Quantum measurement and the infinite regress

Suppose that a quantum system described by some state vector $|\Psi\rangle$ interacts with a measuring instrument whose measurement eigenstates are $|\psi_+\rangle$ and $|\psi_-\rangle$. These eigenstates combine with the macroscopic instrument to reveal one or other of the two possible outcomes, which we can imagine to involve the deflection of a pointer either to the left (+ result) or the right (− result). Following von Neumann and recognizing that the instrument itself consists of quantum particles, we describe the state of the instrument before the measurement in terms of a state vector $|\phi_0\rangle$, corresponding to the central pointer position. The total state of the quantum system plus the measuring instrument before the measurement is made is described by the state vector $|\Phi_0\rangle$, which is given by the product:

$$|\Phi_0\rangle = |\Psi\rangle|\phi_0\rangle = \tfrac{1}{\sqrt{2}}\big(|\psi_+\rangle + |\psi_-\rangle\big)|\phi_0\rangle,$$

$$= \tfrac{1}{\sqrt{2}}\big(|\psi_+\rangle|\phi_0\rangle + |\psi_-\rangle|\phi_0\rangle\big),$$

where we have made use of the expansion theorem to express $|\Psi\rangle$ in terms of the measurement eigenstates, and we have assumed that $\langle\psi_+|\Psi\rangle = \langle\psi_-|\Psi\rangle = 1/\sqrt{2}$ (the results are equally probable).

We now want to know how $|\Phi_0\rangle$ can be expected to evolve in time during the act of measurement. The time-dependent Schrödinger equation is given by (see Appendix 5):

$$i\hbar\frac{\partial|\Psi\rangle}{\partial t} = \hat{H}|\Psi\rangle,$$

which can be integrated to give:

$$|\Psi_t\rangle = e^{-i\hat{H}t/\hbar}|\Psi_0\rangle,$$

where $|\Psi_0\rangle$ is the state vector at some initial time $t = 0$ and $|\Psi_t\rangle$ is the same state vector at some later time $t = t$. If, at first sight, an exponential containing the differential operator $\hat{H}$ in the exponent looks a little strange, recall that the exponential can be expanded as a power series in $\hat{H}$, with terms in $\hat{H}, \hat{H}^2, \hat{H}^3$, and so on, which makes it a little clearer how it operates on the state vector $|\Psi_0\rangle$. In fact, the exponential is often called the *time evolution operator* and given the special symbol $\hat{U}$:

$$|\Psi_t\rangle = \hat{U}|\Psi_0\rangle, \quad \hat{U} = e^{-i\hat{H}t/\hbar}.$$

The operator $\hat{U}$ is said to be *unitary*, meaning that its application does not change the magnitude of the state vector or any relationships (i.e. 'angles') between vectors that it may comprise. Applying the time evolution operator to $|\Phi_0\rangle$ gives:

$$|\Phi_t\rangle = \hat{U}|\Phi_0\rangle = \tfrac{1}{\sqrt{2}}\left(\hat{U}|\psi_+\rangle|\phi_0\rangle + \hat{U}|\psi_-\rangle|\phi_0\rangle\right).$$

It is clear that if the instrument interacts with a quantum system which is already present in one of the measurement eigenstates ($|\psi_+\rangle$, say), then the total system (quantum system plus instrument) *must* evolve into a product quantum state given by $|\psi_+\rangle|\phi_+\rangle$. This is equivalent to saying that this interaction will always produce a + result (the pointer always moves to the left). In this case, the effect of $\hat{U}$ on the initial product quantum state $|\psi_+\rangle|\phi_0\rangle$ must be to yield the result $|\psi_+\rangle|\phi_+\rangle$, that is:

$$\hat{U}|\psi_+\rangle|\phi_0\rangle = |\psi_+\rangle|\phi_+\rangle.$$

Similarly,

$$\hat{U}|\psi_-\rangle|\phi_0\rangle = |\psi_-\rangle|\phi_-\rangle.$$

Substituting these last two expressions into the expression above for $|\Phi_t\rangle$ gives:

$$|\Phi_t\rangle = \tfrac{1}{\sqrt{2}}\left(|\psi_+\rangle|\phi_+\rangle + |\psi_-\rangle|\phi_-\rangle\right).$$

We are now no further forward than before the measurement was made. This last equation suggests that the measuring instrument evolves into a superposition state in which the pointer simultaneously points both to the left and to the right. Reducing the state vector $|\Phi_t\rangle$ of the system-plus-measuring-device would seem to require a further measurement. But then the whole argument can be repeated *ad infinitum*. Are we therefore locked into an endless chain of measuring processes? At what point does the chain stop (at what point is the state vector reduced)? The problem arises because of the unitary characteristics of the time-evolution operator. *The time evolution of a quantum system as described by the time-dependent Schrödinger equation is continuous and deterministic.* This equation *cannot* describe the discontinuous, indeterministic projection of $|\Psi\rangle$ into the measurement eigenstates.

Appendix 16

Von Neumann's 'impossibility proof'

Suppose the state vector of an ensemble of N quantum particles is given by $|\Psi\rangle$. This state vector can produce two possible measurement outcomes, R_+ and R_-, after interaction with a device with measurement operator $\hat{M}$. We know from the formalism of standard quantum theory (i.e. without hidden variables) that to obtain the quantum probabilities for these results, we need to use the expansion theorem and express $|\Psi\rangle$ as a linear combination of the two eigenstates of $\hat{M}$, which we denote $|\psi_+\rangle$ and $|\psi_-\rangle$ (see Appendix 9):

$$|\Psi\rangle = |\psi_+\rangle\langle\psi_+|\Psi\rangle + |\psi_-\rangle\langle\psi_-|\Psi\rangle.$$

The effect of $\hat{M}$ on $|\Psi\rangle$ is, therefore:

$$\hat{M}|\Psi\rangle = \hat{M}|\psi_+\rangle\langle\psi_+|\Psi\rangle + \hat{M}|\psi_-\rangle\langle\psi_-|\Psi\rangle,$$
$$= R_+|\psi_+\rangle\langle\psi_+|\Psi\rangle + R_-|\psi_-\rangle\langle\psi_-|\Psi\rangle.$$

The expectation value $\langle\hat{M}\rangle$ is therefore given by:

$$\langle\hat{M}\rangle = \langle\Psi|\hat{M}|\Psi\rangle = |\langle\psi_+|\Psi\rangle|^2 R_+ + |\langle\psi_-|\Psi\rangle|^2 R_-,$$
$$= P_+R_+ + P_-R_-,$$

where $P_+ = |\langle\psi_+|\Psi\rangle|^2$ is the projection probability for the result R_+ and $P_- = |\langle\psi_-|\Psi\rangle|^2$ is the projection probability for the result R_-. Under these circumstances, the variance, $\langle\hat{M}^2\rangle - \langle\hat{M}\rangle^2$, is given by:

$$\langle\hat{M}^2\rangle - \langle\hat{M}\rangle^2 = (P_+R_+^2 + P_-R_-^2) - (P_+R_+ + P_-R_-)^2,$$

which is clearly non-zero. In this case, the ensemble collectively described by the state vector $|\Psi\rangle$ exhibits dispersion when analysed according to the standard quantum formalism.

If we now suppose that the particles are governed by hidden variables, denoted collectively as λ, then we can identify a sub-ensemble N_+ of N which contains particles with

λ values that on interaction with the measuring device predetermine the result R_+ with unit probability. So, for the particles that form the sub-ensemble N_+ we have:

$$\hat{M}|\Psi\rangle_{N_+} = R_+|\Psi\rangle_{N_+},$$

where we have used the subscript N_+ to indicate that this expression applies only to the sub-ensemble N_+. It follows immediately that under these circumstances:

$$\langle \hat{M}^2 \rangle_{N_+} - \langle \hat{M} \rangle^2_{N_+} = R_+^2 - R_+^2 = 0$$

and the sub-ensemble N_+ is said to be dispersion-free.

As part of his 'impossibility proof', von Neumann considered the simultaneous application of a second measurement, with operator $\hat{L}$ and results S_+ and S_-, on particles in the sub-ensemble N_+. The same arguments that we used above can be used again to identify particles with values that predetermine the results R_+ *and* S_+ with unit probability. These particles form a sub-sub-ensemble N_{++}. Von Neumann postulated that the expectation value of the joint measurement, $\langle \hat{M} + \hat{L} \rangle$, can be calculated as the sum of the expectation values $\langle \hat{M} \rangle + \langle \hat{L} \rangle$. This is von Neumann's 'additivity postulate', which can be demonstrated to apply for all operators—including non-commuting operators—in quantum theory. When applied to the sub-sub-ensemble N_{++}, this postulate yields:

$$\langle \hat{M} + \hat{L} \rangle_{N_{++}} = \langle \hat{M} \rangle_{N_{++}} + \langle \hat{L} \rangle_{N_{++}},$$
$$= R_+ + S_+.$$

Herein lies the difficulty, von Neumann claimed. Note that, unlike the equation for the quantum theoretical expectation value $\langle \hat{M} \rangle$ given above, the expectation value for the combined operator $\langle \hat{M} + \hat{L} \rangle_{N_{++}}$ in this hidden variables formulation is given by the sum of two eigenvalues corresponding to two measurement processes applied simultaneously, each of which must be obtained with unit probability (i.e. with certainty). Whereas the quantum theoretical expression for $\langle \hat{M} \rangle$ above is interpreted to mean that the result R_+ *or* the result R_- may be obtained with quantum probabilities given by P_+ and P_-, respectively, the expression for $\langle \hat{M} + \hat{L} \rangle_{N_{++}}$ can only mean that the results R_+ *and* S_+ *must* each be obtained with unit probability. However, although the expectation values of non-commuting quantum mechanical operators are additive, as von Neumann postulated, their eigenvalues are not. If they were, then an appropriate combination of measurement operators would allow us simultaneously to measure any combination of complementary observables with arbitrary precision. This, experiment tells us, is something we cannot do. Von Neumann therefore concluded that dispersion-free ensembles (and hence hidden variables) are impossible.

Appendix 17

Photon spin correlations

To anticipate or interpret the results of experiments conducted on pairs of correlated photons using the arrangement given in Fig. 8.2, we need to know the initial state vector of the photon pair and the possible measurement eigenstates.

The two photons are emitted with opposite spin orientations or circular polarizations. The total state vector of the pair can therefore be written as a linear combination of the product of the states $|L_A\rangle$ (photon A in a state of left-circular polarization) and $|L_B\rangle$ (photon B in a state of left-circular polarization) and the product of $|R_A\rangle$ (photon A in a state of right-circular polarization) and $|R_B\rangle$ (photon B in a state of right-circular polarization). You might have expected that these products should have been $|L_A\rangle|R_B\rangle$ and $|R_A\rangle|L_B\rangle$ to get the correct left–right symmetry, but remember that the convention for circular polarization given in Chapter 5 (see Fig. 5.4) specifies the direction of rotation for photons propagating *towards* the detector. Left (anticlockwise) rotation with respect to PA_1 corresponds to right (clockwise) rotation with respect to PA_2, and so we interchange the labels for photon B.

We also need to recall that photons are bosons, and from Appendix 11, we note that bosons have two-particle state vectors that are symmetric to the exchange of the particles. The initial state vector of the pair is therefore given by:

$$|\Psi\rangle = \tfrac{1}{\sqrt{2}}\big(|L_A\rangle|L_B\rangle + |R_A\rangle|R_B\rangle\big)$$

The experimental arrangement shown in Fig. 8.2 can produce any one of four possible outcomes for each successfully detected pair. If we denote detection of a photon in a state of vertical polarization as a + result and detection in a state of horizontal polarization as a − result, there are four measurement possibilities (see Table A17.1).

The joint measurement eigenstates are the products of the final state vectors of the individual photons. Denoting these final states as $|v_A\rangle$ (photon A detected in a vertical polarization state with respect to orientation a of PA_1), $|h_A\rangle$ (photon A detected in a horizontal polarization state with respect to orientation a of PA_1), and similarly $|v_B\rangle$ and

Table A17.1 Four measurement possibilities for detection of a photon in a state of vertical and horizontal polarization

PA$_1$	PA$_2$	Measurement eigenstate
+	+	$\lvert\psi_{++}\rangle$
+	−	$\lvert\psi_{+-}\rangle$
−	+	$\lvert\psi_{-+}\rangle$
−	−	$\lvert\psi_{--}\rangle$

$\lvert h_B\rangle$ for photon B, we have

$$\lvert\psi_{++}\rangle = \lvert v_A\rangle\lvert v_B\rangle \qquad \lvert\psi_{+-}\rangle = \lvert v_A\rangle\lvert h_B\rangle,$$

$$\lvert\psi_{-+}\rangle = \lvert h_A\rangle\lvert v_B\rangle \qquad \lvert\psi_{--}\rangle = \lvert h_A\rangle\lvert h_B\rangle.$$

Now we must do something about the fact that the initial state vector $\lvert\Psi\rangle$ is given above in a basis of circular polarization states, whereas the measurement eigenstates are given in a basis of linear polarization states. We therefore use the expansion theorem to express the initial state vector in terms of the possible measurement eigenstates as follows:

$$\lvert\Psi\rangle = \lvert\psi_{++}\rangle\langle\psi_{++}\lvert\Psi\rangle + \lvert\psi_{+-}\rangle\langle\psi_{+-}\lvert\Psi\rangle + \lvert\psi_{-+}\rangle\langle\psi_{-+}\lvert\Psi\rangle + \lvert\psi_{--}\rangle\langle\psi_{--}\lvert\Psi\rangle.$$

We must now find expressions for the individual projection amplitudes. The measurement eigenstate $\lvert\psi_{++}\rangle$ is defined above in terms of the vertical polarization states of photons A and B. Combining this definition with the expression for $\lvert\Psi\rangle$ given above in terms of the initial circular polarization states allows us to deduce:

$$\langle\psi_{++}\lvert\Psi\rangle = \langle v_A\lvert\langle v_B\lvert \cdot \tfrac{1}{\sqrt{2}}\big(\lvert L_A\rangle\lvert L_B\rangle + \lvert R_A\rangle\lvert R_B\rangle\big),$$

$$= \tfrac{1}{\sqrt{2}}\big(\langle v_A\lvert L_A\rangle\langle v_B\lvert L_B\rangle + \langle v_A\lvert R_A\rangle\langle v_B\lvert R_B\rangle\big).$$

The terms collected in the bracket in the above expression are projection amplitudes for this particular combination of circular and linear polarization states. We can look these up in Appendix 12, Table A12.2. We discover:

$$\langle\psi_{++}\lvert\Psi\rangle = \tfrac{1}{\sqrt{2}}\left(\tfrac{1}{\sqrt{2}}\cdot\tfrac{1}{\sqrt{2}} + \tfrac{1}{\sqrt{2}}\cdot\tfrac{1}{\sqrt{2}}\right) = \tfrac{1}{\sqrt{2}}.$$

Repeating this process for the other projection amplitudes in the expansion for $\lvert\Psi\rangle$ gives:

$$\langle\psi_{+-}\lvert\Psi\rangle = 0$$

$$\langle\psi_{-+}\lvert\Psi\rangle = 0$$

and

$$\langle\psi_{--}\lvert\Psi\rangle = -\tfrac{1}{\sqrt{2}}.$$

and so

$$|\Psi\rangle = \tfrac{1}{\sqrt{2}}\big(|\psi_{++}\rangle - |\psi_{--}\rangle\big).$$

This last expression merely reflects the fact that the physics of the two-photon emission process (no net angular momentum) combines with the arrangement of the measuring device as given in Fig. 8.2, in which both polarization analysers have the same orientation, to ensure that detection of a combined $+$ result for photon A and $-$ result for photon B (and *vice versa*) is not possible. We further deduce from this last expression that the joint probability for both photons to produce $+$ results, $P_{++}(a, a) = |\langle\psi_{++}|\Psi\rangle|^2 = 1/2$; similarly the joint probability for both photons to give $-$ results, $P_{--}(a, a) = |\langle\psi_{--}|\Psi\rangle|^2 = 1/2$. The notation (a, a) indicates the orientations of the two analysers.

Finally, assuming $|\Psi\rangle$ is normalized, we can define an expectation value for the joint measurement as:

$$E(a, a) = \langle\Psi|\hat{M}_1^a\hat{M}_2^a|\Psi\rangle,$$

where $\hat{M}_1^a$ is the measurement operator corresponding to PA_1 with orientation a, and $\hat{M}_2^a$ is the measurement operator corresponding to PA_2 also with orientation a. If we further define the results of the operation of $\hat{M}_1^a$ on photon A as R_+^A and R_-^A, depending on whether A is detected in a final vertical or horizontal polarization state, we can derive the following expression for $E(a, a)$:

$$E(a, a) = \tfrac{1}{2}\big(\langle\psi_{++}|\hat{M}_1^a\hat{M}_2^a|\psi_{++}\rangle - \langle\psi_{++}|\hat{M}_1^a\hat{M}_2^a|\psi_{--}\rangle$$
$$- \langle\psi_{--}|\hat{M}_1^a\hat{M}_2^a|\psi_{++}\rangle + \langle\psi_{--}|\hat{M}_1^a\hat{M}_2^a|\psi_{--}\rangle\big),$$

which reduces to:

$$E(a, a) = P_{++}R_+^A R_+^B + P_{--}R_-^A R_-^B$$

or

$$E(a, a) = \tfrac{1}{2}\big(R_+^A R_+^B + R_-^A R_-^B\big)$$

because we can assume the measurement eigenstates $|\psi_{++}\rangle$ and $|\psi_{--}\rangle$ to be orthonormal. We can use the expectation value as a measure of the *correlation* between the measurement results by assigning some 'values' to the results R_+^A and R_-^A, etc. If we set[1] $R_+^A = 1, R_+^B = 1$ and $R_-^A = -1, R_-^B = -1$, then $E(a, a)$ becomes:

$$E(a, a) = +1,$$

that is, the joint results are perfectly correlated.

---

[1] Any actual values and their units associated with the results (such as voltages) can in principle be absorbed into constants associated with the measurement operators.

Appendix 18

Quantum versus local hidden variable correlations

We now want to examine the effects of rotating the analyser PA_2 through some angle with respect to PA_1, as shown in Fig. 8.3. PA_1 is aligned in the same direction as before, which we continue to denote as orientation a, so its measurement eigenstates are $|v_A\rangle$ and $|h_A\rangle$, with eigenvalues R_+^A and R_-^A. We denote the new orientation of PA_2 as b and designate its new measurement eigenstates as $|v_B'\rangle$ and $|h_B'\rangle$, corresponding respectively to polarization along the new vertical v' and horizontal h' directions, with corresponding eigenvalues R'^B_+ and R'^B_-. We denote the angle between the vertical axes of the analysers as $(b - a)$. The eigenstates of the joint measurement in this new arrangement are given by:

$$|\psi'_{++}\rangle = |v_A\rangle|v_B'\rangle, \quad |\psi'_{+-}\rangle = |v_A\rangle|h_B'\rangle,$$
$$|\psi'_{-+}\rangle = |h_A\rangle|v_B'\rangle, \quad |\psi'_{--}\rangle = |h_A\rangle|h_B'\rangle.$$

We must now express the initial state vector $|\Psi\rangle$ in terms of the new joint measurement eigenstates. We can obviously proceed in the same way as before (see Appendix 17):

$$|\Psi\rangle = |\psi'_{++}\rangle\langle\psi'_{++}|\Psi\rangle + |\psi'_{+-}\rangle\langle\psi'_{+-}|\Psi\rangle + |\psi'_{-+}\rangle\langle\psi'_{-+}|\Psi\rangle + |\psi'_{--}\rangle\langle\psi'_{--}|\Psi\rangle$$

in which

$$\langle\psi'_{++}|\Psi\rangle = \langle v_A||v_B'| \cdot \frac{1}{\sqrt{2}}\left(|L_A\rangle|L_B\rangle + |R_A\rangle|R_B\rangle\right),$$

$$= \frac{1}{\sqrt{2}}\left(\langle v_A|L_A\rangle\langle v_B'|L_B\rangle + \langle v_A|R_A\rangle\langle v_B'|R_B\rangle\right),$$

$$= \frac{1}{\sqrt{2}}\left(\frac{1}{\sqrt{2}} \cdot \frac{e^{-i(b-a)}}{\sqrt{2}} + \frac{1}{\sqrt{2}} \cdot \frac{e^{i(b-a)}}{\sqrt{2}}\right),$$

$$= \frac{1}{\sqrt{2}}\cos(b - a).$$

Similarly,

$$\langle \psi'_{+-}|\Psi\rangle = \tfrac{1}{\sqrt{2}}\sin(b-a),$$

$$\langle \psi'_{-+}|\Psi\rangle = \tfrac{1}{\sqrt{2}}\sin(b-a),$$

$$\langle \psi'_{--}|\Psi\rangle = -\tfrac{1}{\sqrt{2}}\cos(b-a).$$

Thus,

$$|\Psi\rangle = \tfrac{1}{\sqrt{2}}\Big[|\psi'_{++}\rangle\cos(b-a) + |\psi'_{+-}\rangle\sin(b-a)$$

$$+\,|\psi'_{-+}\rangle\sin(b-a) - |\psi'_{--}\rangle\cos(b-a)\Big].$$

The consistency of this last expression with the result we obtained in Appendix 17 can be confirmed by setting $b = a$. We can use the expressions derived above for the projection amplitudes to obtain the projection probabilities for each of the four possible joint results:

$$P_{++}(a,b) = |\langle \psi'_{++}|\Psi\rangle|^2 = \tfrac{1}{2}\cos^2(b-a),$$

$$P_{+-}(a,b) = |\langle \psi'_{+-}|\Psi\rangle|^2 = \tfrac{1}{2}\sin^2(b-a),$$

$$P_{-+}(a,b) = |\langle \psi'_{-+}|\Psi\rangle|^2 = \tfrac{1}{2}\sin^2(b-a),$$

$$P_{--}(a,b) = |\langle \psi'_{--}|\Psi\rangle|^2 = \tfrac{1}{2}\cos^2(b-a).$$

If we again ascribe results of $+1$ for v or v' polarization and -1 for h or h' polarization, then the expectation value, $E(a,b)$, is given by:

$$E(a,b) = P_{++}(a,b) - P_{+-}(a,b) - P_{-+}(a,b) + P_{--}(a,b).$$

Using the projection probabilities given above, we deduce:

$$E(a,b) = \cos^2(b-a) - \sin^2(b-a) = \cos 2(b-a).$$

The function $\cos 2(b-a)$ is plotted against $(b-a)$ in Fig. 8.4.

For the simple hidden variable theory described in Chapter 8, the joint probability $P_{++}(a,b)$ will depend on the probability that the λ value for photon B lies within $\pm 45°$ of both the v and v' axes—the doubly shaded area shown in Fig. 8.5. This probability is given by the ratio of the range of angles that determine the area of overlap $(90° - |b-a|)$ to the range of all possible angles $(180°)$. Thus:

$$P_{++}(a,b) = \frac{90° - |b-a|}{180°}.$$

This expression for $P_{++}(a,b)$ is valid for $0° \le |b-a| \le 90°$. We can use a similar line of reasoning to show that

$$P_{+-}(a,b) = \frac{|b-a|}{180°},$$

$$P_{-+}(a,b) = \frac{|b-a|}{180°},$$

$$P_{--}(a,b) = \frac{90° - |b-a|}{180°}.$$

From the expression above for the expectation value in terms of the projection probabilities, it follows that the prediction for $E(a, b)$ based on this simple local hidden variable theory is:

$$E(a, b) = \frac{180° - 4|b - a|}{180°},$$

$$= 1 - \frac{|b - a|}{45°}$$

valid for $0° \leq |b - a| \leq 90°$.

Note that when $|b - a| = 0°, 45°$, and $90°, E(a, b) = +1, 0$, and -1, respectively. This local hidden variable theory is therefore consistent with the quantum theory predictions at these three angles. However, from the comparison of the two correlation functions shown in Fig. 8.4, we can see that the two theories predict different results at all other angles, with maximum divergence between the predictions occurring at angles $(b - a) = 22\frac{1}{2}°$ and $67\frac{1}{2}°$.

Appendix 19

Bell's inequality

Having established the 'space' of possible results obtained from the three different experiments (see Fig. 8.7), Bertlmann denotes the number of socks that pass test a and fail test b as $n[a_+b_-]$. These results fall in the space illustrated to the left of Fig. 8.7(a). It is clear from this figure that $n[a_+b_-]$ can be written as the sum of the numbers of socks which belong to two subsets, one in which the individual socks pass test a, fail b and pass c, denoted $n[a_+b_-c_+]$, and one in which the socks pass test a, fail b and fail c, denoted $n[a_+b_-c_-]$. In other words,

$$n[a_+b_-] = n[a_+b_-c_+] + n[a_+b_-c_-].$$

Similarly, we can deduce that (see Fig. 8.7(b)):

$$n[b_+c_-] = n[a_+b_+c_-] + n[a_-b_+c_-].$$

Adding these expressions together gives:

$$n[a_+b_-] + n[b_+c_-] = n[a_+b_-c_+] + n[a_+b_-c_-]$$
$$+ n[a_+b_+c_-] + n[a_-b_+c_-],$$

which can be simply rearranged to read (see Fig. 8.7(c)):

$$n[a_+b_-] + n[b_+c_-] = n[a_+b_+c_-] + n[a_+b_-c_-]$$
$$+ n[a_+b_-c_+] + n[a_-b_+c_-].$$

The first two terms on the right-hand side of this last expression are simply the number $n[a_+c_-]$, so we can simplify this to give:

$$n[a_+b_-] + n[b_+c_-] = n[a_+c_-] + n[a_+b_-c_+] + n[a_-b_+c_-].$$

The number of socks giving results a_+, b_-, and c_+ and the number giving results a_-, b_+, and c_+, obviously cannot be negative (they must be positive or, at least, zero). It therefore follows that:

$$n[a_+b_-] + n[b_+c_-] \geq n[a_+c_-].$$

Bertlmann notes that subjecting one of the socks in each pair to each test in turn will necessarily change its physical characteristics irreversibly such that, even if it survives the first test, it may not give the result for the second test that might be expected of a brand new sock. And, of course, if the sock fails either of the first two tests, it will simply not be available for the last.

Bertlmann therefore exploits that fact that his socks always come in pairs. He assumes that, apart from differences in colour, the physical characteristics of each sock in a pair are identical. Thus, a test performed on the right sock (sock B) can be used to predict what the result of the same test would be if it were performed on the left sock (sock A), even though the test on A is not actually carried out. He must further assume that whatever test he chooses to perform on B in no way affects the outcome of any other test he might perform on A. Bertlemann assumes that his socks are Einstein-separable.

He now devises three different sets of experiments to be carried out on three samples containing the same total number of pairs of his socks. In experiment 1, for each pair, sock A is subjected to test a, and sock B is subjected to test b. If sock B fails test b, this implies that sock A would also have failed test b had it been performed on A. He denotes the number of pairs of socks for which A passes test a and B fails test b as $N_{+-}(a, b)$. If the socks are assumed to be Einstein-separable, it follows that this number must be equal to the theoretical number of socks A which pass test a and fail test b, that is:

$$N_{+-}(a, b) = n[a_+b_-].$$

In experiment 2, for each pair, sock A is subjected to test b, and sock B is subjected to test c. The same kind of reasoning allows Bertlmann to deduce that:

$$N_{+-}(b, c) = n[b_+c_-].$$

Finally, in experiment 3, for each pair, sock A is subjected to test a, and sock B is subjected to test c. Bertlmann deduces that

$$N_{+-}(a, c) = n[a_+c_-].$$

Applying the inequality derived above gives:

$$N_{+-}(a, b) + N_{+-}(b, c) \geq N_{+-}(a, c).$$

Bertlmann now generalizes this result for any batch of pairs of socks. By dividing each number in the above expression by the total number of pairs of socks (which was the same for each experiment), he arrives at the relative frequencies with which each joint result was obtained. He identifies these relative frequencies with probabilities for obtaining the results for experiments to be performed on any batch of pairs of socks that, statistically, have the same properties. Thus,

$$P_{+-}(a, b) + P_{+-}(b, c) \geq P_{+-}(a, c).$$

This is a form of Bell's inequality.

Appendix 20

Bell's inequality for non-ideal cases

The generalization of Bell's inequality requires that we consider experiments involving four different orientations of the two polarization analysers. We denote these orientations as a, b, c, and d. We suppose that the results of measurements made on photon A and photon B are determined by some local hidden variable (or variables) denoted λ. The λ values are distributed among the photons according to a distribution function $\rho(\lambda)$, which is essentially the ratio of the number of photons with the value λ, N_λ, divided by the total number of photons. We assume that this function is suitably normalized, so that $\int \rho(\lambda)d\lambda = 1$.

The average or expectation values of the results of measurements made on individual photons depend on the particular orientation of the polarization analyser and the λ value. We denote the expectation value for photon A entering PA_1 set up with orientation a as $A(a, \lambda)$. Similarly, the expectation value for photon B entering PA_2 set up with orientation b is $B(b, \lambda)$. The possible result of each measurement is ± 1, corresponding to detection in the vertical or horizontal channels, respectively. It must then follow that the absolute values of the expectation values cannot exceed unity, that is,

$$|A(a, \lambda)| \leq 1, \quad |B(a, \lambda)| \leq 1.$$

We assume that the individual results for A depend on a and on λ, but are independent of b and vice versa (Einstein separability).

The expectation value for the joint measurement of A and B, $E(a, b, \lambda)$ is given by the product $A(a, \lambda)B(b, \lambda)$. We can eliminate λ from this expression by averaging the results over many photon pairs (emphasizing the statistical nature of the hidden variable approach) or by integrating over all λ:

$$E(a, b) = \int A(a, \lambda)B(b, \lambda)\rho(\lambda)\, d\lambda.$$

This follows if we assume that we can perform measurements on a sufficiently large number of photon pairs so that all possible values of λ are sampled. The same reasoning can be used to show that:

$$E(a, b) - E(a, d) = \int [A(a, \lambda)B(b, \lambda) - A(a, \lambda)B(d, \lambda)]\rho(\lambda)\,d\lambda,$$

$$= \int A(a, \lambda)[B(b, \lambda) - B(d, \lambda)]\rho(\lambda)\,d\lambda.$$

And so, since $|A(a, \lambda)| \le 1$,

$$|E(a, b) - E(a, d)| \le \int |B(b, \lambda) - B(d, \lambda)|\rho(\lambda)\,d\lambda.$$

Similarly,

$$|E(c, b) + E(c, d)| \le \int |B(b, \lambda) + B(d, \lambda)|\rho(\lambda)\,d\lambda.$$

If we now add these two expressions together, we get:

$$|E(a, b) - E(a, d)| + |E(c, b) + E(c, d)|$$

$$\le \int [|B(b, \lambda) - B(d, \lambda)| + |B(b, \lambda) + B(d, \lambda)|]\rho(\lambda)\,d\lambda$$

We know from the earlier arguments given above that the expectation values in this last expression must individually be less than or equal to 1. It therefore follows that:

$$|B(b, \lambda) - B(d, \lambda)| + |B(b, \lambda) + B(d, \lambda)| \le 2$$

Thus,

$$|E(a, b) - E(a, d)| + |E(c, b) + E(c, d)| \le 2 \int \rho(\lambda)\,d\lambda$$

and since, by definition $\int \rho(\lambda)\,d\lambda = 1$, we have:

$$|E(a, b) - E(a, d)| + |E(c, b) + E(c, d)| \le 2.$$

Note that nowhere in this derivation have we needed to assume that we will obtain a perfect correlation between the measured results for any combination of the analyser orientations. This generalized inequality is therefore valid for non-ideal cases in which limitations in the experimental apparatus prevent the observation of perfect correlation.

Appendix 21

Three-photon GHZ states

Suppose we can create a polarization-entangled state comprising of three photons with an overall initial state vector given by:

$$|\psi\rangle = \tfrac{1}{\sqrt{2}}(|v_A\rangle|v_B\rangle|v_C\rangle + |h_A\rangle|h_B\rangle|h_C\rangle),$$

where v and h denote vertical and horizontal polarization and the subscripts A, B, C denote the photons. We set up an experiment to make a linear-polarization measurement on photon A using an analyser oriented at an angle φ to the v/h axis, projecting the photon into $|v'\rangle$ or $|h'\rangle$ measurement states. At the same time, we subject photons B and C to circular-polarization measurements, with measurement states $|L\rangle$ or $|R\rangle$. To anticipate the results of such experiments, we recast the above expression for $|\psi\rangle$ in a basis of $|v'\rangle/|h'\rangle$ and $|L\rangle/|R\rangle$ states using the projection amplitudes given in Appendix 12. In general,

$$|v_i\rangle = |v_i'\rangle\cos\varphi + |h_i'\rangle\sin\varphi \quad |h_i\rangle = |h_i'\rangle\cos\varphi - |v_i'\rangle\sin\varphi,$$

$$|v_i\rangle = \frac{1}{\sqrt{2}}(|L_i\rangle + |R_i\rangle) \quad |h_i\rangle = \frac{i}{\sqrt{2}}(|R_i\rangle - |L_i\rangle),$$

where the index i can be any of A, B, or C. We can simplify the next step considerably by setting the linear-polarization analyser to an angle $\varphi = 45°$, in which case the expressions for $|v_i\rangle$ and $|h_i\rangle$ reduce to:

$$|v_i\rangle = \tfrac{1}{\sqrt{2}}(|v_i'\rangle + |h_i'\rangle) \quad |h_i\rangle = \tfrac{1}{\sqrt{2}}(|h_i'\rangle - |v_i'\rangle).$$

In the expression for $|\psi\rangle$, we can now substitute for $|v_A\rangle$ and $|h_A\rangle$, transforming these states into a $|v_A'\rangle/|h_A'\rangle$ basis. Similarly, we can substitute $|v_B\rangle$ and $|h_B\rangle$ and $|v_C\rangle$ and $|h_C\rangle$, transforming these into $|L_B\rangle/|R_B\rangle$ and $|L_C\rangle/|R_C\rangle$ bases, respectively. After a little algebra, we get:

$$|\psi\rangle = \tfrac{1}{2}(|v_A'\rangle|L_B\rangle|L_C\rangle + |v_A'\rangle|R_B\rangle|R_C\rangle + |h_A'\rangle|L_B\rangle|R_C\rangle + |h_A'\rangle|R_B\rangle|L_C\rangle).$$

Table A21.1 Combination states predicted by local hidden variables

λ parameters	Measurement			Predicted measurement state
	A	B	C	
$\lambda_A\binom{+1}{-1}\lambda_B\binom{-1}{+1}\lambda_C\binom{-1}{+1}$	$|v'\rangle/|h'\rangle$	$|L\rangle/|R\rangle$	$|L\rangle/|R\rangle$	$|v'_A\rangle|L_B\rangle|L_C\rangle$
	$|L\rangle/|R\rangle$	$|v'\rangle/|h'\rangle$	$|L\rangle/|R\rangle$	$|R_A\rangle|h'_B\rangle|L_C\rangle$
	$|L\rangle/|R\rangle$	$|L\rangle/|R\rangle$	$|v'\rangle/|h'\rangle$	$|R_A\rangle|L_B\rangle|h'_C\rangle$
	$|v'\rangle/|h'\rangle$	$|v'\rangle/|h'\rangle$	$|v'\rangle/|h'\rangle$	$|v'_A\rangle|h'_B\rangle|h'_C\rangle$

Terms involving combination states such as $|v'_A\rangle|L_B\rangle|R_C\rangle$ *cancel* and so do not appear in this final expression for $|\psi\rangle$.

The nature of the correlations between the three particles is now clear. Whenever we detect photon A to be in a vertical $|v'_A\rangle$ state, we expect that photons B and C will be detected to be in *identical* circular-polarization states, *both* $|L\rangle$ *or* both $|R\rangle$. Whenever we detect photon A to be in a horizontal $|h'_A\rangle$ state, we expect photons B and C to be detected in *opposite* states of circular polarization, *either* $|L_B\rangle/|R_C\rangle$ *or* $|R_B\rangle/|L_C\rangle$. According to quantum theory, there are no other possibilities.

We can arrive at exactly analogous results by choosing to perform the linear-polarization measurement on either photons B or C. In any of the three cases, detection of one of the photons in a $|v'_i\rangle$ state means that the other two must be detected in identical states of circular polarization, and detection in a $|h'_i\rangle$ state means that the other two photons must be detected in opposite states of circular polarization.

The challenge now is to account for these preferred states using a simple local hidden-variables theory. We can presume that the $|v'\rangle/|h'\rangle$ and $|L\rangle/|R\rangle$ measurement outcomes for all three photons are governed by two-valued hidden variables. We denote these as $\lambda_i\binom{\pm1}{\pm1}$, where the index i is again any of A, B, or C, the upper value in the parentheses predisposes the photon to a $|v'\rangle(+1)$ or $|h'\rangle(-1)$ result, and the lower value predisposes the photon to an $|L\rangle(+1)$ or $|R\rangle(-1)$ result. For example, the combination of values for the λ parameters given in Table A21.1 reproduces the combination states predicted by quantum theory in the situations where we make linear-polarization measurements on A, B, or C and circular polarization measurements on the other two. We can easily find other combinations of values that will satisfactorily predict all the other combination states predicted by quantum theory. So far, so good.

We now recall that the values of the λ parameters must be fixed at the moment the three photons become entangled and physically separate. Their values remain fixed as they propagate towards their respective analysers. We conclude therefore that if we change the arrangement so that we now make only linear-polarization measurements on all three photons, then we assume that the particular combination of values of the λ parameters cannot change in response to this.[1] The particular combination of λ parameters given in Table A21.1 lead us to expect therefore that the photons will be detected in the combination state $|v'_A\rangle|h'_B\rangle|h'_C\rangle$. Remember that we need this combination of values for the λ parameters to explain the measured combination states predicted by quantum

[1] It is possible to relax this assumption—see the discussion on the 'locality' loophole in Chapter 9.

theory. If we relax this requirement, we cannot use local hidden variables to explain the quantum theory correlations in experiments where one photon is subjected to a linear-polarization measurement, and the other two are subjected to circular-polarization measurements. We can go through similar arguments to deduce that when we make only linear-polarization measurements on all three photons, we expect to see only the combination states $|v'_A\rangle|v'_B\rangle|v'_C\rangle$, $|v'_A\rangle|h'_B\rangle|h'_C\rangle$, $|h'_A\rangle|v'_B\rangle|h'_C\rangle$ and $|h'_A\rangle|v'_B\rangle|v'_C\rangle$. We can summarize these results as follows: our local hidden-variable theory says that when photon A, B, or C is measured in a $|v'\rangle$ state, the other two must be detected in identical states of linear polarization, both $|v'\rangle$ or both $|h'\rangle$. When one of the photons is detected in a $|h'\rangle$ state, the other two must be detected in opposite states of linear polarization, either $|v'\rangle/|h'\rangle$ or $|h'\rangle/|v'\rangle$.

What does quantum theory predict? To answer this, we must simply recast the initial three-photon state vector in a basis of $|v'\rangle/|h'\rangle$ states only. After some further algebra, we get:

$$|\psi\rangle = \tfrac{1}{2}(|v'_A\rangle|v'_B\rangle|h'_C\rangle + |v'_A\rangle|h'_B\rangle|v'_C\rangle + |h'_A\rangle|v'_B\rangle|v'_C\rangle + |h'_A\rangle|h'_B\rangle|h'_C\rangle).$$

In other words, quantum theory predicts the *exact opposite set of combination states*. Quantum theory says that when we measure one photon in a $|v'\rangle$ state, the other two must be detected in opposite states of linear polarization, either $|v'\rangle/|h'\rangle$ or $|h'\rangle/|v'\rangle$. When one of the photons is detected in a $|h'\rangle$ state, the other two must be detected in identical states of linear polarization, both $|v'\rangle$ or both $|h'\rangle$.

We can find no way around this. If we use a local hidden-variable theory to account for the preferred combination states in which we make a linear-polarization measurement on one photon and circular-polarization measurements on the other two, as we have done above, then we find we can no longer successfully predict the preferred states when we make only linear-polarization measurements on all three. If we try to use a local hidden-variable theory to reproduce the quantum theory predictions for the experiment where we make only linear-polarization measurements, we will find that we can no longer reproduce the preferred combination states in the experiments where we measure the linear polarization of one photon and the circular polarization of the other two.

We are presented with a simple, direct test. We can set up an experiment to establish the correlations between the photons for the case where we make mixed linear- and circular-polarization measurements. Having confirmed the quantum theory predictions for this arrangement, we switch to experiments in which we make only linear-polarization measurements. If the observed combination states are those predicted by our local hidden-variables theory, then reality is local, and quantum theory is incomplete. If the observed combination states are those predicted by quantum theory, all classes of locally realistic hidden variables theories are effectively ruled out.

Would these conclusions be different if we had set up a different initial three-photon state vector? Certainly, the precise nature of the correlations changes when we start with a different initial state vector, but the overall conclusion is the same. Consider, for example, the following three-photon state vector as an alternative:

$$|\psi\rangle = \tfrac{1}{\sqrt{2}}(|v_A\rangle|v_B\rangle|h_C\rangle + |h_A\rangle|h_B\rangle|v_C\rangle).$$

In the experiment in which we make a linear polarization measurement on A and circular-polarization measurements on B and C, this state vector is recast as:

$$|\psi\rangle = \frac{i}{2}(|v'_A\rangle|L_B\rangle|R_C\rangle - |v'_A\rangle|R_B\rangle|L_C\rangle - |h'_A\rangle|L_B\rangle|L_C\rangle + |h'_A\rangle|R_B\rangle|R_C\rangle),$$

which reverses the pattern of correlations compared with our first example considered above. However, when we now make only linear-polarization measurements on all three photons, we find that the combination states predicted by quantum theory are the same as the previous example: $|v'_A\rangle|v'_B\rangle|h'_C\rangle$, $|v'_A\rangle|h'_B\rangle|v'_C\rangle$, $|h'_A\rangle|v'_B\rangle|v'_C\rangle$, and $|h'_A\rangle|h'_B\rangle|h'_C\rangle$ (there are some differences in phase factors, but these will not be measurable). It turns out that we *can* devise a set of values for the local hidden variables which will be fully compatible with the results of $|v'\rangle/|h'\rangle$ measurements made on either photon A or B and circular polarization measurements made on either photons B and C or A and C and the linear-only polarization measurements. For example, the hidden variable set $\lambda_A\binom{+1}{-1} \lambda_B\binom{-1}{+1} \lambda_C\binom{+1}{-1}$ will predict the same results as quantum theory for all these measurements. But it will *not* cope with the situation in which we make circular polarization measurements on photons A and B and $|v'\rangle/|h'\rangle$ measurements on photon C, as we can see immediately from the corresponding three-photon state vector:

$$|\psi\rangle = \frac{1}{2}(|L_A\rangle|R_B\rangle|h'_C\rangle - |L_A\rangle|L_B\rangle|v'_C\rangle - |R_A\rangle|R_B\rangle|v'_C\rangle + |R_A\rangle|L_B\rangle|h'_C\rangle)$$

The hidden variable set $\lambda_A\binom{+1}{-1} \lambda_B\binom{-1}{+1} \lambda_C\binom{+1}{-1}$ predicts that detection of photon C in a vertical polarization state should correspond to detection of photons A and B in opposite states of circular polarization, in contrast to the quantum theory prediction above. We are again forced to conclude that for this initial state vector, there is no combination of values of the hidden variables which can predict the same results as quantum theory.

Appendix 22

The Clauser–Horne–Shimony–Holt form of Bell's inequality

In an experiment involving a pair of polarization-entangled photons, we position two polarization analysers, PA_1 and PA_2, and detect those photons transmitted by the analysers (+ results) and those reflected (− results). If the orientations of the analysers are a and b, respectively, then from Appendix 18, we know that the expectation value for the joint measurement, $E(a, b)$ is given by:

$$E(a, b) = P_{++}(a, b) - P_{+-}(a, b) - P_{-+}(a, b) + P_{--}(a, b),$$

where $P_{++}(a, b)$ represents the probability of a joint ++ result and so on. To obtain experimental values for $E(a, b)$, it would seem that we are obliged to measure all four probabilities (in fact, coincidence *rates* are measured for each combination), implying that we need at least two photon detectors for each polarization analyser. As the experiments become more complex (cf. the experiments of Aspect, Dalibard, and Roger involving time-varying analysers described in Chapter 9), this requirement starts to impose some practical limitations.

If we seek to avoid these limitations, perhaps by detecting only transmitted photons, then only quantities related to $P_{++}(a, b)$ can be measured. However, consider an experiment in which we remove PA_2 completely. We use the symbol ∞ instead of b to define a probability for joint detection, $P_{++}(a, \infty)$ in these circumstances. Provided that the removal of PA_2 in no way affects the behaviour of either photon A or B, then $P_{++}(a, \infty)$ should include the probabilities of all possible joint results in which photon A is detected, that is, it includes the possible joint results in which photon B is detected (+) *and* not detected (−):

$$P_{++}(a, \infty) = P_{++}(a, b) + P_{+-}(a, b).$$

Similarly,

$$P_{++}(\infty, b) = P_{++}(a, b) + P_{-+}(a, b)$$

and

$$P_{++}(\infty, \infty) = P_{++}(a, b) + P_{+-}(a, b) + P_{-+}(a, b) + P_{--}(a, b).$$

We can now combine these expressions to give

$$E(a, b) = 4P_{++}(a, b) - 2P_{++}(a, \infty) - 2P_{++}(\infty, b) + P_{++}(\infty, \infty),$$

an expression first derived by John F. Clauser, Michael A. Horne, Abner Shimony, and Richard A. Holt.

This expression allows us to calculate the expectation value using only the probabilities (or coincidence rates) of joint $++$ results, for which related quantities can be obtained directly from experiment. Expectation values for other combinations of orientations— (a, d), (c, b), and (c, d)—can be obtained in a similar manner. These can then be used in the generalized form of Bell's inequality (Appendix 20).

Appendix 23

'Which Way' versus interference: testing complementarity

The interference effects arising from the overlap of beams of rubidium atoms passing through two slits can be readily understood in terms of the spatial (or centre-of-mass) components of the atomic state vectors, which we denote $|\psi(r)\rangle$, where r represents a generalized set of spatial coordinates. The state vectors are diffracted by the slits, spread out, and overlap constructively or destructively. In this interference region, the spatial component of the resultant state vector is given by:

$$|\Psi(r)\rangle = \tfrac{1}{\sqrt{2}}(|\psi_1(r)\rangle + |\psi_2(r)\rangle),$$

where $|\psi_1(r)\rangle$ and $|\psi_2(r)\rangle$ are the secondary waves produced by diffraction through slits 1 and 2, respectively. The probability of finding an atom at a point R on a screen lying beyond the two slits is related to the modulus-square of this state vector at position R:

$$
\begin{aligned}
P(R) &= ||\Psi(R)\rangle|^2, \\
&= \tfrac{1}{2}|(|\psi_1(R)\rangle + |\psi_2(R)\rangle)|^2, \\
&= \tfrac{1}{2}(\langle\psi_1(R)|\psi_1(R)\rangle + \langle\psi_2(R)|\psi_1(R)\rangle + \langle\psi_1(R)|\psi_2(R)\rangle \\
&\quad + \langle\psi_2(R)|\psi_2(R)\rangle).
\end{aligned}
$$

We recognize the origin of interference fringes in the cross-terms $\langle\psi_2(R)|\psi_1(R)\rangle$ and $\langle\psi_1(R)|\psi_2(R)\rangle$.

Now imagine that we have inserted micromaser cavities in front of each of the two slits (see Chapter 10, Fig. 10.5). As a rubidium atom passes through one of the cavities, it spontaneously emits a microwave photon before continuing on through the slit, leaving the spatial component of its state vector intact but also leaving behind a 'telltale' signal from which we learn which slit the atom subsequently passed through. If we describe the initial state of the excited $63p_{3/2}$ rubidium atom in terms of the state vector $|a\rangle$ and the lower maser state $61d_{3/2}$ in terms of the vector $|b\rangle$, then this process can be described as follows:

$$|a\rangle \rightarrow |b\rangle + |h\nu\rangle,$$

where $|h\nu\rangle$ denotes the emitted microwave photon.

But this is only half the story. In emitting the photon, one of the micromaser cavities initially in an 'off' state is transformed to an 'on' state. If we denote the initial state of the cavities as $|00\rangle$ (both cavities 'off'), the result of emission will be either of the states $|10\rangle$ (photon in cavity 1) or $|01\rangle$ (photon in cavity 2), which we can represent in terms of the following process:

$$|00\rangle + |h\nu\rangle \rightarrow |10\rangle \text{ or } |01\rangle.$$

Now let us put all of this together. The combination of excited rubidium atoms, unexcited cavities, and the two slits gives the state vector given by the product:

$$|\Psi'(r)\rangle = \tfrac{1}{\sqrt{2}}(|\psi_1(r)\rangle + |\psi_2(r)\rangle)|a\rangle|00\rangle.$$

A rubidium atom passes through one or other of the two cavities and emits a microwave photon. If it passes through cavity 1, the initial product state $|\psi_1(r)\rangle|a\rangle|00\rangle$ is transformed to $|\psi_1(r)\rangle|b\rangle|10\rangle$ (remember that the spatial component of the atomic state vector is unaffected by the emission of the photon), and if it passes through cavity 2, the initial product state $|\psi_2(r)\rangle|a\rangle|00\rangle$ is transformed to $|\psi_2(r)\rangle|b\rangle|01\rangle$. The resulting state vector is then given by the superposition:

$$|\Psi''(r)\rangle = \tfrac{1}{\sqrt{2}}(|\psi_1(r)\rangle|10\rangle + |\psi_2(r)\rangle|01\rangle)|b\rangle.$$

As before, the probability of finding an atom at location R on the screen is given by the modulus-square of the state vector at position R:

$$
\begin{aligned}
P''(R) &= ||\Psi''(R)\rangle|^2, \\
&= \tfrac{1}{2}|(|\psi_1(R)\rangle|10\rangle + |\psi_2(R)|01\rangle)|b\rangle|^2, \\
&= \tfrac{1}{2}(\langle 10|10\rangle\langle\psi_1(R)|\psi_1(R)\rangle + \langle 01|10\rangle\langle\psi_2(R)|\psi_1(R)\rangle \\
&\quad + \langle 10|01\rangle\langle\psi_1(R)|\psi_2(R)\rangle + \langle 01|01\rangle\langle\psi_2(R)|\psi_2(R)\rangle)\langle b|b\rangle, \\
&= \tfrac{1}{2}(\langle\psi_1(R)|\psi_1(R)\rangle + \langle\psi_2(R)|\psi_2(R)\rangle).
\end{aligned}
$$

And herein lies the answer. The states $|10\rangle$ and $|01\rangle$ are *orthogonal*. The presence of the terms $\langle 01|10\rangle$ and $\langle 10|01\rangle$ in the expression for the probability therefore has the effect of destroying the cross-terms and eliminating interference between the spatial components of the state vectors. In seeking to acquire 'which way' information characteristic of particle-like behaviour, we are quite simply *denied* the possibility of observing interference through the operation of a mechanism that has nothing to do with the uncertainty principle.

Appendix 24

The quantum eraser

If we separate the two micromaser cavities using a central, two-sided photo-detector and two shutters, as shown in Fig. 10.6(a), we must include the state vector of this detector in our expression for the combined state vector $|\Psi''(r)\rangle$ described in Appendix 23. We denote the lowest energy unexcited (ground) state of the detector by the state vector $|\alpha\rangle$ and the excited state formed by absorption of a microwave photon by the state vector $|\beta\rangle$. Initially, the two shutters are closed. The combined state vector for this arrangement is therefore (see Appendix 23):

$$|\Psi''(r)\rangle = \tfrac{1}{\sqrt{2}}(|\psi_1(r)\rangle|10\rangle + |\psi_2(r)\rangle|01\rangle)|b\rangle|\alpha\rangle,$$

where, as before, $|\psi_1(r)\rangle$ and $|\psi_2(r)\rangle$ represent the spatial components of the secondary waves formed by passage of the atom through slits 1 and 2, $|10\rangle$ and $|01\rangle$ are the state vectors describing the microwave photon present in cavity 1 and cavity 2, respectively, and $|b\rangle$ is the lower maser state $61d_{3/2}$ of ^{85}Rb.

We allow the rubidium atom to traverse the apparatus and be detected. The atom is registered as a spot on the screen. Then we open the shutters. The photodetector can be excited only when the shutters are opened and the states $|10\rangle$ and $|01\rangle$ combine constructively. Suppose we can form such a constructive superposition as the following linear combination:

$$|+\rangle = \tfrac{1}{\sqrt{2}}(|10\rangle + |01\rangle).$$

It follows therefore that forming the destructive combination

$$|-\rangle = \tfrac{1}{\sqrt{2}}(|10\rangle - |01\rangle)$$

leads to non-detection of the microwave photon. We can readily introduce these combinations into the expression for $|\Psi''(r)\rangle$ above if we simultaneously introduce the following linear combinations of the spatial components of the atomic state vectors:

$$|\psi_+(r)\rangle = \tfrac{1}{\sqrt{2}}(|\psi_1(r)\rangle + |\psi_2(r)\rangle)$$

and

$$|\psi_-(r)\rangle = \tfrac{1}{\sqrt{2}}(|\psi_1(r)\rangle - |\psi_2(r)\rangle).$$

It then follows that:

$$|\Psi''(r)\rangle = \tfrac{1}{\sqrt{2}}(|\psi_+(r)\rangle|+\rangle + |\psi_-(r)\rangle|-\rangle)|b\rangle|\alpha\rangle.$$

Only the constructive superposition, $|+\rangle$, results in detection of the photon by the photodetector, which we can describe in terms of the transitions $|+\rangle \rightarrow |00\rangle$ as $|\alpha\rangle \rightarrow |\beta\rangle$.

After detection, the state vector $|\Psi''(r)\rangle$ is given by:

$$|\Psi''(r)\rangle = \tfrac{1}{\sqrt{2}}(|\psi_+(r)\rangle|00\rangle|\beta\rangle + |\psi_-(r)\rangle|-\rangle|\alpha\rangle)|b\rangle.$$

As in Appendix 23, the probability of finding an atom at location R on the screen is given by the modulus-square of the state vector at position R:

$$
\begin{aligned}
P''(R) &= ||\Psi''(R)\rangle|^2, \\
&= \tfrac{1}{2}|(|\psi_+(R)\rangle|00\rangle|\beta\rangle + |\psi_-(R)\rangle|-\rangle|\alpha\rangle)|b\rangle|^2, \\
&= \tfrac{1}{2}(\langle\psi_+(R)|\psi_+(R)\rangle + \langle\psi_-(R)|\psi_-(R)\rangle)\langle b|b\rangle, \\
&= \tfrac{1}{2}(\langle\psi_1(R)|\psi_1(R)\rangle + \langle\psi_2(R)|\psi_2(R)\rangle),
\end{aligned}
$$

which is the same result obtained in Appendix 23 for the case where we *do* obtain 'which way' information. We conclude that no interference can be observed, even though we have chosen to open the shutters before we looked to see through which cavity (and hence through which slit) the atom went.

But now suppose we code each spot recorded by detection of an atom which is correlated with detection of a photon using a red colour, and we code detection of an atom which is correlated with non-detection of the photon using a blue colour. This allows us to differentiate between the constructive and destructive superpositions of the cavity states $|10\rangle$ and $|01\rangle$. We can split the state vector $|\Psi''(R)\rangle$ into two components, $|\Psi''_{\text{red}}(R)\rangle$ and $|\Psi''_{\text{blue}}(R)\rangle$, each given by:

$$|\Psi''_{\text{red}}(R)\rangle = \tfrac{1}{\sqrt{2}}(|\psi_+(R)\rangle|00\rangle|\beta\rangle)|b\rangle$$

and

$$|\Psi''_{\text{blue}}(r)\rangle = \tfrac{1}{\sqrt{2}}(|\psi_-(r)\rangle|-\rangle|\alpha\rangle)|b\rangle$$

and so the probability of observing *only* red spots is given by:

$$
\begin{aligned}
P''_{\text{red}}(R) &= ||\Psi''_{\text{red}}(R)\rangle|^2, \\
&= \tfrac{1}{2}|(|\psi_+(R)\rangle|00\rangle|\beta\rangle)|b\rangle|^2, \\
&= \tfrac{1}{2}||\psi_+(R)\rangle|^2\langle 00|00\rangle\langle\beta|\beta\rangle\langle b|b\rangle, \\
&= \tfrac{1}{4}((\langle\psi_1(R)|\psi_1(R)\rangle + \langle\psi_2(R)|\psi_1(R)\rangle + \langle\psi_1(R)|\psi_2(R)\rangle \\
&\quad + \langle\psi_2(R)|\psi_2(R)\rangle).
\end{aligned}
$$

Lo and behold! The interference pattern has returned! The corresponding expression for $P''_{\text{blue}}(R)$ is given by:

$$
\begin{aligned}
P''_{\text{blue}}(R) &= ||\Psi''_{\text{blue}}(R)\rangle|^2 \\
&= \tfrac{1}{2}|||\psi_-(R)\rangle|^2 \langle -|-\rangle \langle \alpha|\alpha\rangle \langle b|b\rangle \\
&= \tfrac{1}{4}(\langle \psi_1(R)|\psi_1(R)\rangle - \langle \psi_2(R)|\psi_1(R)\rangle - \langle \psi_1(R)|\psi_2(R)\rangle \\
&\quad + \langle \psi_2(R)|\psi_2(R)\rangle)
\end{aligned}
$$

giving rise to interference fringes formed by the blue spots that are 180° out of phase with the interference fringes formed by the red spots (see Fig. 10.6(b)). This explains the fact that when the combined probability, $P''_{\text{red}}(R) + P''_{\text{blue}}(R)$, is studied, the interference pattern disappears.

Appendix 25

Beam me up, Scotty

To understand what's involved in the quantum teleportation of a photon, we need to imagine that we first set up a two-photon entangled state:

$$|\Psi_{AB}\rangle = \tfrac{1}{\sqrt{2}}(|v_A\rangle|h_B\rangle - |h_A\rangle|v_B\rangle),$$

where the indices A and B denote the photons, and v and h denote vertical and horizontal polarization with respect to some arbitrary frame. Photon A is sent to Alice, and photon B is sent to Bob. Alice then carefully entangles photon A with another photon of unknown state, X, using a semi-reflecting mirror or a Köster prism. Photons A and X become entangled, lose their separate identity and interfere. The photons are detected using photomultipliers placed on either side of the semi-reflecting surface. Clearly, four resultant combinations are possible, described by the following measurement eigenstates:

$$|\psi_{++}\rangle = \tfrac{1}{\sqrt{2}}(|v_X\rangle|v_A\rangle + |h_X\rangle|h_A\rangle) \quad |\psi_{--}\rangle = \tfrac{1}{\sqrt{2}}(|v_X\rangle|v_A\rangle - |h_X\rangle|h_A\rangle),$$

$$|\psi_{+-}\rangle = \tfrac{1}{\sqrt{2}}(|v_X\rangle|h_A\rangle + |h_X\rangle|v_A\rangle) \quad |\psi_{-+}\rangle = \tfrac{1}{\sqrt{2}}(|v_X\rangle|h_A\rangle - |h_X\rangle|v_A\rangle).$$

These should not be confused with the measurement states described in Appendix 18, in which the subscripts $+$, $-$ referred to detection of vertical versus horizontal polarization. No polarization measurements are being made here (the photons are simply being detected in one, other, or both photomultipliers)—the subscripts are a convenient way of differentiating the different possible outcomes and have no deeper significance. Of these combinations, the first three result in photons being detected by only one or other of the photomultipliers or no detection at all. Only the state $|\psi_{-+}\rangle$ results in one photon being detected by each photomultiplier.

However, the polarization state of X is initially unknown. For simplicity, we express it as a simple linear combination of vertical and horizontal states:

$$|\psi_X\rangle = a|v_X\rangle + b|h_X\rangle,$$

where the coefficients a and b are suitably normalized so that $|a|^2 + |b|^2 = 1$.

The complete, three-photon state is then given by:

$$\begin{aligned}
|\Psi_{XAB}\rangle &= |\psi_X\rangle|\Psi_{AB}\rangle, \\
&= \tfrac{1}{\sqrt{2}}|\psi_X\rangle(|v_A\rangle|h_B\rangle - |h_A\rangle|v_B\rangle)), \\
&= \frac{a}{\sqrt{2}}(|v_X\rangle|v_A\rangle|h_B\rangle - |v_X\rangle|h_A\rangle|v_B\rangle)) + \frac{b}{\sqrt{2}}(|h_X\rangle|v_A\rangle|h_B\rangle \\
&\quad - |h_X\rangle|h_A\rangle|v_B\rangle)).
\end{aligned}$$

We can recast this last expression for $|\Psi_{XAB}\rangle$ in terms of the measurement eigenstates of the X–A pair. After a little algebra, we get:

$$|\Psi_{XAB}\rangle = \tfrac{1}{2}\big[|\psi_{++}\rangle|\psi_B^1\rangle - |\psi_{+-}\rangle|\psi_B^2\rangle - |\psi_{-+}\rangle|\psi_B^3\rangle + |\psi_{--}\rangle|\psi_B^4\rangle\big],$$

where the four possible states of photon B are:

$$|\psi_B^1\rangle = a|h_B\rangle - b|v_B\rangle,$$

$$|\psi_B^2\rangle = a|v_B\rangle - b|h_B\rangle,$$

$$|\psi_B^3\rangle = a|v_B\rangle + b|h_B\rangle,$$

and

$$|\psi_B^4\rangle = a|h_B\rangle + b|v_B\rangle.$$

Of these possible photon B states, $|\psi_B^3\rangle$ is an exact replica of the initial (still unknown) state of photon X, $|\psi_X\rangle$. The other three possible states of photon B differ from the initial state of X only by 180° rotations of the plane of polarization around the x-, y-, and z-axes.

We are now ready to teleport photon X. Once Alice has successfully entangled photon X with photon A, she notes by which photomultipliers they were ultimately detected and communicates this information to Bob by *conventional* means. If Alice observes that both photomultipliers detected a photon, then Bob understands that the three-photon state vector $|\Psi_{XAB}\rangle$ collapsed or reduced to the product state $|\psi_{-+}\rangle|\psi_B^3\rangle$, and so he knows that photon B is an exact replica of the initial state of X. If Alice observes any other result, she can also communicate this to Bob. All Bob has to do is pass photon B through an appropriate combination of half waveplates to again replicate the initial state of X. Either way, photon X has been teleported from Alice to Bob.

Note once again that it is necessary for a conventional communication (which takes place no faster than the speed of light) for Bob to know whether or not to rotate the polarization of B and in what way. Without this communication, all Bob receives is randomly polarized photons from which it is impossible to replicate the state of X.

Appendix 26

The de Broglie–Bohm theory

In his redevelopment of a variation of quantum theory based on general non-local hidden variables, Bohm started with the basic, non-relativistic Schrödinger wave equation (see Appendix 5):

$$i\hbar \frac{\partial \Psi}{\partial t} = -\frac{\hbar^2}{2m}\nabla^2\Psi + V\Psi,$$

where $\nabla^2 = \partial^2/\partial x^2 + \partial^2/\partial y^2 + \partial^2/\partial z^2$ is the Laplacian operator, and V is the (classical) potential energy. Bohm assumed that the general solutions for this equation could take the form:

$$\Psi = R \cdot e^{iS/\hbar},$$

where R is a real amplitude function, S is a real phase function, and $\hbar = h/2\pi$.

Starting with the time-dependent form of Schrödinger's equation and substituting for Ψ gives, after some rearrangement:

$$i\hbar \frac{\partial \Psi}{\partial t} = i\hbar e^{iS/\hbar}\frac{\partial R}{\partial t} - R \cdot e^{iS/\hbar}\frac{\partial S}{\partial t}.$$

Similarly for the time-independent form:

$$-\frac{\hbar^2}{2m}\nabla^2\Psi + V\Psi$$

$$= -\frac{e^{iS/\hbar}}{2m}\left[\hbar^2\nabla^2 R - R(\nabla S)^2 + 2i\hbar\nabla R\nabla S + i\hbar R\nabla^2 S\right] + VR \cdot e^{iS/\hbar},$$

where $\nabla = \partial/\partial x + \partial/\partial y + \partial/\partial z$ is the gradient operator. Equating the real and imaginary parts of both sides of this modified Schrödinger equation allows us to deduce the following coupled partial differential equations:

$$\frac{\partial S}{\partial t} = \frac{\hbar^2}{2m}\frac{\nabla^2 R}{R} - \frac{(\nabla S)^2}{2m} - V$$

and

$$\frac{\partial R}{\partial t} = -\frac{\nabla R \nabla S}{m} - \frac{R}{2m}\nabla^2 S.$$

The appearance of R in the expression for the time dependence of S and *vice versa* means that the two fields R and S co-determine one another.

For a reason that will become apparent below, we now rearrange the expression for $\partial S/\partial t$ and apply the gradient operator ∇ to the result:

$$\nabla\left[\frac{\partial S}{\partial t} + \frac{(\nabla S)^2}{2m}\right] = -\nabla\left[V - \frac{\hbar^2}{2m}\frac{\nabla^2 R}{R}\right],$$

which becomes:

$$\frac{\partial}{\partial t}\nabla S + \frac{\nabla S}{m}\cdot\nabla\cdot\nabla S = -\nabla(V + U),$$

where

$$U = -\frac{\hbar^2}{2m}\frac{\nabla^2 R}{R}.$$

So far, the discussion has centred on manipulation of the conventional Schrödinger wave equation, and aside from the assumption of a specific form for the wave function in terms of the real functions R and S, nothing new has been added. The de Broglie–Bohm theory departs from conventional quantum theory by now assuming the existence of a real particle of mass m, following a real trajectory through space. We can define a particle velocity, $\mathbf{v}$, and momentum $\mathbf{p}$ ($= m\mathbf{v}$ in this non-relativistic version of the theory) on which we impose a 'guidance condition':

$$\mathbf{p} = m\mathbf{v} = \nabla S,$$

which effectively ties the particle trajectories to be orthogonal to surfaces of constant phase of the wave function, S. The *total* derivative of momentum with respect to time is given by $d\mathbf{p}/dt = d(\nabla S)/dt$ which we evaluate along a 'flow' or 'stream' line using the so-called convective derivative[1]:

$$\frac{d}{dt} = \frac{\partial}{\partial t} + \mathbf{v}\cdot\nabla,$$

where $\mathbf{v}$ is the velocity vector. Applying this derivative to ∇S and identifying the velocity with $\nabla S/m$ gives:

$$\frac{d}{dt}\mathbf{p} = \frac{d}{dt}\nabla S = \frac{\partial}{\partial t}\nabla S + \frac{\nabla S}{m}\cdot\nabla\cdot\nabla S.$$

---

[1] And, indeed there are strong similarities between these equations and equations of classical hydrodynamics, with $\mathbf{v}$ corresponding to the velocity of a fluid and m the mass of a dust particle being carried by the fluid—see Holland, Peter R. (1993). *The quantum theory of motion.* Cambridge University Press, Cambridge, p. 120.

Equating this with the result we obtained above yields:

$$\frac{d}{dt}\mathbf{p} = m\frac{d}{dt}\mathbf{v} = -\nabla(V + U),$$

which is a modified form of Newton's second law. Here, the force (equal to mass times acceleration) is given as the negative gradient of the classical potential V plus a 'quantum potential' U.

Given that the particle position and its trajectory are 'defined' at all times, it has not been necessary so far to resort to probabilities. When we consider a large number (ensemble) of particles all describable in terms of the same wave function, the above reasoning can still be applied—that is, there is nothing in principle preventing us from following the trajectories of each particle. However, in practice, we do not usually have access to a complete specification of all the particle initial conditions and, just as in Boltzmann's statistical mechanics, we calculate probabilities (using in this case $|\Psi|^2 = R^2$) as a practical necessity.

All of the quantum 'weirdness', including entanglement and 'spooky' action at a distance, is accounted for in the de Broglie–Bohm theory by the quantum potential, U, which is in turn dependent on the amplitude of the wave function, R.

Appendix 27

Neutron worlds

In order to analyse the neutron interferometer experiment described in Fig. 14.2 in terms of the many-worlds interpretation, it is first necessary to set up the definitions of the state vectors required and the actions of the mirrors M_1 and M_2 and the beamsplitters BS_1 and BS_2. We start by defining the state of the world before we perform the experiment in terms of a state vector, denoted $|W\rangle$. The action of BS_1 is to create a superposition of worlds in which the neutron is transmitted and reflected. These worlds are defined by the state vectors $|w_1\rangle$ and $|w_2\rangle$, and the neutrons in each world may follow upwards (u) or downwards (d) trajectories with respect to the central axis of the apparatus. This transformation can be written:

$$|W\rangle \rightarrow \tfrac{1}{\sqrt{2}}(|w_1^u\rangle + |w_2^d\rangle)$$

implying that $|W\rangle$ is transformed into a superposition of $|w_1\rangle$ (with the neutron following an upwards trajectory) and $|w_2\rangle$ (with the neutron following a downward trajectory) in which the worlds have equal measures of existence.

The action of the mirrors is to transform an upwards/downwards trajectories into downwards/upwards trajectories:

$$\text{At mirror } M_1 : |w_1^u\rangle \rightarrow |w_1^d\rangle \quad M_2 : |w_2^d\rangle \rightarrow |w_2^u\rangle.$$

The effect of BS_2 is to further split the world $|w_1\rangle$ into $|w_{1a}\rangle$ and $|w_{1b}\rangle$ and $|w_2\rangle$ into $|w_{2a}\rangle$ and $|w_{2b}\rangle$, but now we must acknowledge the need to impose a *phase convention* which yields superpositions as follows:

$$|w_1^d\rangle \rightarrow \tfrac{1}{\sqrt{2}}(|w_{1a}^u\rangle - |w_{1b}^d\rangle)$$

and

$$|w_2^u\rangle \rightarrow \tfrac{1}{\sqrt{2}}(|w_{2a}^u\rangle + |w_{2b}^d\rangle).$$

Now we must further acknowledge that the neutron in either world has no 'memory' of the path it has just taken. So, the states describing the trajectory from BS_2 to D_1, for example, cannot be distinguished on the basis of the path taken prior to BS_2. The states $|w^u_{1a}\rangle$ and $|w^u_{2a}\rangle$ therefore describe the same world (which we will denote using the state vector $|w^u_a\rangle$). Similarly, $|w^d_{1b}\rangle$ and $|w^d_{2b}\rangle$ describe the same world, which we denote $|w^d_b\rangle$. Making the substitutions gives:

$$|w^d_1\rangle \rightarrow \tfrac{1}{\sqrt{2}}(|w^u_a\rangle - |w^d_b\rangle)$$

and

$$|w^u_2\rangle \rightarrow \tfrac{1}{\sqrt{2}}(|w^u_a\rangle + |w^d_b\rangle).$$

Merging these worlds results in interference, and we can trace the history of events as follows:

$$|W\rangle \rightarrow \tfrac{1}{\sqrt{2}}(|w^u_1\rangle + |w^d_2\rangle) \rightarrow \tfrac{1}{\sqrt{2}}(|w^d_1\rangle + |w^u_2\rangle) \rightarrow |w^u_a\rangle.$$

The sequence of transformations results in detection of the neutron just by detector D_1, as expected.

If we now try to 'tag' the neutron somehow and follow its actual trajectory through the apparatus in order to distinguish between the worlds $|w^u_{1a}\rangle$ and $|w^u_{2a}\rangle$ and, similarly, $|w^d_{1b}\rangle$ and $|w^d_{2b}\rangle$ we can see immediately that there would be no cancellation of terms. All four worlds would have equal measures of existence. In two of these worlds, the neutron is detected by D_1, and in the other two, it is detected by D_2. With repeated measurements the observer in any one of these worlds would conclude that the neutrons are detected by D_1 and D_2 with equal probability, and interference is lost.

Bibliography

Enough material has been published on the subject of quantum theory and its interpretations to fill whole libraries. The following, then, is hardly a comprehensive list of sources of information on the subject matter of this book. Rather, it is a list of publications that I have found particularly helpful in crafting this edition, and which I can heartily recommend to readers in search of further enlightenment.

Advanced texts

Atkins, P. W. (1983). *Molecular quantum mechanics* (2nd edn.). Oxford University Press, Oxford.

Bohm, David (1951). *Quantum theory*. Prentice-Hall, Englewood Cliffs, NJ.

Bohm, David (1980). *Wholeness and the implicate order*. Routledge, London.

Bohm, D. and Hiley, B. J. (1993). *The undivided universe*. Routledge, London.

Dirac, P. A. M. (1958). *The principles of quantum mechanics* (4th edn.). Clarendon Press, Oxford.

Dodd, J. E. (1984). *The ideas of particle physics*. Cambridge University Press, Cambridge.

d'Espagnat, Bernard (1989). *The conceptual foundations of quantum mechanics* (2nd edn.). Addison-Wesley, New York.

Feynman, Richard P., Leighton, Robert B., and Sands, Matthew (1965). *The Feynman lectures on physics*, Vol. III. Addison-Wesley, Reading MA.

French, A. P. (1968). *Special relativity*. Van Nostrand Reinhold, Wokingham, UK.

French, A. P. and Taylor, E. F. (1978). *An introduction to quantum physics*. Van Nostrand Reinhold, Wokingham, UK.

Heisenberg, Werner (1930). *The physical principles of the quantum theory*. Dover, New York.

Holland, Peter R. (1993). *The quantum theory of motion*. Cambridge University Press, Cambridge.

Isham, Chris J. (1995). *Lectures on quantum theory*. Imperial College Press, London.

Jammer, Max (1974). *The philosophy of quantum mechanics*. Wiley, New York.

Merzbacher, Eugene (1970). *Quantum mechanics* (2nd edn.). Wiley, New York.

Omnès, Roland (1994). *The interpretation of quantum mechanics*. Princeton University Press, Princeton, NJ.

Omnès, Roland (1999). *Understanding quantum mechanics*. Princeton University Press, Princeton, NJ.

Prigogine, Ilya (1980). *From being to becoming*. W. H. Freeman, San Francisco, CA.

Rae, Alastair I. M. (1986). *Quantum mechanics* (2nd edn.). Adam Hilger, Bristol, UK.

Rindler, Wolfgang (1982). *Introduction to special relativity*. Oxford University Press, Oxford.

Thaler, Bernd (2000). *Visual quantum mechanics*. Springer-Verlag, New York.

von Neumann, John (1955). *Mathematical foundations of quantum mechanics*. Princeton University Press, Princeton, NJ.

Young, N. (1988). *An introduction to Hilbert space*. Cambridge University Press, Cambridge.

Biographies

Bernstein, Jeremy (1991). *Quantum profiles*. Princeton University Press, Princeton, NJ. (Contains biographical sketches of John Bell and John Wheeler.)

Cassidy, David C. (1992). *Uncertainty: the life and science of Werner Heisenberg*. W. H. Freeman, New York.

Enz, Charles P. (2002). *No time to be brief: a scientific biography of Wolfgang Pauli*. Oxford University Press, Oxford.

Feynman, Richard P. (1985). *'Surely you're joking, Mr. Feynman!'*. Unwin, London.

Gleick, James (1992). *Genius. Richard Feynman and modern physics*. Little Brown, London.

Goodchild, Peter (1980). *J. Robert Oppenheimer*. BBC, London.

Hoffmann, Banesh (1975). *Albert Einstein*. Paladin, St. Albans, UK.

Horgan, John (1993). Last words of a quantum heretic (a biographical sketch of David Bohm). *Scientific American*, February, 38.

Klein, Martin J. (1985). *Paul Ehrenfest: the making of a theoretical physicist*, Vol. 1 (3rd edn.). North-Holland, Amsterdam.

Kragh, Helge S. (1990). *Dirac: a scientific biography*. Cambridge University Press, Cambridge.

Mehra, Jagdish (1994). *The beat of a different drum: the life and science of Richard Feynman*. Oxford University Press, Oxford.

Moore, Walter (1989). *Schrödinger: life and thought*. Cambridge University Press, Cambridge.

Nasar, Sylvia (1998). *A beautiful mind*. Faber & Faber, London. (A biography of the mathematician John Nash, containing a biographical sketch of John von Neumann.)

Pais, Abraham (1982). *Subtle is the Lord: the science and the life of Albert Einstein*. Oxford University Press, Oxford.

Pais, Abraham (1991). *Niels Bohr's times*. Oxford University Press, Oxford.

Peat, David F. (1997). *Infinite potential: the life and times of David Bohm*. Addison-Wesley, Reading, MA.

Popper, Karl (1976). *Unended quest: an intellectual autobiography*. Fontana, London.

Poundstone, William (1992). *Prisoner's dilemma: John von Neumann, game theory and the puzzle of the bomb*. Random House, New York.

Rogers, Ben (2000). *A. J. Ayer: a life*. Vintage, London.

Wheeler, John Archibald, with Ford, Kenneth (1998). *Geons, black holes and quantum foam: a life in physics*. W. W. Norton, New York.

Anthologies and collections

Bell, J. S. (1987). *Speakable and unspeakable in quantum mechanics*. Cambridge University Press, Cambridge.

Born, Max (1969). *Physics in my generation* (2nd edn.). Springer, New York.

DeWitt, B. S. and Graham, N. (eds.) (1975). *The many worlds interpretation of quantum mechanics*. Pergamon, Oxford.

French, A. P. and Kennedy, P. J. (eds.) (1985). *Niels Bohr: a centenary volume*. Harvard University Press, Cambridge, MA.

Heisenberg, Werner (1983). *Encounters with Einstein*. Princeton University Press, Princeton, NJ.

Hiley, B. J. and Peat, F. D. (eds.) (1987). *Quantum implications*. Routledge & Kegan Paul, London.

Kilmister, C. W. (ed.) (1987). *Schrödinger: centenary celebration of a polymath*. Cambridge University Press, Cambridge.

Oppenheimer, Robert J. (1989). *Atom and void*. Princeton University Press, Princeton, NJ.

Schilpp, P. A. (ed.) (1949). *Albert Einstein. philosopher–scientist*. The Library of Living Philosophers, Open Court, La Salle, IL.

Wheeler, John Archibald (1994). *At home in the universe*. AIP Press, New York.

Wheeler, John Archibald and Zurek, Wojciech Hubert (eds.) (1983). *Quantum theory and measurement*. Princeton University Press, Princeton, NJ.

History and philosophy of science

Albert, David Z. (1992). *Quantum mechanics and experience*. Harvard University Press, Cambridge, MA.

Ayer, A. J. (1936). *Language, truth and logic*. Penguin, London.

Ayer, A. J. (1956). *The problem of knowledge*. Penguin, London.

Ayer, A. J. (1976). *The central questions of philosophy*. Penguin, London.

Beller, Mara (1999). *Quantum dialogue*. University of Chicago Press, Chicago.

Bohm, David (1957). *Causality and chance in modern physics*. Routledge, London.

Broad, William and Wade, Nicholas (1982). *Betrayers of the truth*. Oxford University Press, Oxford.

Cartwright, Nancy (1983). *How the laws of physics lie*. Oxford University Press, Oxford.

Collins, Harry and Pinch, Trevor (1993). *The Golem*. Cambridge University Press, Cambridge.

Cushing, James T. (1994). *Quantum mechanics: historical contingency and the Copenhagen hegemony*. University of Chicago Press, Chicago.

Cushing, James T. (1998). *Philosophical concepts in physics*. Cambridge University Press, Cambridge.

Descartes, René (1968). *Discourse on method and the meditations*. Penguin, London.

Duhem, Pierre (1954). *The aim and structure of physical theory*, translated by Wiener, Philip P. Princeton University Press, Princeton, NJ.

Duhem, Pierre (1996). *Essays in the history and philosophy of science*, translated and edited by Ariew, Roger and Barker, Peter. Hackett, Indianapolis, IN.

d'Espagnat, Bernard (1989). *Reality and the physicist*. Cambridge University Press, Cambridge.

Feyerabend, Paul (1987). *Farewell to reason*. Verso, London.

Fine, Arthur (1996). *The shaky game* (2nd edn.). University of Chicago Press, Chicago.

Gamow, George (1966). *Thirty years that shook physics*. Dover, New York.

Gillies, Donald (1993). *Philosophy of science in the twentieth century*. Blackwell, Oxford.

Hacking, Ian (1983). *Representing and intervening*. Cambridge University Press, Cambridge.

Harré, R. (1984). *The philosophies of science* (2nd edn.). Oxford University Press, Oxford.

Heisenberg, Werner (1989). *Physics and philosophy*. Penguin, London.

Hermann, Armin (1971). *The genesis of quantum theory (1899–1913)*. MIT Press, Cambridge, MA.

Hume, David (1948). *Dialogues concerning natural religion*. Hafner Press, New York.

Hume, David (1969). *A treatise of human nature*. Penguin, London.

Jeans, James (1981). *Physics and philosophy*. Dover, New York.

Kant, Immanuel (1934). *Critique of pure reason*. J. M. Dent, London.

Kragh, Helge (1999). *Quantum generations*. Princeton University Press, Princeton, NJ.

Kuhn, Thomas S. (1970). *The structure of scientific revolutions* (2nd edn.). University of Chicago Press, Chicago.

Kuhn, Thomas S. (1978). *Black-body theory and the quantum discontinuity 1894–1912*. University of Chicago Press, Chicago.

Leibniz, Gottfried Wilhelm (1973). *Philosophical writings*. J. M. Dent, London.

Lockwood, Michael (1990). *Mind, brain and the quantum: the compound 'I'*. Blackwell, Oxford.

Losee, John (2001). *A historical introduction to the philosophy of science* (4th edn.). Oxford University Press, Oxford.

Medawar, Peter (1984). *The limits of science*. Oxford University Press, Oxford.

Mehra, Jagdish (1999). *Einstein, physics and reality*. World Scientific, London.

Murdoch, Dugald (1987). *Niels Bohr's philosophy of physics*. Cambridge University Press, Cambridge.

Nagel, Ernest and Newman, James R. (1958). *Gödel's proof*. Routledge, London.

Omnès, Roland (1999). *Quantum philosophy*. Princeton University Press, Princeton, NJ.

Popper, Karl R. (1959). *The logic of scientific discovery*. Hutchinson, London.

Popper, Karl R. (1982). *Quantum theory and the schism in physics*. Unwin Hyman, London.

Popper, Karl R. (1990). *A world of propensities*. Thoemmes, Bristol.

Psillos, Stathis (1999). *Scientific realism*. Routledge, London.

Russell, Bertrand (1967). *The problems of philosophy*. Oxford University Press, Oxford.

Ryle, Gilbert (1963). *The concept of mind*. Penguin, London.

Schacht, Richard (1984). *Classical modern philosophers*. Routledge & Kegan Paul, London.

Scruton, Roger (1986). *Spinoza*. Oxford University Press, Oxford.

Tomonaga, Sin-itiro (1997). *The story of spin*. University of Chicago Press, Chicago.

Van Fraassen, Bas C. (1980). *The scientific image*. Oxford University Press, Oxford.

Popular science

Aczel, Amir D. (2003). *Entanglement*. Wiley, London.

Barbour, Julian (1999). *The end of time*. Weidenfeld & Nicholson, London.

Barrow, John D. (1991). *Theories of everything*. Vintage, London.

Damasio, Antonio R. (1994). *Descartes' error*. Picador, London.

Davies, Paul (1984). *God and the new physics*. Penguin, London.

Davies, Paul (1988). *Other worlds*. Penguin, London.

Davies, Paul (1995). *About time*. Penguin, London.

Davies, Paul (2001). *How to build a time machine*. Penguin, London.

Davies, P. C. W. and Brown, J. R. (eds.) (1986). *The ghost in the atom*. Cambridge University Press, Cambridge.

Dennett, Daniel C. (1991). *Consciousness explained*. Penguin, London.

Dennett, Daniel C. (1996). *Kinds of minds*. Weidenfeld & Nicholson, London.

Deutsch, David (1997). *The fabric of reality*. Penguin, London.

Edelman, Gerald (1992). *Bright air, brilliant fire*. Penguin, London.

Feynman, Richard (1967). *The character of physical law*. MIT Press, Cambridge, MA.

Feynman, Richard P. (1985). *QED: the strange theory of light and matter*. Penguin, London.

Feynman, Richard P. (1998a). *Six easy pieces*. Perseus, Cambridge, MA.

Feynman, Richard P. (1998b). *Six not-so-easy pieces*. Allen Lane, London.

Gamow, George (1965). *Mr. Tompkins in paperback*. Cambridge University Press, Cambridge.

Gell-Mann, Murray (1994). *The quark and the jaguar*. Little Brown, London.

Gleick, James (1988). *Chaos. Making a new science*. Heinemann, London.

Gott, Richard J. (2001). *Time travel in Einstein's universe*. Weidenfeld & Nicholson, London.

Greenfield, Susan A. (1995). *Journey to the centers of the mind*. W. H. Freeman, New York.

Greenfield, Susan A. (2000). *The private life of the brain*. Penguin, London.

Gregory, Bruce (1988). *Inventing reality: physics as language*. Wiley, New York.

Gregory, Richard L. (1981). *Mind in science*. Penguin, London.

Gribbin, John (1995). *Schrödinger's kittens*. Penguin, London.

Gribbin, John (1998). *Q is for quantum: particle physics from A to Z*. Weidenfeld & Nicholson, London.

Hawking, Stephen W. (1988). *A brief history of time*. Bantam, London.

Koestler, Arthur (1964). *The sleepwalkers*. Penguin, London.

LeDoux, Joseph (2002). *Synaptic self*. Viking Penguin, London.

Lindley, David (1996). *Where does the weirdness go?* Basic Books, New York.

Milburn, Gerard J. (1998). *The Feynman processor*. Perseus, Cambridge, MA.

Penrose, Roger (1990). *The emperor's new mind*. Vintage, London.

Penrose, Roger (1995). *Shadows of the mind*. Vintage, London.

Penrose, Roger (1997). *The large, the small and the human mind*. Cambridge University Press, Cambridge.

Pinker, Steven (1997). *How the mind works*. Penguin, London.

Polkinghorne, J. C. (1984). *The quantum world*. Penguin, London.

Prigogine, Ilya and Stengers, Isabelle (1985). *Order out of chaos*. Fontana, London.

Rae, Alastair (1986). *Quantum physics: illusion or reality?* Cambridge University Press, Cambridge.

Rohrlich, Fritz (1987). *From paradox to reality*. Cambridge University Press, Cambridge.

Rose, Stephen (1992). *The making of memory*. Bantam Books, London.

Sachs, Mendel (1988). *Einstein versus Bohr: the continuing controversies in physics*. Open Court, La Salle, IL.

Sacks, Oliver (1985). *The man who mistook his wife for a hat*. Picador, London.

Singh, Simon (2000). *The science of secrecy*. Fourth Estate, London.

Smolin, Lee (2000). *Three roads to quantum gravity*. Weidenfeld & Nicholson, London.

Snow, C. P. (1981). *The physicists*. Macmillan, London.

Squires, Euan (1986). *The mystery of the quantum world*. Adam Hilger, Bristol, UK.

Stewart, Ian (1989). *Does God play dice?—the mathematics of chaos*. Blackwell, Oxford.

Thorne, Kip S. (1994). *Black holes and time warps*. W. W. Norton, New York.

Treiman, Sam (1999). *The odd quantum*. Princeton University Press, Princeton, NJ.

Warwick, Kevin (1998). *In the mind of the machine*. Arrow Books, London.

Zukav, Gary (1980). *The dancing Wu Li masters*. Bantam, New York.

Research papers and review articles

Arndt, Markus, Nairz, Olaf, Vos-Andreae, Julian, Keller, Claudia, van der Zouw, Gerbrand, and Zeilinger, Anton (1999). Wave–particle duality of C_{60} molecules. *Nature*, **401**, 680.

[1] Aspect, Alain (1976). Proposed experiment to test the nonseparability of quantum mechanics. *Physical Review D*, **14**, 1944.

Aspect, Alain, Grangier, Philippe, and Roger, Gérard (1981). Experimental tests of realistic local theories via Bell's theorem. *Physical Review Letters*, **47**, 460.

Aspect, Alain, Grangier, Philippe, and Roger, Gérard (1982a). Experimental realization of Einstein–Podolsky–Rosen–Bohm *Gedankenexperiment*: a new violation of Bell's inequalities. *Physical Review Letters*, **49**, 91.

Aspect, Alain, Dalibard, Jean, and Roger, Gérard (1982b). Experimental test of Bell's inequalities using time-varying analyzers. *Physical Review Letters*, **49**, 1804.

Baggott, Jim (1990). Quantum mechanics and the nature of physical reality. *Journal of Chemical Education*, **67**, 638.

[1] Bell, J. S. (1964). On the Einstein–Podolsky–Rosen paradox. *Physics*, **1**, 195.

[1] Bell, J. S. (1966). On the problem of hidden variables in quantum mechanics. *Reviews of Modern Physics*, **38**, 447.

Bell, John (1990). Against 'measurement'. *Physics World*, **3**, 33. (See also the reply from Gottfried, Kurt (1991). *Physics World*, **4**, 34.)

Belousek, Darrin W. (1996). Einstein's 1927 unpublished hidden-variable theory: its background, context and significance. *Studies in the History and Philosophy of Modern Physics*, **4**, 437.

Bennett, Charles H., Brassard, Gilles, Crépeau, Claude, Josza, Richard, Peres, Asher, and Wooters, William K. (1993). Teleporting an unknown quantum state via dual classical and Einstein–Podolsky–Rosen channels. *Physical Review Letters*, **70**, 1895.

[1] Bohm, David (1952). A suggested interpretation of the quantum theory in terms of 'hidden' variables, I and II. *Physical Review*, **85**, 166.

Bohm, D. and Bub, J. (1966). A proposed solution of the measurement problem in quantum mechanics by a hidden variable theory. *Reviews of Modern Physics*, **38**, 453.

[1] Bohr, N. (1935). Can quantum-mechanical description of physical reality be considered complete? *Physical Review*, **48**, 696.

Bouwmeester, Dik, Pan, Jian-Wei, Daniell, Matthew, Weinfurter, Harald, and Zeilinger, Anton (1999). Observation of three-photon Greenberger–Horne–Zeilinger entanglement. *Physical Review Letters*, **82**, 1345.

[1] Clauser, John F., Horne, Michael A., Shimony, Abner, and Holt, Richard A. (1969). Proposed experiment to test local hidden-variable theories. *Physical Review Letters*, **23**, 880.

Clauser, John F. and Shimony, Abner (1978). Bell's theorem: experimental tests and implications. *Reports on Progress in Physics*, **41**, 1881.

Deutsch, David and Lockwood, Michael (1994). The quantum physics of time travel. *Scientific American*, March, 68.

DeWitt, Bryce S. (1970). Quantum mechanics and reality. *Physics Today*, **23**, 155.

---

[1] These papers are reproduced in Wheeler, J. A. and Zurek, W. H. (eds.) (1983). *Quantum theory and measurement*. Princeton University Press, Princeton, NJ.

[2]Diósi, Lajos and Kiefer, Claus (2001). Robustness and diffusion of pointer states. arXiv:quant-ph/0005071 v2 10 February.

Dürr, S., Nonn, T., and Rempe, G. (1998). Origin of quantum-mechanical complementarity probed by a "which-way" experiment in an atom interferometer. *Nature*, **395**, 33.

[1]Einstein, A., Podolsky, B., and Rosen, N. (1935). Can quantum-mechanical description of physical reality be considered complete? *Physical Review*, **47**, 777.

Englert, Berthold-Georg, Scully, Marlan O., and Walther, Herbert (1994). The duality in matter and light. *Scientific American*, December, 86.

Englert, Bethold-Georg, Scully, Marlan O., and Walther, Herbert (1999). Quantum erasure in double-slit interferometers with which-way detectors. *American Journal of Physics*, **67**, 325.

d'Espagnat, Bernard (1979). The quantum theory and reality. *Scientific American*, **241**, 128.

[1]Everett III, Hugh (1957). 'Relative state' formulation of quantum mechanics. *Reviews of Modern Physics*, **29**, 454.

Franson, J. D. (1989). Bell inequality for position and time. *Physical Review Letters*, **62**, 2205.

[1]Freedman, Stuart J. and Clauser, John F. (1972). Experimental test of local hidden-variable theories. *Physical Review Letters*, **28**, 938.

Friedman, Jonathan R., Patel, Vijay, Chen, W., Tolpygo, S. K., and Lukens, J. E. (2000). Quantum superposition of distinct macroscopic states. *Nature*, **406**, 43.

Ghiradi, G. C., Rimini, A., and Weber, T. (1986). Unified dynamics for microscopic and macroscopic systems. *Physical Review D*, **34**, 470.

Gisin, N. and Gisin, B. (1999). A local hidden variable model of quantum correlation exploiting the detection loophole. *Physics Letters A*, **260**, 323.

Gisin, Nicholas, Ribordy, Gregoire, Tittel, Wolfgang, and Zbinden, Hugo (2002). Quantum cryptography. *Reviews of Modern Physics*, **74**, 145.

Greeberger, Daniel M., Horne, Michael A., Shimony, Abner, and Zeilinger, Anton (1990). Bell's theorem without inequalities. *American Journal of Physics*, **58**, 1131.

[2]Griffiths, Robert B. (1996). Consistent histories and quantum reasoning. arXiv:quant-ph/9606004 v1 4 June.

[2]Griffiths, Robert B. (1998). Choice of consistent families, and quantum incompatibility. ArXiv:quant-ph/9708028 v2 31 December.

[2]Hagan, S., Hameroff, S. R., and Tuszynski, J. A. (2000). Quantum computation in brain microtubles. arXiv:quant-ph/0005025 v1 4 May.

Haroche, Serge (1998). Entanglement, decoherence and the quantum/classical boundary. *Physics Today*, July, 36.

[1]Heisenberg, Werner (1927). The physical content of quantum kinematics and mechanics. *Zeitschrift für Physik*, **43**, 172.

Herzog, Thomas J., Kwiat, Paul G., Weinfurther, Harald, and Zeilinger, Anton (1995). Complementarity and the quantum eraser. *Physical Review Letters*, **75**, 3034.

[2]Isham, C. J. and Butterfield, J. (1999). Some possible roles for topos theory in quantum theory and quantum gravity. arXiv:quant-ph/9910005 v1 2 October.

[2]Joos, Erich (1999). Elements of environmental decoherence. arXiv:quant-ph/9908008 v1 2 August.

[2]Kent, Adrian (1997). Against many-worlds interpretations. arXiv:gr-qc/9703089 v1 31 March.

[2]Kiefer, Claus and Joos, Erich (1998). Decoherence: concepts and examples. arXiv:quant-ph/9803052 v1 19 March.

Kurtsiefer, Christian, Oberparleiter, Markus, and Weinfurter, Harald (2001). High-efficiency entangled photon pair collection in type-II parametric fluorescence. *Physical Review A*, **64**, 023802.

Kwiat, Paul G., Mattle, Klaus, Weinfurter, Harald, and Zeilinger, Anton (1995). New high-intensity source of polarization-entangled photons. *Physical Review Letters*, **75**, 4337.

Kwiat, Paul, Weinfurter, Harald, and Zeilinger, Anton (1996). Quantum seeing in the dark. *Scientific American*, November, 72.

---

[2] Preprints dating from 1994 are available on the arXiv.org e-print archive website, which is supported by the US National Science Foundation with Cornell University: http://arXiv.org/. Select the links for quantum physics (quant-ph) or general relativity and quantum cosmology (gr-qc), as appropriate.

Leggett, A. J. (2002). Testing the limits of quantum mechanics: motivation, state of play, prospects. *Journal of Physics: Condensed Matter*, **14**, R415.

[2]Marshall, William, Simon, Christoph, Penrose, Roger, and Bouwmeester, Dik (2002). Towards quantum superpositions of a mirror. arXiv:quant-ph/0210001 v1 30 September.

Mermin, David N. (1990). Quantum mysteries revisited. *American Journal of Physics*, **58**, 731.

Monroe, C., Meekhof, D. M., King, B. E., and Wineland, D. J. A. (1996). 'Schrödinger cat' superposition state of an atom. *Science*, **272**, 1131.

Ou, Z. Y. and Mandel, L. (1988). Violation of Bell's inequality and classical probability in a two-photon correlation experiment. *Physical Review Letters*, **61**, 50.

Pan, Jian-Wei, Bouwmeester, Dik, Daniell, Matthew, Weinfurter, Harald, and Zeilinger, Anton (2000). Experimental test of quantum nonlocality in three-photon Greenburger–Horne–Zeilinger entanglement. *Nature*, **403**, 515.

Rowe, M. A., Kielpinski, D., Meyer, V., Sackett, C. A., Itano, W. M., Monroe, C., and Wineland, D. J. (2001). Experimental violation of a Bell's inequality with efficient detection. *Nature*, **409**, 791.

[1]Schrödinger, Erwin (1980). The present situation in quantum mechanics: a translation of Schrödinger's 'cat paradox' paper (trans. John D. Trimmer). *Proceedings of the American Philosophical Society*, **124**, 323.

Scully, Marlan O., Englert, Berthold-Georg, and Walther, Herbert (1991). Quantum optical tests of complementarity. *Nature*, **351**, 111.

Shimony, Abner (1986). The reality of the quantum world. *Scientific American*, **258**, 36.

Simmonds, R. W., Marchenkov, A., Hoskinson, E., Davis, J. C., and Packard, R. E. (2001). Quantum interference of superfluid ^{3}He. *Nature*, **412**, 55.

Simon, Jonathan Z. (1994). The physics of time travel. *Physics World*, December, 27.

[2]Tegmark, Max (1997). The interpretation of quantum mechanics: many worlds or many words? arXiv:quant-ph/9709032 v1 15 September.

[2]Tegmark, Max (1999). The importance of quantum decoherence in brain processes. arXiv:quant-ph/9907009 v2 10 November.

Tegmark Max and Wheeler, John Archibald (2001). 100 years of quantum mysteries. *Scientific American*, February, 68.

[2]Tipler, Frank (2000). Does quantum non-locality exist? Bell's theorem and the many-worlds interpretation. arXiv:quant-ph/0003146 v1 30 March.

Tittel, W., Brendel, J., Gisin, B., Herzog, T., Zbinden, H., and Gisin, N. (1998). Experimental demonstration of quantum correlations over more than 10 km. *Physical Review A*, **57**, 3229.

Tittel, W., Brendel, J., Gisin, N., and Zbinden, H. (1999). Long-distance Bell-type tests using energy-time entangled photons. *Physical Review A*, **59**, 4150.

Tittel, W., Brendel, J., Zbinden, H., and Gisin, N. (1998). Violation of Bell inequalities by photons more than 10 km apart. *Physical Review Letters*, **81**, 3563.

Tittel, Wolfgang and Weihs, Gregor (2001). Photonic entanglement for fundamental tests and quantum communication. *Quantum information and computation*, **1**, 3.

[2]Vaidman, Lev (1996). On schizophrenic experiences of the neutron or why we should believe in the many-worlds interpretation of quantum theory. arXiv:quant-ph/9609006 v1 7 September.

Weihs, Gregor, Jennewein, Thomas, Simon, Christoph, Weinfurter, Harald, and Zeilinger, Anton (1998). Violation of Bell's inequality under strict Einstein locality conditions. *Physical Review Letters*, **81**, 5039.

[1]Wigner, Eugene (1961). Remarks on the mind–body question. In I. J. Good (ed.), *The scientist speculates: an anthology of partly-baked ideas*. Heinemann, London.

Wiltshire, David L. (1996). Introduction to quantum cosmology. In Robson B., Visvanathan N., and Woolcock W. S. (eds.), *Cosmology, the physics of the universe*. World Scientific, Singapore.

Wooters, William and Zurek, Wojciech H. (1979). Complementarity in the double-slit experiment: quantum nonseparability and a quantitative statement of Bohr's principle. *Physical Review D*, **19**, 473.

Zeilinger, Anton (2000). Quantum teleportation. *Scientific American*, April, 50.

[2]Zurek, Wojciech H. (1998). Decoherence, einselection and the existential interpretation (the rough guide). arXiv:quant-ph/9805065 v1 21 May.

Name index

The symbol *f* indicates footnote

Subject index

Printed in the United States
By Bookmasters